The Urban Book Series

The Urban Book Series is a resource for urban studies and geography research worldwide. It provides a unique and innovative resource for the latest developments in the field, nurturing a comprehensive and encompassing publication venue for urban studies, urban geography, planning and regional development.

The series publishes peer-reviewed volumes related to urbanization, sustainability, urban environments, sustainable urbanism, governance, globalization, urban and sustainable development, spatial and area studies, urban management, transport systems, urban infrastructure, urban dynamics, green cities and urban landscapes. It also invites research which documents urbanization processes and urban dynamics on a national, regional and local level, welcoming case studies, as well as comparative and applied research.

The series will appeal to urbanists, geographers, planners, engineers, architects, policy makers, and to all of those interested in a wide-ranging overview of contemporary urban studies and innovations in the field. It accepts monographs, edited volumes and textbooks.

Indexed by Scopus.

Robert Goodspeed • Esra Suel
Huanfa Chen • Joana Barros • Christopher Pettit
Editors

Digital-Era Urban Transformations

Advancements in Data Science, Analytics and Technology

Springer

Editors
Robert Goodspeed
University of Michigan
Ann Arbor, MI, USA

Esra Suel
University College London
London, UK

Huanfa Chen
University College London
London, UK

Joana Barros
University College London
London, UK

Christopher Pettit
UNSW Sydney
Sydney, NSW, Australia

ISSN 2365-757X ISSN 2365-7588 (electronic)
The Urban Book Series
ISBN 978-3-031-98299-6 ISBN 978-3-031-98300-9 (eBook)
https://doi.org/10.1007/978-3-031-98300-9

This Springer imprint is published by the registered company Springer Nature Switzerland AG
The registered company address is: Gewerbestrasse 11, 6330 Cham, Switzerland

Preface

The Computational Urban Planning and Urban Management (CUPUM) international conference is one of the premier international conferences for the exchange of ideas and applications of computer technologies to address a range of social and environmental problems relating to urban areas. The first conference took place in 1989 in Hong Kong. Since then, this biennial conference has been hosted in cities across Asia, Australia, Europe, North America, and South America (Table 1). This volume is published in conjunction with the 19th CUPUM conference, which was hosted by The Bartlett Centre for Advanced Spatial Analysis at the University College London in London, UK, in June 2025.

Table 1 Past CUPUM conferences

Number	Year	Place	Country
I	1989	Hong Kong	China
II	1991	Oxford	United Kingdom
III	1993	Atlanta	USA
IV	1995	Melbourne	Australia
V	1997	Mumbai	India
VI	1999	Venice	Italy
VII	2001	Honolulu	USA
VIII	2003	Sendai	Japan
IX	2005	London	United Kingdom
X	2007	Iguazu Falls	Brazil
XI	2009	Hong Kong	China
XII	2011	Lake Louise (Calgary/Banff)	Canada
XIII	2013	Utrecht	The Netherlands
XIV	2015	Boston	USA
XV	2017	Adelaide	Australia
XVI	2019	Wuhan	China
XVII	2021	Helsinki	Finland
XVIII	2023	Montreal	Canada
XIV	2025	London	United Kingdom

Table 2 Board of directors of CUPUM

Name	Region	Institution	Country	E-mail
Chris Pettit (Chair)	Australia	University of New South Wales	AUS	c.pettit@unsw.edu.au
Joanna Badach	Europe	Gdańsk Tech	PL	joanna.badach@pg.edu.pl
Sharon Biermann	Australia	University of Western Australia	AUS	sharon.biermann@uwa.edu.au
Adam Dennett	Europe	University College London	GBR	a.dennett@ucl.ac.uk
Robert Goodspeed	North America	University of Michigan	USA	rgoodspe@umich.edu
Marketta Kyttä	Europe	Aalto University	FIN	marketta.kytta@aalto.fi
Wenfeng Li	Asia	University of Hong Kong	CHN	wfli@hku.hk
Achilleas Psyllidis	Europe	TU Delft	NL	A.Psyllidis@tudelft.nl
Renee Sieber	North America	McGill University	CAN	renee.sieber@mcgill.ca
Atsushi Suzuki	Asia	Meijo University	JPN	atsuzuki@meijo-u.ac.jp
Sarah Williams	North America	Massachusetts Institute of Technology	USA	sew@mit.edu
Qingming Zhan	Asia	Wuhan University	CHN	qmzhan@whu.edu.cn

CUPUM is led by a board of directors (Table 2). Supporting them in organizing the 2025 conference was the Local Academic Committee, which included Adam Dennett, Susannah Cramer-Greenbaum, Chen Zhong, Andy Hudson-Smith, Zhixuan (Willow) Liu, Maio Zeng Liu, Duncan Smith, Michael Batty, and Susan Francis.

The editors produced the volume with support from these two groups, as well as the CUPUM advisers to the board (Table 3). After an initial call for submissions, the book editors received 31 draft book chapters, 16 of which appear in the final volume. All published chapters underwent a double-blind peer-review process and was presented at the CUPUM 2025 conference in London, June 23–27, 2025. Many of the submitted chapters that were not included in this volume are to be presented as short papers and are likely to appear elsewhere in the scholarly literature.

As editors, we wish to thank the many people who made this volume possible. The authors submitted very high-quality chapters, and we also thank them for their diligent and careful revisions in response to feedback from the reviewers and editors. All of the peer reviewers for this book, which include members of the CUPUM boards as well as other scholars involved in the CUPUM community, provided invaluable feedback to strengthen the overall quality. In addition, Springer has remained a supportive publishing partner, and this marks the seventh volume published in conjunction with a CUPUM event. In particular, we thank Andrea Brody-Barre, the editor of the *Urban Book Series*, for her support of this project, and Kirithiga Gnanasekaran for professionally managing the final production. At the University of Michigan, Mei Yuan provided editorial support. The other volumes

Table 3 Advisers to the CUPUM board

Name	Institution	Country	E-mail
Richard Klosterman (chair)	University of Akron	USA	dick.klosterman@gmail.com
Michael Batty	University College London	GBR	m.batty@ucl.ac.uk
Karl Kim	University of Hawaii	USA	karlk@hawaii.edu
Kazuaki Miyamoto	Tokyo City University	JPN	miyamoto@tcu.ac.jp
Paola Rizzi	Università degli Studi di Sassari	BRA	rizzi@uniss.it
John Stillwell	University of Leeds	GBR	j.c.h.stillwell@leeds.ac.uk
Anthony G.O. Yeh	University of Hong Kong	CHN	hdxugoy@hkucc.hku.hk
Ray Wyatt	University of Melbourne	AUS	raywyatt3@gmail.com
Antônio Nélson Rodrigues da Silva	University of São Paulo	BRA	anelson@sc.usp.br
Stan Geertman	Utrecht University	NLD	s.c.m.geertman@uu.nl
Andrew Allan	University of South Australia	AUS	andrew.allan@unisa.edu.au
Joseph Ferreira	Massachusetts Institute of Technology	USA	jf@mit.edu

published in conjunction with prior CUPUM conferences are *Planning Support Systems for Sustainable Urban Development* (2013), *Planning Support Systems and Smart Cities* (2015), *Planning Support Science for Smarter Urban Futures* (2017), *Computational Urban Planning and Management for Smart Cities* (2019), *Urban Informatics and Future Cities* (2021), and *Intelligence for Future Cities: Planning Through Big Data and Urban Analytics* (2023).

Given that CUPUM returned to London after 20 years and was hosted by the renowned Centre for Advanced Spatial Analysis led by CUPUM co-founder Michael Batty, we knew interest in CUPUM 2025 would run high. However, the outpouring of interest in the conference has confirmed that despite the era of global environmental and political change, CUPUM remains a vital community deeply engaged in the improvement of urban life worldwide. It is our pleasure to steward this latest volume in the CUPUM series, propelling our field forward with fresh ideas and perspectives.

Ann Arbor, MI, USA — Robert Goodspeed
London, UK — Esra Suel
London, UK — Huanfa Chen
London, UK — Joana Barros
Sydney, NSW, Australia — Christopher Pettit
2025

Contents

Introduction

Robert Goodspeed, Joana Barros, Huanfa Chen, Christopher Pettit, and Esra Suel

Abstract This chapter introduces the contents of the book and summarizes the chapters it contains. This seventh volume published in partnership between Computational Urban Planning and Urban Management (CUPUM) and Springer heralds a new phase of research in a new digital era. In reflecting on the development and growth of planning support systems (PSSs) in the 1990s and early 2000s, we argue that the broad digitalization of cities and society has set the stage for a new era of urban planning and management research. Containing diverse contributions exploring AI, urban sensing, satellite and platform data, and visualization tools, this volume illustrates a CUPUM community in motion, engaging with the latest technologies in service of long-standing goals. Much less than displacing the concept of a PSS, the new digital era multiplies the possibilities for achieving transformative planning support.

Keywords Planning support system · Urban planning · Urban analytics · CUPUM

R. Goodspeed (✉)
Taubman College of Architecture and Urban Planning, University of Michigan, Ann Arbor, MI, USA
e-mail: rgoodspe@umich.edu

J. Barros
Centre for Advanced Spatial Analysis, University College London, London, UK

School of Social Sciences, Birbeck, University of London, London, UK

H. Chen
Centre for Advanced Spatial Analysis, University College London, London, UK

C. Pettit
City Futures Research Centre, University of New South Wales, Sydney, NSW, Australia

E. Suel
Centre for Advanced Spatial Analysis, University College London, London, UK

Department of Geography, University of Zürich, Zürich, Switzerland

R. Goodspeed et al. (eds.), *Digital-Era Urban Transformations*, The Urban Book Series, https://doi.org/10.1007/978-3-031-98300-9_1

The world has changed profoundly in the 35 years since CUPUM was inaugurated in Hong Kong in 1989, the same year the concept of PSS first appeared in the scholarly literature (Harris, 1989). During that era, even the most advanced forms of urban analysis generally involved an expert working with government data on a personal computer. Although this work was often limited to the use of static, limited data, such as census records or administrative property records, the outpouring of exciting new models, visualizations, and techniques that occurred under the PSS banner is a testament to the creativity of our field and the power of the computational tools of that time.

This chapter first places the present volume in the context of this broader evolution, highlighting how it demonstrates both a continuation of the development of long-standing concepts like PSS, but also an exciting engagement with the many facets of today's digital era. Next, we provide an overview of the chapters contained in this book, briefly introducing their focus and contributions, before offering concluding remarks.

1 CUPUM Research in an Evolving World

Since the introduction of the PSS concept in 1989, the field has achieved significant advancements toward the realization of improved tools for urban planning. The advancements of this era include the invention and dissemination of new PSS such as CommunityViz (Kwartler & Bernard, 2001), CUF (Landis, 1994), Index (Allen, 2001), LEAM (Deal & Pallathucheril, 2008), UrbanSim (Waddell, 2002), and What If? (Klosterman, 1999, Chap. 2), and the development of urban models (Hunt et al., 2005) and geographic information system (GIS)-based analyses and models. Many of these pioneering PSS were highlighted in key edited volumes, which raised the visibility of the work beyond journal articles and professional projects to an international audience (Brail & Klosterman, 2001; Geertman & Stillwell, 2003; Brail, 2008). Work on complementary topics such as public participation GIS (PPGIS), e-participation, scenario planning, Geodesign, smart cities, and platform urbanism provided a rich intellectual backdrop during this era and constituted important intellectual strands complementing the pivotal role played by PSS.

Importantly, the era of technical development in PSS—the 1990s and early 2000s—was accompanied by a maturity of the PSS concept itself. In the initial, tentative discussions of Harris (1989) and Harris and Batty (1993), PSS emerged as an alternative to the top-down, technocratic use of urban models. Later, Klosterman (1997) introduced important conceptual developments by placing PSS in the context of long-standing planning debates and aligning it with the emerging emphasis on collaboration, pointing out the rich set of operational PSS tools and the growing body of knowledge that had come into existence (Klosterman & Pettit, 2005). Finally, Geertman (2006) brought a nuanced socio-technical perspective to PSS that highlighted the essential influence of factors to explain PSS use, such as the planning style, political context, and policy model operative in any given application context. Geertman's work was complemented by the pioneering research of Dutch collaborators who were also focusing on social questions like PSS adoption, performance, and usefulness, such as Vonk (Vonk & Geertman, 2008), te Brömmelstroet (2012), and Pelzer (2017). This progress was commemorated through the landmark publication of Geertman and Stillwell's (2020) *Handbook of Planning Support Science*, which synthesized this voluminous body of work, much of which appeared in CUPUM books over the years (Geertman et al., 2013, 2015, 2017, 2019, 2021; Goodspeed et al. 2023).

However, to the eyes of contemporary researchers, the early generation of PSS development must seem antiquated indeed. Data were often years out of date, hindering the potential for real-time analysis of urban change. Knowledge of human behavior or nuanced environmental conditions was limited to details gleaned from limited social surveys and rigid governmental monitoring programs. Visualization remained limited to GIS-derived maps or crude three-dimensional (3D) representations made possible with limited computing power. Artificial intelligence (AI)—tools that would allow us to go beyond the narrow analysis of existing data to mimic the human ability to generate new ideas and images—remained a long-unfulfilled academic dream.

The acceleration of the diffusion of information technology to nearly all facets of our society means we are living fully within a new *digital era* that could only be imagined in decades past. The digital mediation of a wide range of communications and transactions, the rise of new global digital infrastructures, even the proliferation of low-cost sensors and satellites, are fundamentally reshaping the landscape of what can be known about cities—and therefore what possibilities are unlocked for new PSS and urban analysis. In fact, these trends began in 2021, where we noted the term *urban informatics* was used to describe a book where big data was a major theme, given that 12 of the chapters involved big data analysis (Goodspeed et al., 2021). However, we noted a lack of engagement with then-new AI tools, which had not yet found their way into the field, let alone generated a more expansive engagement with new technologies that transcended traditional analysis.

We are proud to report that the body of research contained in this book—and reflected in the broader CUPUM community, which has played such an important role in the development of the global field of urban planning, management, and analytics—shows we are not standing still. In fact, most of the chapters bear the fingerprints of this new reality, demonstrating engagement with some of the most important aspects of the digital era in cities:

- Powerful new GenAI tools creating the potential for new planning and design tools (chapters "Rapid Development of a Cloud-Based Chatbot for Housing Policy", "Urban Design Analytics in the Age of Data Science and AI: Discovery, Measurement, Design, and Prediction", and "Daytime to Nighttime Street View Image Generation for 24-Hour Safety Perception Mapping");
- The creation of OpenStreetMap, a unique global crowdsourced geospatial dataset (used in chapters "Virtual Reality as a Collaborative Tool in Road Planning: Insights from a Case Study in Perth, Western Australia", "Towards Net Zero:

Converting Car Commutes to Bicycling in Australia", "Increasing Urban Greenery Through Reconfiguring Street Spaces to Mitigate Urban Heat Island Effects in Metropolitan Adelaide, Australia", and "Do Good Schools Influence Property Prices? Disentangling School and Neighbourhood Effects with Empirical Evidence from Brighton");

- The advent of Google Street View, a unique database of imagery of urban streets in many cities worldwide, has opened new avenues for urban analysis and design (chapter "Daytime to Nighttime Street View Image Generation for 24-Hour Safety Perception Mapping");
- The proliferation of low-cost satellite data introduces new forms of urban analysis (chapter "Gentrification from the Sky: Using Remote Sensing and Machine Learning for Urban Change Detection");
- The adoption of platform firms, such as ride-sharing services, generates unprecedented fine-grain data about urban mobility (chapter "A City's Response to Platformized Mobility: Emerging Interactions of Ride-Hailing Beyond Automobile Travel");
- Innovative sensor technologies allowing for flexible, nuanced analysis of urban environments (chapter "Sensing and Analysing Urban Heat Islands Using Internet of Things: A Real Time Microclimate Network for London"); and
- Ever-improving visualization tools allow for advancements in visual simulation (chapter "Virtual Reality as a Collaborative Tool in Road Planning: Insights from a Case Study in Perth, Western Australia").

Although these chapters contain diverse contributions and applications, in sum, they illustrate a transformation of urban life unfolding that has been much more expansive and profound than the sometimes-simplistic smart cities concept, which has faded from view. Nevertheless, some readers may be disappointed. Where are the whizzy control rooms, the magical PSS, the ubiquitous sensing, and the autonomous infrastructure systems that seemed just over the horizon for so many years? Of course, the sheer complexity and scale of urban systems, and the ways planning practice is deeply embedded in slowly changing institutions and policy frameworks, have made the pace of transformation iterative and often imperceptibly incremental.

For this reason, we decided to lead off the book with a section we call "Planning Support Systems in Evolution," which features chapters that are deeply engaged with the real world of practice and urban planning and decision-making, in many cases using technology to push practice forward. At its heart, the central question tackled by many PSS is how to not only provide *intelligence*, but also support *collective design*—"social interaction, interpersonal communication, and community debate that attempts to achieve collective goals and deal with common concerns" (Klosterman, 1997, p. 51). The remainder of the book is organized into two thematic sections, one on Urban Streets and Mobility Systems and the other on Neighborhoods and Communities, each containing a set of methodologically diverse investigations and applications related to these topical areas.

As much as this introduction has emphasized the connections between the chapters in this volume and the new technologies and PSS, we wish to also make note of the value of contributions that lack a strong link to either. Noteworthy are the papers whose innovation lies in substantive analysis of pressing problems. Improving cities does not always require new digital tools, but often rather the innovative application of existing methods to push the field forward toward greater scholarly understanding and to inform practice. This volume includes chapters on school quality, tsunamis, urban heat, and bicycling, each of which uses largely traditional methodological approaches to ask innovative and timely questions. We are therefore pleased to spotlight these contributions as an important strand in the broader field.

2 Chapter Overviews

Having set the context of CUPUM and the broader field of scholarship our chapters connect with, next we turn to a more detailed discussion of the specific chapters contained within. As noted, we chose to lead off with a section on PSS, given the importance of this topic to our scholarly community. Next, we turn to sections defined loosely around two topical categories: urban streets and mobility systems, and neighborhoods and communities.

2.1 Planning Support Systems in Evolution

This section contains five chapters, each contributing toward innovations and advancements in PSSs or planning practice. As noted, What If? was a pioneering PSS developed by Dick Klosterman that has stood the test of time. Chapter "A 20-Year Review of the What If? Planning Support System" by Yi Lu, Mohammad Qadeer Hussnain, Richard Klosterman, and Christopher Pettit, provides a review of the academic and professional use of this tool. This chapter describes how the tool found its way into a digital urban planning and research infrastructure in Australia and was eventually used in Pakistan, Iran, and Malaysia. Chapter "Bridging the Gap Between Tsunami Risk Awareness and Physical Vulnerabilities: A Data-Driven Approach for Disaster-Resilient Urban Planning" by Yuki Akiyama, Takuma Yamada, Kento Kokubo, Jin Horiuchi, and Chiaki M. Akiyama, reports on a creative project to integrate physical vulnerability with survey data about risk awareness, sharpening the perspective of which communities should be prompted to take action. Chapter "Virtual Reality as a Collaborative Tool in Road Planning: Insights from a Case Study in Perth, Western Australia" by Jonathan Pillai, Parisa Izadpanahi, and Shohreh Nematollahi, describes an innovative project to experiment with virtual reality as an engagement tool. Finally, chapter "Rapid Development of a Cloud-Based Chatbot for Housing Policy Enquiries" by Wayne Williamson, reports on a project that partnered with a planning agency to explore the creation of a chatbot to answer citizen questions. In sum, although these works are highly diverse, they all demonstrate the sustained engagement in PSS and planning practice by the CUPUM community to pioneer new technology-enabled forms of practice in the service of planning goals.

2.2 Urban Streets and Mobility Systems

This section contains six chapters that illustrate the increasing global interest in the analysis of urban streets and mobility systems. In chapter "A City's Response to Platformized Mobility: Emerging Interactions of Ride-Hailing Beyond Automobile Travel" Si Qiao, Fang Bian, and Anthony Gar-On Yeh report a novel analysis of ride-hailing users and general mobility data to explore nuanced questions of what types of destinations are especially influencing those modes. Parisa Izadpanahi, Scott N. Lieske, and Simone Z. Leao, in chapter "Towards Net Zero: Converting Car Commutes to Bicycling in Australia" provide a thoughtful and rigorous analysis that explores the possible greenhouse gas (GHG) reduction benefits of shifting short car commutes to bicycling in the Australian context, quantifying the benefits of mode shift. A chapter titled "Inferring On-Street Parking Occupancy with Smart Meter Data" by Daniel Jordon, Robert C. Hampshire, and Tayo Fabusuyi, demonstrates insights that can be gleaned from new sources of smart infrastructure system data. In chapter "Urban Design Analytics in the Age of Data Science and AI: Discovery, Measurement, Design, and Prediction" Stephen Law, Demin Hu, Xuhui Lin, Xiaowei Gao, Zicheng Fan, Yao Shen, and Tao Yang not only present an overarching framework for the potential use of analytical tools in the urban design process, but also provide a concrete example of how AI tools can be used to simulate changes to urban streets. Similarly, in chapter "Fingerprinting NYC's Scaffolding Problem with Longitudinal Dashcam Data" Dorin Ruinsky Shapira, Matt Franchi, and Wendy Ju report on an innovative analysis of dashcam data to identify the locations of construction scaffolding on New York City streets, informing a current debate about how such structures can be better regulated and managed. Finally, chapter "Development of a Method for Identifying the Presence of Front Roads Based on Building Coverage Ratio" by Takahito Shimizu, Kotaro Mizutani, Yuki Akiyama, Hiroki Baba, and Chiaki M. Akiyama, reports a clever analysis of urban data to glean new insights into urban form.

2.3 Neighborhoods and Communities

This section contains six chapters that analyze different facets of urban neighborhoods and communities. The chapter "Gentrification from the Sky: Using Remote Sensing and Machine Learning for Urban Change Detection" by Javier Alfaro, Sanja Šćepanović, Stephen Law, and Daniele Quercia, explores the use of remote sensing data to create new insights into urban change. The chapter "Increasing Urban Greenery Through Reconfiguring Street Spaces to Mitigate Urban Heat

Island Effects in Metropolitan Adelaide, Australia" by Jinhui Yu, Raziyeh Teimouri, and Andrew Allan, reports on a project that analyzed urban form to find opportunities to increase urban greenery to mitigate against urban heat. In the chapter "Sensing and Analysing Urban Heat Islands Using Internet of Things: A Real Time Microclimate Network for London" Dongyi Ma, Andrew Hudson-Smith, Martin De Jode, and Leah Lovett also focus on the problem of urban heat, this time through the use of novel real-time sensors that provide new microclimate data. The chapter "Daytime to Nighttime Street View Image Generation for 24-Hour Safety Perception Mapping" by Jiajing Dai, Zhiyi Liu, Tingting Li, Tianyi Ren, Waishan Qiu, Da Chen, and Wenjing Li, applies new visual AI tools to convert street view images (taken in daytime) to create simulations of nighttime scenes, to explore safety perceptions. The chapter "Spatial Variation Analysis of the Local Population Distribution During Urban Expansion and Contraction" by Kiichiro Kumagai, Kohei Uda, and Ayano Shiraishi, examines the phenomenon of *urban spongification* in Japan, a critical issue arising from the country's population decline. The chapter presents a method to identify such areas, analyzes how they have changed over time, and explores their economic impacts. Finally, the chapter "Do Good Schools Influence Property Prices? Disentangling School and Neighbourhood Effects with Empirical Evidence from Brighton" by Josiah Tana and Adam Dennett, reports on an analysis of a unique natural experiment where school enrollment policies were changed in Brighton, England, to improve our understanding of the relationship between school quality and property values.

3 Conclusions

Reflecting on the chapters, it is noteworthy to observe that the digital era has not heralded a conceptual transformation for our field, neither through the wholescale adoption of alternative rubrics like smart cities, urban science, urban analytics, or urban informatics, nor through the development of Geertman's proposed Planning Support Science. The chapters show that although we may be living in a digital era, it is also an era beset with many acute urban challenges, both new and old. Whether seeking to boost resilience to urban heat (chapters "Increasing Urban Greenery Through Reconfiguring Street Spaces to Mitigate Urban Heat Island Effects in Metropolitan Adelaide, Australia" and "Sensing and Analysing Urban Heat Islands Using Internet of Things: A Real Time Microclimate Network for London"), prepare for natural disasters like tsunamis (chapter "Bridging the Gap Between Tsunami Risk Awareness and Physical Vulnerabilities: A Data-Driven Approach for Disaster-Resilient Urban Planning"), grapple with urban decline (chapter "Spatial Variation Analysis of the Local Population Distribution During Urban Expansion and Contraction Processes"), or reduce GHG emissions (chapter "Towards Net Zero: Converting Car Commutes to Bicycling in Australia"), the chapters illustrate how CUPUM researchers are pioneering the use of digital tools and methods to help solve a wide number of urban challenges. Furthermore, we are heartened to see how the broadest concept of PSSs—as new tools for intelligence and collective design—remains broadly relevant as a unifying concept across so many cultural and national contexts. Although conceptually diverse, in sum the chapters illustrate a vibrant CUPUM community of researchers, who are eagerly exploring the possibilities created by the digital era to realize improvements to urban life.

References

Allen, E. (2001). INDEX: Software for community indicators. In R. Brail & R. Klosterman (Eds.), *Planning support systems: Integrating geographic information systems, models, and visualization tools* (pp. 229–261). ESRI Press.

Brail, R. K. (2008). *Planning support systems for cities and regions*. Lincoln Institute of Land Policy.

Brail, R. K., & Klosterman, R. E. (2001). *Planning support systems: Integrating geographic information systems, models, and visualization tools*. ESRI Press.

Deal, B., & Pallathucheril, V. (2008). Simulating regional futures: The land-use evolution and impact assessment model (LEAM). In R. Brail (Ed.), *Planning support systems for cities and regions* (pp. 61–84). Lincoln Institute for Land Policy.

Geertman, S. (2006). Potentials for planning support: A planning-conceptual approach. *Environment and Planning B Planning and Design, 33*(6), 863.

Geertman, S., & Stillwell, J. C. H. (2003). *Planning support systems in practice*. Springer.

Geertman, S., & Stillwell, J. (2020). *Handbook of planning support science*. Edward Elgar Publishing Limited.

Geertman, S., Toppen, F., & Stillwell, J. C. H. (2013). *Planning support systems for sustainable urban development* (Lecture notes in geoinformation and cartography). Springer.

Geertman, S., Ferreira, J., Goodspeed, R., Stillwell, J. (2015). Planning support systems and smart cities. Springer International Publishing Cham.

Geertman, S., Allan, A., Pettit, C., Stillwell, J. (2017). Planning support science for smarter urban futures. Springer International Publishing Cham.

Geertman, S., Zhan, Q., Allan, A., Pettit, C. (2019). Computational urban planning and management for smart cities. Springer International Publishing Cham.

Geertman S. C. M., Pettit, C., Goodspeed, R., Staffans, A. (2021). Urban informatics and future cities. Springer International Publishing Cham.

Goodspeed, R., Pettit, C., Staffans, A., & Geertman, S. (2021). Introduction. In S. Geertman et al. (Eds.), *Urban informatics and future cities*. Springer.

Goodspeed, R., Sengupta, R., Kyttä M., Pettit, C. (2023). Intelligence for future cities. Planning through big data and urban analytics. Springer Nature Switzerland Cham.

Harris, B. (1989). Beyond geographic information systems. *Journal of the American Planning Association, 55*(1), 85–90. http://www.informaworld.com/10.1080/01944368908975408

Harris, B., & Batty, M. (1993). Locational models, geographic information and planning support systems. *Journal of Planning Education and Research, 12*(3), 184.

Hunt, J. D., Kriger, D. S., & Miller, E. J. (2005). Current operational urban land-use–transport modelling frameworks: A review. *Transport Reviews, 25*(3), 329–376.

Klosterman, R. E. (1997). Planning support systems: A new perspective on computer-aided planning. *Journal of Planning Education and Research, 17*(1), 45–54. http://jpe.sagepub.com/content/17/1/45.abstractN2

Klosterman, R. E. (1999). The what if? Collaborative planning support system. *Environment and Planning B Planning and Design, 26*(3), 393–408.

Klosterman, R. E., & Pettit, C. J. (2005). An update on planning support systems. *Environment and Planning B: Planning and Design, 32*(4), 477–484.

Kwartler, M., & Bernard, R. N. (2001). CommunityViz: An integrated planning support system. In R. K. Brail & R. E. Klosterman (Eds.), *Planning support systems: Integrating geographic information systems, models, and visualization tools*. ESRI Press.
Landis, J. D. (1994). The California urban futures model—A new-generation of metropolitan simulation-models. *Environment and Planning B: Planning and Design, 21*(4), 399–420.
Pelzer, P. (2017). Usefulness of planning support systems: A conceptual framework and an empirical illustration. *Transportation Research Part A Policy and Practice, 104*, 84–95. https://doi.org/10.1016/j.tra.2016.06.019
te Brömmelstroet, M. (2012). Transparency, flexibility, simplicity: From buzzwords to strategies for real PSS improvement. *Computers, Environment and Urban Systems, 36*(1), 96–104. https://doi.org/10.1016/j.compenvurbsys.2011.06.002
Vonk, G., & Geertman, S. (2008). Improving the adoption and use of planning support systems in practice. *Applied Spatial Analysis and Policy, 1*(3), 153–173.
Waddell, P. (2002). UrbanSim—Modeling urban development for land use, transportation, and environmental planning. *Journal of the American Planning Association, 68*(3), 297–314.

Planning Support Systems in Evolution

A 20-Year Review of the What if? Planning Support System

Yi Lu, Mohammad Qadeer Hussnain, Richard Klosterman, and Christopher Pettit

Abstract This paper reviews the 20-year history of What if?, a widely used planning support system (PSS) in the field of intelligent urban planning and management. It reviews the literature on and features of several What if? PSS versions and reports on an online survey of What if? PSS users. This research concludes by analysing users' experience with the What if? PSS, and considering the implications for the future development of PSS.

Keywords What if? · Planning support systems · Qualitative analysis · Australian Housing Data Analytics Platform

1 Background and Introduction

Planning support systems (PSS) harness the power of big urban data through data-driven modelling approaches to support community-based land-use and transport planning (Pettit et al., 2018). In the field of urban planning and management, PSS help urban planners and policymakers simulate the planning of their cities more rigorously and reliably. Initially integrated with urban modelling approaches and visualization tools, PSS have been further expanded as a more comprehensive platforms with additional intelligent approaches and features. Along with the advancement of computer science (CS), geographic information system (GIS) and remote sensing (RS), the last two decades have witnessed an immense growth in PSS development (Klosterman & Pettit, 2005; Pelzer et al., 2014; Newton et al., 2024).

Y. Lu · C. Pettit
University of New South Wales, Sydney, NSW, Australia
e-mail: c.pettit@unsw.edu.au

M. Q. Hussnain
Better Cotton, Lahore, Pakistan

R. Klosterman (✉)
University of Akron, Akron, OH, USA

R. Goodspeed et al. (eds.), *Digital-Era Urban Transformations*, The Urban Book Series, https://doi.org/10.1007/978-3-031-98300-9_2

The What if? PSS is one of the earliest and most widely used GIS-based PSS. It operates with a rule-based model that projects future land use patterns by balancing the supply of and demand for different land uses at various temporal and spatial scales. Alternative visions for a city's future can be explored by considering different land use strategies and constraints. For example, a 'Business as Usual' scenario could assess the effects of continuing current development policies with relatively stable population and employment trends (Pettit et al., 2020a). In comparison, a 'Sustainable Development' scenario could incorporate sustainable development goals into the proposed urban development scenarios (Zhou et al., 2022; Martellozzo et al., 2018).

The What if? PSS adopts a 'bottom-up' modelling structure that begins with homogeneous land units called uniform analysis zones (UAZs), in contrast to traditional 'top-down' approaches that incorporate larger spatial units (Pettit et al., 2020a; Lu et al., 2024). UAZs are GIS-generated polygons that are uniform with respect to all selected factors. For instance, all points within a UAZ have identical slopes, are in the same administrative district, and have similar accessibility to a selected point of interest (POI).

A typical application of the What if? PSS consists of three modules: suitability, demand, and allocation. The suitability module uses a set of factors to consider the overall suitability of all UAZs for accommodating different land use demands. The demand module calculates the amount of land required to accommodate projected population and employment changes at a macro level. The allocation module generates future land use patterns by allocating the projected land use demands to the most suitable UAZs determined by the suitability module. The allocation module can also incorporate specified infrastructure plans and urban planning schemes by allocating projected land use demands to specific UAZs with permitted land use planning controls, and necessary infrastructure in a given projection year.

Three versions of the What if? PSS have been developed over the past 20 years, with a fourth version released in the year 2023 (Fig. 1). The first version, What if? 1.0 (Wi1), was developed in 1999 by Richard Klosterman (Klosterman, 1999), and released at the beginning of this century. What if? 1.0 used suitability, growth, and allocation procedures to project residential, commercial, and preservation land uses. Its output included reports and GIS maps that displaying current land use map, suitability outcomes, and projected land use layouts (Klosterman, 2001).

Based on this initial version, What if? 2.0 (Wi2) was subsequently developed by Richard Klosterman in 2008 (Klosterman, 2008). It was initially designed as a commercial package but was later distributed as a free online software package. Wi2 included all the functionality of Wi1, and was also capable of projecting the residential population, number of households, and sector-based employment for subregions within the study area. Wi2 has been downloaded by more than 1000 users from 20 countries. Wi1 and Wi2 were desktop applications that were programmed in Visual Basic and could be downloaded along with extensive user guides. Both

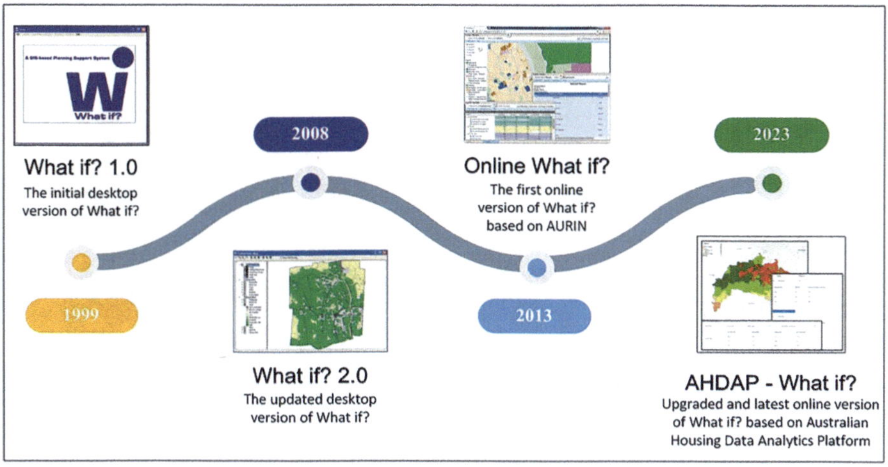

Fig. 1 History of the What if? PSS

versions were designed, documented, and funded by Richard Klosterman, with the assistance of a part-time professional programmer.

With the advancement of computer science, cloud- or web-based spatial data processing, management, and modelling have gradually become a major trend in information systems, including PSS (Xu et al., 2022). Under this trend, the Online What if? (OWI) was released in 2011. OWI was developed by Chris Pettit and Richard Klosterman, with the support of the Australian Urban Research Infrastructure Network (Sinnott et al., 2011).

Inheriting the policy-oriented characteristics of the What if? PSS series, OWI provides easier access to the application through cloud services, with efficient data storage and computing power, eliminating the need for local infrastructure (Russo et al., 2018). Furthermore, it imports multi-source spatial datasets, urban policy schemes, and explicit rules for determining the suitability of UAZs, projecting regional land use demands, and allocating these demands to UAZs efficiently (Pettit et al., 2015). Based on an online survey undertaken by the authors, OWI has been used by more than 250 international users in over 22 countries by the year of 2022.

The fourth and latest version of What if? has been developed as part of the 'Housing Analytics Workbench Tools' on the Australian Housing Data Analytics Platform (AHDAP). Following the 'bottom-up' structure of previous versions, this AHDAP-What if? Employs an optimized workflow and more user-friendly interfaces (Fig. 2, Table 1). The AHDAP-What if? also integrates with related analytical and visualization tools such as Colouring Australia (Hecht et al., 2023) and other Housing Analytics Workbench tools, providing a more comprehensive understanding of planning scenarios from different perspectives.

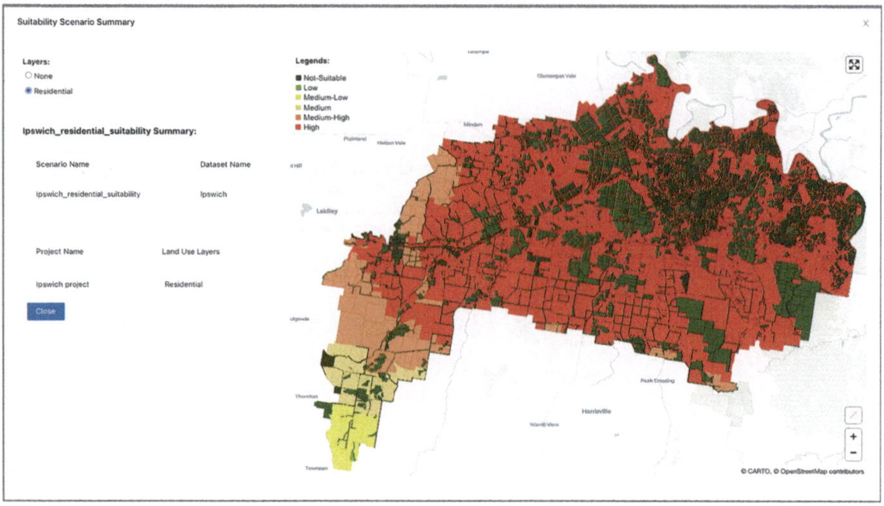

Fig. 2 Suitability evaluation outcome generated by the AHDAP-What if?

Table 1 Differences between the What if? PSS

	What if? 1.0 and 2.0	Online What if? (OWI)	AHDAP—What if?
System platform	Desktop	Web and cloud-based	
Data storage	ESRI shapefile	PostGIS database	
Function	Factor rating and weights used for suitability analyses		Only factor weights used for suitability analysis
	Driving factors are recorded in different units		Driving factors are normalized
Programming language	Visual Basic	Front-End: JavaScript Back-End: Java	Front-End: JavaScript Back-End: Python

2 Review of the What if? Scholarly Literature

The What if? related literature was identified with an initial search conducted on Google Scholar, using the keywords 'What if?', 'Whatif?', 'What if? Planning Support System', 'OWI', and 'Online What if?'; 980 publications were identified. After filtering duplicates and manual cleaning, 33 publications were identified that are directly related to the What if? PSS theme. The first What if? Publication (Klosterman, 1999) has been cited more than 500 times in Google Scholar and downloaded more than 2000 times on ResearchGate. What if? Papers have been published in Australia (ten), the United States (seven), Pakistan (six), the United Kingdom (three), Iran (three), and Mainland China, Hong Kong, Singapore, Italy, and Malaysia (one each) by 2023.

Additionally, a series of use cases are described in the Wi1- and Wi2-related literature. Eight of the cases are derived from Australia, four are from Pakistan, and the remaining ones are reported by researchers in Iran, Malaysia, and the United States. The first use case (Klosterman, 1999) describes the application of Wi2 in a declining region of the United States. Case studies of Wi2 in Australia include Hervey Bay (Pettit & Pullar, 1999, 2000; Pettit et al., 2002), Melbourne metropolitan area (Buxton et al., 2011), Mitchell Shire (Pettit et al., 2008), as well as a comparison between different visualizations for What if? Generated land use planning scenarios (Bishop et al., 2009). One of the three What if? Papers published in Iran were originally written in Persian (Ashnedar, 2000), while another Iranian application is described by Asgari et al. (2007) for evaluating alternative growth policies for sustainable urban growth. A Malaysian application (Adnan et al., 2014) projected the growth of Seberang Prai Utara under low, medium, and high growth scenarios.

Several OWI applications have been described in the literature, including Australian case studies in Western Sydney, New South Wales (Pettit et al., 2020b), Hervey Bay (Pettit et al., 2013), and Ipswich City (Lu et al., 2022) in Queensland, and the Perth-Peel region of Western Australia (Pettit et al., 2015, 2020a), respectively. Additionally, OWI applications have been reported in Pakistan. Hussnain et al. (2019) describe its use to prepare the Muzaffargarh Peri-Urban Plan and Dadu Master Plans, and Hussnain et al. (2020) developed the EasyUAZ package that facilitates data pre-processing and cleaning for OWI applications. OWI has also been applied at a regional scale to project the allocation of new residential dwellings under the 'Five Million Housing' program for the City of Faisalabad (Hussnain et al., 2021). These case studies have demonstrated the capability of the What if? PSS to produce objective and reliable projections of land use changes under various circumstances, highlighting its importance for urban planning and policy making.

This literature review illustrates different versions of the What if? PSS have been applied in both developed countries, like Australia and the United States, and developing countries such as Iran and Pakistan. It further reveals the generality of What if? PSS at different geographic scales, from a small, rapidly developing city to metropolitan regions.

3 What if? User Survey Results and Analysis

3.1 Respondents Profile

An online survey of What if? Users were conducted from October to December 2020 to better understand users' experience with multiple What if? PSS. The survey was sent to 1034 registered users of Wi2 or OWI; 79 responses were received. Both Wi2 and OWI have been operating for over a decade, so much of the users' contact information may be outdated. After removing duplicate entries from the same IP addresses and performing necessary data cleaning work, 43 valid responses were identified.

The survey respondents were distributed in 20 cities from 14 countries: Argentina, Australia, Canada, Croatia, Egypt, Ecuador, Iran, Italy, Mexico, Pakistan, Switzerland, Tanzania, the Netherlands, and the United States. Twenty-nine percent of the respondents had used Wi2, 33% had used OWI, and 38% had used both versions. Fifty-eight percent of the respondents worked in academic or research institutions, 16% worked in government or other public sectors, and 26% were from industry and private consulting organizations. More than half (58%) of the respondents were spatial planners, 21% worked as GIS professionals, 12% were engineers or worked in other domains, and 9% were researchers.

3.2 User Motivation and Model Uses

According to these responses, 32% and 35% of respondents were drawn to Wi2 and OWI, respectively, because they were open-source and freely available online. Additionally, 27% and 20% of Wi2 and OWI respondents selected the programs for their applicability to strategic planning. Another 27% and 20% of respondents, separately, chose the applications due to the availability of extensive documentation or frequent references in academic literature. Specifically, ease of online access was the main reason for 20% of OWI users. Smaller groups (14% and 5%) mentioned other attraction factors in their responses.

The largest proportion (37%) of the respondents accessed OWI via the AURIN website (https://whatif.aurin.org.au/whatif/). Twenty-six percent of respondents accessed Wi2 via the What if? Website. For the remaining respondents, 21% accessed What if? Through the Planning Support Systems for Pakistan website (https://citypulse.com.pk/pss); and 16% of the respondents accessed it through academic articles or industry publications. These results demonstrate the importance of providing an official website for both current and potential What if? Users.

As Fig. 3 indicates, the uses are comparable for Wi2 and OWI. Regarding Wi2, 39% of the respondents have explored Wi2's functions and options, compared to 38% for OWI. Fifteen percent and 21% of the Wi2 and OWI respondents used What if? to evaluate the tool. More of the respondents used Wi2 for teaching and class assignments than OWI (23% vs. 16%). On the other hand, OWI was used for more real-world projects (25%) compared to Wi2 (23%).

3.3 User Satisfaction with What if?

Despite the importance of evaluating PSS, relatively little attention has been paid to establishing clear evaluation standards. Pelzer (2017) has identified four proposed evaluation measurements: performance (Norton et al., 2015), effectiveness (Bujari

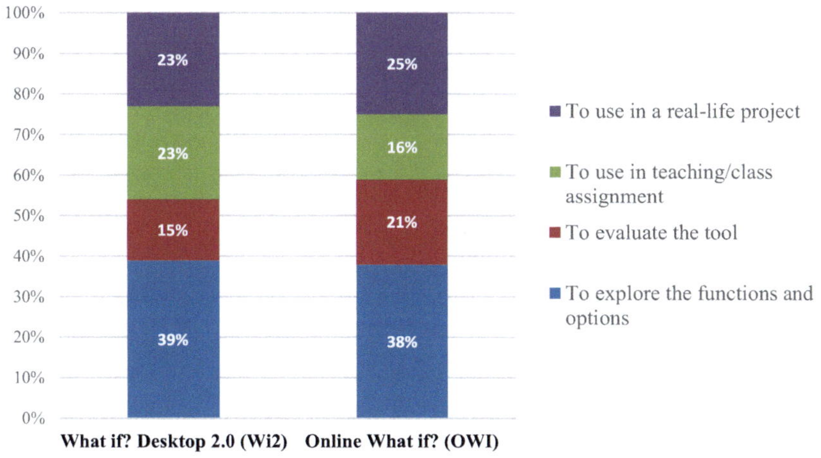

Fig. 3 Purpose for using the What if? PSS

et al., 2021), added value (Pettit et al., 2018), and usefulness (Zhang et al., 2019). The indicators for evaluating users' satisfaction with What if? PSS are based on these measurements. The respondents were also asked to provide their level of satisfaction on several dimensions of the What if? PSS on a 5-point scale, 'Highly satisfied', 'Satisfied', 'Neutral', 'Dissatisfied', and 'Highly Dissatisfied'.

As shown in Fig. 4, Wi2 was evaluated on seven indicators. The results illustrate that 75% of respondents were satisfied or highly satisfied with the 'Documentation (D)', and a quarter of them were neutral (15%) or highly dissatisfied (10%). In addition, 58% of respondents were neutral on 'User Support (U)', 16% were satisfied, and 26% were dissatisfied or highly dissatisfied. Forty percent of the respondents were satisfied or highly satisfied with the 'Graphical User Interface (G)', 39% were neutral, and 21% were dissatisfied or highly dissatisfied. Similarly, 48% of respondents were satisfied or highly satisfied with the 'Workflow Logic (W)', 31% neutral, and 21% were dissatisfied or highly dissatisfied. Forty-four percent of the respondents were satisfied or highly satisfied with the tool's 'Flexibility for Different Planning Tasks (F).' Forty-four percent were neutral, and 12% were dissatisfied or highly dissatisfied with the 'Calculation Time (C).' Sixty-three percent of the respondents were satisfied or highly satisfied with the 'Level of Detail' (L) provided in the system outputs, 21% were neutral, and 26% were dissatisfied or highly dissatisfied.

OWI was evaluated on six indicators (Fig. 5). Seventy-eight percent of the respondents were satisfied or highly satisfied with the 'Documentation (D)', 15% were neutral, and only 7% were dissatisfied or highly dissatisfied. Forty-eight percent of the respondents were either satisfied or highly satisfied with the OWI's 'User Support (U)', 39% hold a neutral attitude, and 13% were dissatisfied or highly dissatisfied. Respondents (75%) were satisfied or highly satisfied with the 'Graphical User Interface (G), 21% of respondents were neutral, and only 4% of the

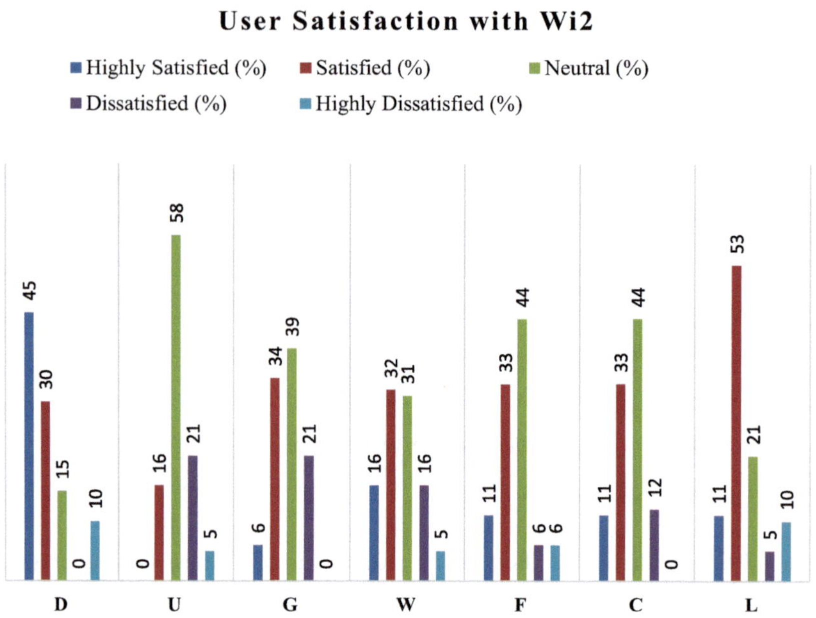

*D: Documentation, U: User Support, G: Graphical User Interface, W: Workflow Logic, F: Flexibility for Different Planning Tasks, C: Calculation Time, L: Level of Detail.

Fig. 4 User satisfaction with Wi2

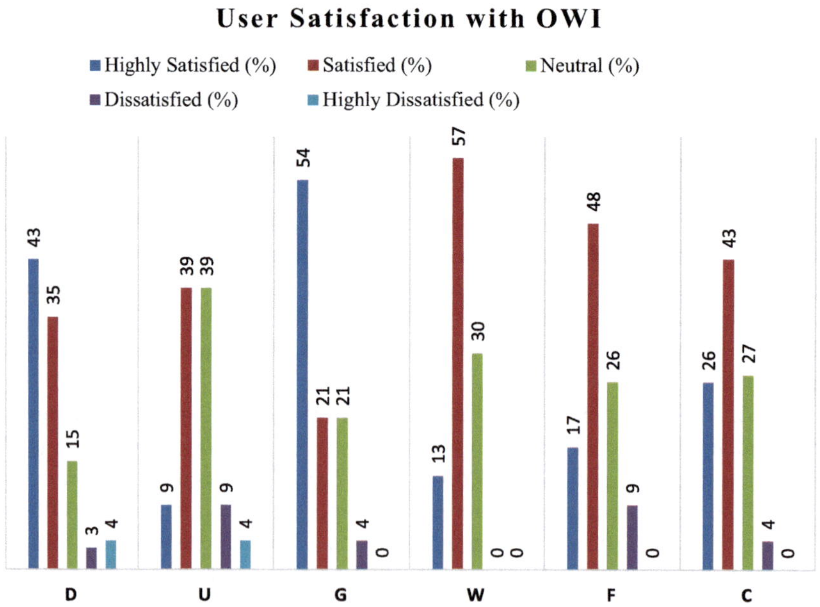

*D: Documentation, U: User Support, G: Graphical User Interface, W: Workflow Logic, F: Flexibility for Different Planning Tasks, C: Calculation Time.

Fig. 5 User satisfaction with OWI

respondents were not satisfied at all. Respondents (70%) were satisfied or highly satisfied with the 'Workflow Logic (W)', 30% were neutral, and none of the respondents were dissatisfied or highly dissatisfied. Respondents (65%) were satisfied or highly satisfied with the application's 'Flexibility for Different Planning Tasks (F)' 26% had a neutral opinion, and the remaining 9% of the respondents were dissatisfied. Respondents (69%) were either highly satisfied or satisfied with the 'Calculation Time' (C), 27% were neutral, and 4% were dissatisfied.

3.4 Analysis of the Survey Results

The survey results indicate that the respondents had generally positive perceptions of Wi2 and OWI, the two most widely used What if? Applications. Wi2 was recognized for its strong theoretical foundations, a reasonably large user community, and its support for personalized scenario planning. Its convenience for applied research with citizen groups was also highlighted as a major strength of Wi2. On the other hand, OWI was noted for its more efficient working procedure, ease of access, user-friendly GUI, and its focus on problem-solving capability for urban land management issues.

The survey results also identify several areas where What if? Could be improved. The main areas for enhancement in both Wi2 and OWI include their 'User Support (U)' and 'Flexibility for Different Planning Tasks (F)'. Several respondents called for more responsive and accessible support for Wi2 and suggested that the Wi2 workflow logic could be simplified. OWI was criticized for a lack of formal training programs and its low awareness in professional practice. Suggestions for improving OWI included enhancing its reliability, providing better error identification, and developing more detailed tutorials. Both Wi2 and OWI were criticised for their relatively limited application to different planning tasks, suggesting that expanding the What if? PSS to different application areas with various characteristics and features could further enhance their generality and utility.

4 Discussion and Conclusions

The 20-year history of the What if? PSS indicates that it has generated significant academic interest, with more than 30 research projects and publications. The literature review indicates that different versions of the What if? PSS series have been applied to study areas around the world at multiple spatial scales. The user survey reveals relatively high user satisfaction levels with the What if? PSS on several key indicators. This experience highlights the attractiveness of a theoretically sound, freely available, and widely applied PSS. However, the survey results also suggest that there is room for further improvements.

This history reveals that while the What if? PSS have proven to be valuable tools for academic research over the last 20 years; their broader adoption in professional practice remains a challenge. Two factors help explain its limited usage outside academia. The first is the 'technology fit' issue recognized for PSS (Pelzer et al., 2015). From this perspective, the What if? PSS have been designed primarily for study areas experiencing growth and development, making them less applicable for regions with stable or declining population and employment. In addition, the What if? PSS have been developed in academic environments with limited funding and have generally been available at no cost. This contrasts with commercial PSS such as CommunityViz (Kazak et al., 2013; Walker, 2017; Flacke et al., 2020), Envision Tomorrow Plus (ET+) (Abou Jaoude et al., 2022), UrbanFootPrint (Goodspeed, 2019), and UrbanSim (Jin & Lee, 2018) that received substantial startup grants and continued revenue streams that promoted their use in professional practice.

This experience also illustrates that while public organizations such as AURIN can provide the funding for the development of What if?, there remains a substantial funding gap for subsequent maintenance and user support for free, open-source PSS, especially for professional practice. The lack of continued funding will limit the continued development and application of PSS that bridge the gap from research and practice. As a result, further work is required to address issues of sustainable funding, technical fit, and user support.

References

Abou Jaoude, G., Mumm, O., & Carlow, V. M. (2022). An overview of scenario approaches: A guide for urban design and planning. *Journal of Planning Literature, 37*(3), 467–487. https://doi.org/10.1177/08854122221083546

Adnan, N. A., Omar, N. S. C., & Ismail, H. F. (2014). Utilizing What-if? 2.0 GIS planning support system for future potential land use development area simulations. In *2014 IEEE 10th International Colloquium on Signal Processing and Its Applications* (pp. 189–194). IEEE. https://doi.org/10.1109/CSPA.2014.6805745

Asgari, A., Klosterman, R., & Razani, A. (2007). Sustainable urban growth management using what-if? *International Journal of Environmental Research, 1*(3), 218–230.

Ashnedar, M. (2000). The "What If" System of Planning Support. *Scientific-Research Quarterly of Geographical Data, 9*(35), 51–52. https://www.sepehr.org/article_28622.html

Bishop, I., Pettit, C., Van, B., & Lynch, J. (2009). Visualising What if? Generated land use planning scenarios. In *11th International Conference on Computers in Urban Planning and Urban Management (CUPUM), Hong Kong, China*, 2009 (pp. 16–18).

Bujari, A., Calvio, A., Foschini, L., Sabbioni, A., & Corradi, A. (2021). A digital twin decision support system for the urban facility management process. *Sensors, 21*(24), 8460. https://doi.org/10.3390/s21248460

Buxton, M., Butt, A., Farrell, S., & O'Neill, D. J. (2011). *Scenario planning for Melbourne's peri-urban region* (Vol. 180). RMIT University.

Flacke, J., Shrestha, R., & Aguilar, R. (2020). Strengthening participation using interactive planning support systems: A systematic review. *ISPRS International Journal of Geo-Information, 9*(1), 49. https://doi.org/10.3390/ijgi9010049

Goodspeed, R. (2019). *Scenario planning: Embracing uncertainty to make better decisions.* Lincoln Institute of Land Policy. Policy Brief. JSTOR.

Hecht, R., Danke, T., Herold, H., Hudson, P., Munke, M., & Rieche, T. (2023). Colouring cities: A citizen science platform for knowledge production on the building stock-potentials for urban and architectural history. In *Workshop on research and education in urban history in the age of digital libraries* (pp. 145–164). Springer.
Hussnain, M. Q., Waheed, A., Wakil, K., Tahir, A., Pettit, C., Pelizaro, C., & Jabbar, J. A. (2019). *A planning support system to aid spatial planning in Pakistan.* Paper presented at the 15th International Conference on Computers in Urban Planning and Urban Management, Adelaide, Australia.
Hussnain, M. Q., Waheed, A., Wakil, K., Jabbar, J. A., Pettit, C. J., & Tahir, A. (2020). Evaluating a workflow tool for simplifying scenario planning with the online whatif? Planning support system. *ISPRS International Journal of Geo-Information, 9*(12), 706. https://doi.org/10.3390/ijgi9120706
Hussnain, M. Q., Wakil, K., Waheed, A., Ahmad, M., & Pettit, C. J. (2021). *Where to build the five million houses? A regional-scale application of the What If? Planning support system.* Paper presented at the the 17th International Conference on CUPUM—Computational Urban Planning and Urban Management, Helsinki, Finland
Jin, J., & Lee, H.-Y. (2018). Understanding residential location choices: An application of the UrbanSim residential location model on Suwon, Korea. *International Journal of Urban Sciences, 22*(2), 216–235.
Kazak, J., Szewrański, S., & Decewicz, P. (2013). Indicator-based assessment of land use planning in Wrocław Region with CommunityViz. In *Proceedings REAL CORP* (pp. 1248–1251).
Klosterman, R. E. (1999). The what if? Collaborative planning support system. *Environment and Planning B: Planning and Design, 26*(3), 393–408. https://doi.org/10.1068/b260393
Klosterman, R. E. (2001). *Paper 10 the what if? Planning support system. Planning support systems-integrating geographic information systems, models, and visualization tools.* ESRI Press.
Klosterman, R. E. (2008). *A new tool for a new planning: The What If? TM planning support system. Planning support systems for cities and regions.* Lincoln Institute of Land Policy.
Klosterman, R. E., & Pettit, C. J. (2005). *An update on planning support systems.* SAGE Publications Sage UK.
Lu, Y., Laffan, S., & Pettit, C. (2022). The integration of cellular automata and what if? For scenario planning: Future residential expansion in the City of Ipswich. *ISPRS Annals of the Photogrammetry, Remote Sensing and Spatial Information Sciences, 10,* 151–158. https://doi.org/10.5194/isprs-annals-X-4-W3-2022-151-2022
Lu, Y., Laffan, S., & Pettit, C. (2024). Coupling cellular automata and what if? Models for residential expansion simulation: A case study of Southwest Sydney, Australia. *Transactions in GIS, 28*(6), 1465–1485. https://doi.org/10.1111/tgis.13198
Martellozzo, F., Amato, F., Murgante, B., & Clarke, K. (2018). Modelling the impact of urban growth on agriculture and natural land in Italy to 2030. *Applied Geography, 91,* 156–167. https://doi.org/10.1016/j.apgeog.2017.12.004
Newton, P., Pettit, C., Barr, S., & Bruns, L., Jr. (2024). Digital innovations for city sustainability analysis and decision-making. In *Future cities making: Mission-oriented research for urban sustainability transitions in Australia* (pp. 215–240). Springer. https://doi.org/10.1007/978-981-97-7671-9_10
Norton, B. A., Coutts, A. M., Livesley, S. J., Harris, R. J., Hunter, A. M., & Williams, N. S. (2015). Planning for cooler cities: A framework to prioritise green infrastructure to mitigate high temperatures in urban landscapes. *Landscape and Urban Planning, 134,* 127–138. https://doi.org/10.1016/j.landurbplan.2014.10.018
Pelzer, P. (2017). Usefulness of planning support systems: A conceptual framework and an empirical illustration. *Transportation Research Part A: Policy and Practice, 104,* 84–95. https://doi.org/10.1016/j.tra.2016.06.019

Pelzer, P., Geertman, S., van der Heijden, R., & Rouwette, E. (2014). The added value of planning support systems: A practitioner's perspective. *Computers, Environment and Urban Systems, 48*, 16–27. https://doi.org/10.1016/j.compenvurbsys.2014.05.002
Pelzer, P., Arciniegas, G., Geertman, S., & Lenferink, S. (2015). Planning support systems and task-technology fit: A comparative case study. *Applied Spatial Analysis and Policy, 8*, 155–175. https://doi.org/10.1007/s12061-015-9135-5
Pettit, C., & Pullar, D. (1999). Planning scenarios for the growth of Hervey Bay. In *6th International Conference on Geocomputation, Brisbane, Queensland*, 1999.
Pettit, C., & Pullar, D. (2000). Towards incorporating dynamic consequences into regional planning scenarios. In *AURISA*.
Pettit, C., Pullar, D., & Stimson, R. (2002). An integrated multi-scaled decision support framework used in the formulation and evaluation of land-use planning scenarios for the growth of Hervey Bay.
Pettit, C., Keysers, J., Bishop, I., & Klosterman, R. (2008). Applying the what if? Planning support system for better understanding urban fringe growth. In *Landscape analysis and visualisation* (pp. 435–454). Springer. https://doi.org/10.1007/978-3-540-69168-6_21
Pettit, C. J., Klosterman, R. E., Nino-Ruiz, M., Widjaja, I., Russo, P., Tomko, M., Sinnott, R., & Stimson, R. (2013). The online what if? Planning support system. In *Planning support systems for sustainable urban development* (pp. 349–362). Springer. https://doi.org/10.1007/978-3-642-37533-0_20
Pettit, C. J., Klosterman, R. E., Delaney, P., Whitehead, A. L., Kujala, H., Bromage, A., & Nino-Ruiz, M. (2015). The online what if? Planning support system: A land suitability application in Western Australia. *Applied Spatial Analysis and Policy, 8*(2), 93–112. https://doi.org/10.1007/s12061-015-9133-7
Pettit, C., Bakelmun, A., Lieske, S. N., Glackin, S., Thomson, G., Shearer, H., Dia, H., & Newman, P. (2018). Planning support systems for smart cities. *City, Culture and Society, 12*, 13–24. https://doi.org/10.1016/j.ccs.2017.10.002
Pettit, C., Biermann, S., Pelizaro, C., & Bakelmun, A. (2020a). A data-driven approach to exploring future land use and transport scenarios: The online what if? Tool. *Journal of Urban Technology, 27*(2), 21–44.
Pettit, C., Shi, Y., Han, H., Rittenbruch, M., Foth, M., Lieske, S., van den Nouwelant, R., Mitchell, P., Leao, S., & Christensen, B. (2020b). A new toolkit for land value analysis and scenario planning. *Environment and Planning B: Urban Analytics and City Science, 47*(8), 1490–1507. https://doi.org/10.1177/2399808320924676
Russo, P., Lanzilotti, R., Costabile, M. F., & Pettit, C. J. (2018). Towards satisfying practitioners in using planning support systems. *Computers, Environment and Urban Systems, 67*, 9–20. https://doi.org/10.1016/j.compenvurbsys.2017.08.009
Sinnott, R., Galang, G., Tomko, M., & Stimson, R. (2011). *Australian urban research infrastructure network*. Paper presented at the eResearch Australasia Conference 2011, Sebel Albert Park, Melbourne, Australia.
Walker, D. (2017). *The planners guide to CommunityViz: The essential tool for a new generation of planning*. Routledge.
Xu, C., Du, X., Fan, X., Giuliani, G., Hu, Z., Wang, W., Liu, J., Wang, T., Yan, Z., & Zhu, J. (2022). Cloud-based storage and computing for remote sensing big data: A technical review. *International Journal of Digital Earth, 15*(1), 1417–1445. https://doi.org/10.1080/17538947.2022.2115567
Zhang, L., Geertman, S., Hooimeijer, P., & Lin, Y. (2019). The usefulness of a web-based participatory planning support system in Wuhan, China. *Computers, Environment and Urban Systems, 74*, 208–217. https://doi.org/10.1016/j.compenvurbsys.2018.11.006
Zhou, M., Ma, Y., Tu, J., & Wang, M. (2022). SDG-oriented multi-scenario sustainable land-use simulation under the background of urban expansion. *Environmental Science and Pollution Research, 29*(48), 72797–72818. https://doi.org/10.1007/s11356-022-20904-9

Bridging the Gap Between Tsunami Risk Awareness and Physical Vulnerabilities: A Data-Driven Approach for Disaster-Resilient Urban Planning

Yuki Akiyama, Takuma Yamada, Kento Kokubo, Jin Horiuchi, and Chiaki M. Akiyama

Abstract This study introduces a novel approach to integrating tsunami risk awareness (TRA) and physical risk assessments by tsunami at the subdivision (Oaza) level, offering a comprehensive spatial analysis for disaster-resilient urban and regional planning. Using detailed tsunami inundation data and TRA data from a large-scale questionnaire survey, it quantifies the gap between perceived tsunami risk and actual physical vulnerabilities in parts of Japan's Pacific and Seto Inland Sea coasts. Geographic information systems (GIS) were employed to spatially aggregate these datasets, identifying regions where low TRA and high physical risk indicate critical weaknesses in disaster preparedness. Notable areas, such as parts of the Nōbi Plain and Kojima Bay, show high tsunami risk but low resident awareness, necessitating urgent interventions. This study supports disaster risk reduction by providing actionable insights for policymakers and urban planners to enhance structural defenses and public awareness, emphasizing the integration of community awareness into spatial risk assessments.

Keywords Tsunami risk awareness (TRA) · Physical vulnerability · Disaster preparedness · Spatial analysis · Risk visualization

---

Y. Akiyama (✉) · T. Yamada
Department of Urban and Civil Engineering, Faculty of Engineering, Tokyo City University, Setagaya-ku, Tokyo, Japan
e-mail: akiyamay@tcu.ac.jp; g2132101@tcu.ac.jp

K. Kokubo · J. Horiuchi
Graduate School of Integrative Science and Engineering, Tokyo City University, Setagaya-ku, Tokyo, Japan
e-mail: g2481618@tcu.ac.jp; g2481642@tcu.ac.jp

C. M. Akiyama
Faculty of Engineering, Reitaku University, Kashiwa, Chiba, Japan
e-mail: chakiyam@reitaku-u.ac.jp

R. Goodspeed et al. (eds.), *Digital-Era Urban Transformations*, The Urban Book Series, https://doi.org/10.1007/978-3-031-98300-9_3

1 Introduction

Tsunamis, caused by earthquakes or volcanic activity, inflict catastrophic damage on coastal regions and pose significant challenges for urban and regional planning. Major recent examples include the 2004 Indian Ocean Earthquake (Mw9.1) and the 2011 Great East Japan Earthquake (Mw9.0). The 2004 tsunami devastated coastal cities in Indonesia, Sri Lanka, and Thailand, killing over 220,000 people (UNESCO, 2020). Similarly, the 2011 tsunami severely impacted Japan's Tohoku region, with approximately 18,000 fatalities (Fire and Disaster Management Agency of Japan, 2021).

Historically, tsunamis have caused extensive destruction. The 1896 Meiji Sanriku Earthquake (Mw8.5–8.6) killed around 22,000 people in Japan's Tohoku region. The 1946 Aleutian Islands Earthquake (Mw8.6) led to 159 deaths in Hawaii (Mori et al., 2022). The 1960 Chile Earthquake (Mw9.5) generated tsunamis across the Pacific, resulting in 57 deaths in Hawaii and 139 fatalities in Japan. These events demonstrate the recurring nature and destructive potential of tsunamis globally, emphasizing their persistent threat to human life.

1.1 Efforts in Urban and Regional Planning to Mitigate Tsunami Damage

Tsunami disasters caused by seismic or volcanic events underscore the need for advanced prevention strategies. These disasters result in significant human losses and demand revisions in urban and regional planning. In response, measures such as enhanced evacuation plans, early warning systems (implemented in Japan, Taiwan, Chile, and parts of the USA), and improved tsunami-resistant infrastructure like seawalls, levees, forests, and evacuation towers have been adopted (Takemoto et al., 2021; Santos & Isidro, 2020). These strategies are key to disaster prevention and recovery efforts.

Urban and regional planning in vulnerable coastal areas is crucial in reducing tsunami risk. Developing mitigation infrastructure and evacuation routes is vital for post-disaster reconstruction (Koshimura & Shuto, 2014). Strengthening land-use regulations and reorganizing urban plans, as seen after the 1946 Aleutian and 1960 Chile earthquakes, has proven effective (Dengler, 1998). Following the 2011 Great East Japan Earthquake, Japan relocated residential areas to higher ground. In anticipation of the Nankai Trough Megaquake, pre-disaster relocations and office relocations have begun (Takemura et al., 2023). These initiatives are essential to regional planning and contribute to building tsunami-resistant, resilient cities.

1.2 Limits of Modern Infrastructure Exposed by Tsunami Disasters and Lessons Learned

Efforts to reduce tsunami damage have increased globally, especially after the 2004 Indian Ocean and 2011 Great East Japan earthquakes. These disasters exposed the limits of modern infrastructure in combating tsunamis, with Japan's advanced defenses like seawalls failing against the massive waves, leading to significant casualties. These events demonstrate that technology alone cannot fully prevent natural disaster risks (Koshimura et al., 2014).

The key lesson from these disasters is that infrastructure alone is insufficient for disaster prevention. Urban and regional planning must also prioritize non-structural measures, such as improved evacuation planning and early warning systems. Tools like risk assessments and hazard mapping have proven critical in minimizing damage (Wood & Schmidtlein, 2016). Following the 2011 disaster, predicting tsunami inundation zones using GIS and optimizing evacuation routes became central to disaster planning in affected areas (Yamashita & Fujita, 2012). These approaches are essential for enabling rapid and effective evacuations.

Therefore, tsunami risk reduction requires not just infrastructure but also societal efforts, including raising public awareness, regular evacuation drills, and community-led disaster activities. These are essential for creating more resilient and sustainable cities (Kitagawa, 2019; Santos & Isidro, 2020).

1.3 The Role of Public Awareness in Reducing Tsunami Damage

Strengthening tsunami resilience requires more than infrastructure; a comprehensive, societal-level approach to disaster prevention is essential. Public awareness of disaster preparedness is critical, as it ensures residents understand local risks and take appropriate actions, directly impacting casualty rates.

In areas like Aceh, Indonesia, and Japan's Tohoku region, increasing public awareness has proven effective. After the 2004 Indian Ocean Earthquake, regular evacuation drills in Aceh significantly improved disaster resilience (Sufri et al., 2020). Similarly, post-2011 education in Japan has enhanced tsunami risk awareness and early evacuation behaviors (Matsumaru & Kawaguchi, 2015).

In high-risk areas, understanding tsunami threats: Tsunami Risk Awareness (TRA), is crucial for timely evacuation. Individual recognition of danger, alongside warning systems, is key to saving lives (Kitagawa, 2019). As seen in the 2004 and 2011 earthquakes, advanced infrastructure alone cannot prevent disaster without prompt evacuation. Assessing TRA is vital to reducing casualties (Amri et al., 2024).

1.4 Objective of This Study

Tsunami inundation depths and resulting building damage have been extensively simulated, enabling quantitative predictions (Koshimura, 2016). Open data on tsunami inundation depths in Japan is publicly available, providing insight into physical risk. However, measuring Tsunami Risk Awareness (TRA) quantitatively remains challenging. TRA data, gathered through interviews with residents (Salah & Sasaki, 2021) and surveys (Cerase et al., 2019), has been limited in scope due to technical and financial constraints.

Understanding TRA on a large scale is crucial for governments. High TRA regions are likely to take appropriate action, reducing damage, whereas areas with low TRA and high physical risk may face delayed evacuations and increased casualties. Identifying such gaps before a disaster and implementing measures can efficiently mitigate overall damage (Fig. 1).

This study assesses these gaps by evaluating both physical risk and TRA, enabling governments to focus on regions with low TRA for non-structural interventions such as awareness campaigns and evacuation drills. This comprehensive TRA analysis also contributes to improving urban and regional planning for tsunami prevention.

To achieve the objectives of this study, data covering a wide area, including anticipated tsunami inundation depths, building distribution, and residents' TRA, is required. As mentioned earlier, obtaining wide-area TRA data has been extremely

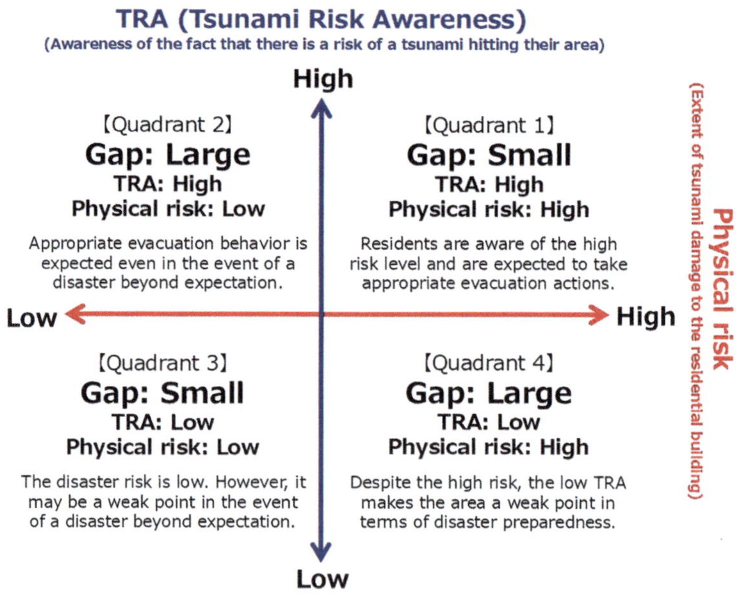

Fig. 1 Illustration of regional classification based on the magnitude of physical risk (horizontal axis) and TRA (vertical axis)

challenging. However, this study breaks through this barrier by utilizing the "livability data", introduced in Sect. 3, marking the first time that wide-area TRA has been captured in Japan.

2 Study Area

The study area comprises Shizuoka, Aichi, Mie, Hyogo, and Okayama Prefectures in Japan (Fig. 2). The Pacific coastal regions of Shizuoka, Aichi, and Mie are predicted to face the greatest impact from the Nankai Trough Megaquake, with tsunamis up to 20 m high expected, threatening national economic stability (Nanjo & Yoshida, 2018). Surveys show that residents in these areas tend to have higher levels of Tsunami Risk Awareness (TRA) (Harada et al., 2014), although low-TRA pockets could still exist, posing disaster preparedness challenges.

In addition, Inland areas, despite perceiving lower risks, can also face significant threats. For instance, during the Great East Japan Earthquake, the tsunami reached 4.5 km inland, greatly increasing casualties (Goto et al., 2011), showing that even areas with low TRA may still be vulnerable.

Hyogo and Okayama Prefectures, which face inland seas, are expected to have lower TRA compared to the national average (Hirano et al., 2023). During the 2011 earthquake, despite tsunami advisories, few residents in these areas evacuated

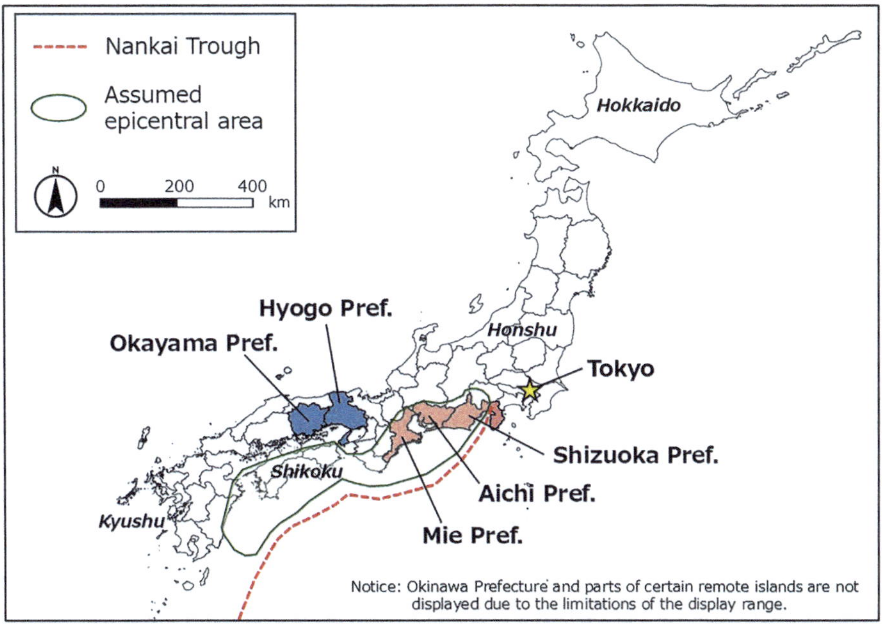

Fig. 2 Study area of this study and the assumed source region of the Nankai Trough Megaquake

(Lewis, 2011). Although located farther from the earthquake's epicenter, some coastal regions still face potential tsunami risks up to 5 m, indicating that low-TRA areas with significant physical risk could exist.

Given these considerations, these prefectures were deemed suitable for demonstrating the methods explored in this study.

3 Data and Study Flow

This chapter introduces the data used in this study and explains the study flow using these data.

3.1 Data

3.1.1 TRA Data

The key data for this study is residents' Tsunami Risk Awareness (TRA), which measures awareness of local tsunami risks and directly affects evacuation behavior and casualty reduction. However, collecting comprehensive TRA data across wide regions poses technical and financial challenges.

This study addresses these challenges by using data from the "Good Room Net Livability Ranking" survey, conducted by Daito Trust Construction Co., Ltd., in collaboration with Macromill, Inc. Covering all 47 prefectures of Japan and collecting approximately 800,000 responses by 2023, this survey evaluates resident satisfaction and well-being. It publishes annual rankings of the most livable municipalities in Japan (Daito Trust Construction, 2023).

The survey's strength lies in gathering subjective data on perceptions such as "familiarity," "safety," and "public services," which are aggregated to assess overall "livability." Due to the large sample size, the data is statistically reliable and widely used in academic research (So, 2021).

Among these subjective evaluations is a measure of disaster preparedness, including perceptions of tsunami risk. This study uses the "Subjective low concern for tsunami risk" data as TRA, evaluated on a five-point scale. In this study, each region's TRA score is calculated using the average of these evaluations as follows: very satisfied (5), satisfied (4), neutral (3), dissatisfied (2), and very dissatisfied (1).

3.1.2 Tsunami Inundation Depth Data

This study utilized the Tsunami Inundation Assumption Data from the National Land Numerical Information (MLIT, 2024) to analyze tsunami inundation areas and depths. This data is based on past tsunami records and predicted earthquake scenarios, including the Nankai Trough Megaquake and the Tokyo Inland Earthquake. It provides polygon data for each prefecture, showing assumed inundation areas and maximum depths from tsunami simulations. These simulations consider local topography, geology, and the extent of inundation. The estimated tsunami inundation depths are classified into the following six categories:

- ~0.3 m
- 0.3–1.0 m
- ~2.0 m
- 2.0–3.0 m
- 3.0–5.0 m (Only Okayama Prefecture: 3.0–4.0 m)
- 5.0–10.0 m (Only Shizuoka, Aichi, and Mie prefectures)

These classifications of runup heights are determined based on fragility curves proposed in existing studies that evaluate the potential impacts on infrastructure and buildings. As described later in Sect. 4.2, fragility curves have been developed to assess the damage to infrastructure and buildings based on tsunami inundation depths. By integrating these fragility curves with the estimated runup heights, a quantitative framework is provided to evaluate the magnitude of loss and damage caused by tsunamis.

3.1.3 Building Polygon Data

In this study, detailed building data were essential for assessing tsunami impacts on structures. We utilized building polygon data from the "Residential Map" provided by Zenrin Co., Ltd. This map is created through on-site surveys, enabling precise identification of individual buildings, their locations, shapes, and usage. It is considered one of the most detailed local maps in Japan. Although a paid resource, it is widely used in postal services, real estate, logistics, and disaster prevention. For this study, we only used buildings designated for residential use, including detached houses and apartment buildings, to assess the damage to residential buildings caused by tsunamis.

3.2 Analytical Unit

The TRA data used in this study includes residents' home locations, not just at the municipality level but also at the more granular "Oaza" level. Oaza is a subdivision within Japanese municipalities, similar to townships, districts, or parishes in other countries. Typically, Oaza names derive from historical villages, reflecting regional culture and history, making them crucial for local geographic awareness and disaster planning (Mori, 2013). While their use has declined in urban areas, Oaza divisions remain significant in rural regions and are key for developing evacuation routes and localized risk assessments (Kim et al., 2015). They play a vital role in urban development and disaster management, especially when accounting for geographic features and historical land use (Nakatsuka et al., 2009).

Despite their importance, spatial data like Oaza polygons is not freely available. However, open data for smaller units, known as "Chōchōaza" (subregions), is accessible. This study merged Chōchōaza units using GIS to create Oaza polygon data for analysis.

3.3 *Study Flow*

The study process is outlined in Fig. 3. First, the TRA data (Sect. 3.1.1) is aggregated at the Oaza level using the Oaza polygon data from Sect. 3.2 to assess the TRA of each Oaza (Sect. 4.1). This result will quantify the qualitative TRA data shown in Fig. 1. Next, the tsunami inundation depth data (Sect. 3.1.2) and building polygon data (Sect. 3.1.3) are integrated to estimate the inundation depth for each building. These results are then aggregated at the Oaza level to evaluate the proportion of buildings affected by tsunami inundation (Sect. 4.2). This will serve as an indicator to assess the physical risk shown in Fig. 1. Finally, combining these results for each Oaza will reveal areas with preparedness gaps and weak points (Sect. 5).

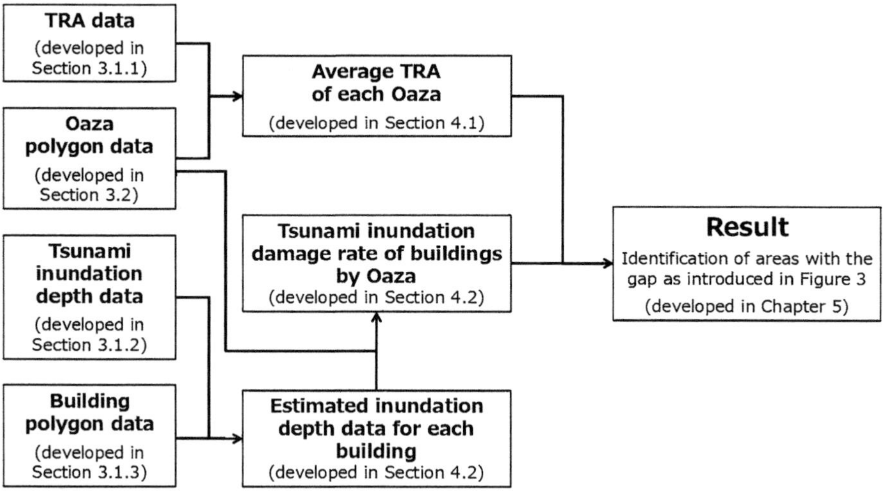

Fig. 3 Study flow

4 Data Preparation for Analysis

This chapter introduces the method for preparing the data used in the analysis.

4.1 Evaluation of TRA at the Oaza Level

We developed a dataset to assess TRA at the Oaza level by calculating the average score of respondents in each Oaza. The average TRA was classified into four categories, with higher numbers of categories indicating greater awareness of tsunami risk:

- ≥0 and <1: High concern (Category 4)
- ≥1 and <2: Moderately high concern (Category 3)
- ≥2 and <3: Moderately low concern (Category 2)
- ≥3: Low concern (Category 1)

In areas with fewer than five respondents, data reliability issues arise due to small sample sizes and privacy concerns (Toronto Metropolitan University, 2015). Therefore, we limited our analysis to Oaza with at least five respondents.

Figure 4 shows the average TRA at the Oaza level across the study area's prefectures. In Shizuoka, where significant damage from the Nankai Trough Megaquake is expected, many Oaza along the Pacific coast were categorized as Category 4, indicating high concern. Similar trends were found along the Pacific coasts of Aichi and Mie Prefectures, as well as in areas along Ise Bay. These results suggest that residents in these regions have a heightened awareness of tsunami risks. Inland areas of Shizuoka and Aichi also had Oaza classified as Category 3 or higher (areas a, b, and c in Fig. 4), reflecting the risk of tsunami inundation through plains and rivers.

Conversely, in Hyogo and Okayama Prefectures, even coastal Oaza along the Seto Inland Sea, where tsunamis up to 5 m are predicted, showed fewer Category 4 classifications. Most areas were categorized as Category 3, and some coastal Oaza along the Seto Inland Sea were classified as Category 2 (area d in Fig. 4). These findings reveal significant differences in TRA depending on whether regions directly face the Pacific Ocean.

4.2 Evaluation of Physical Risk at the Oaza Level

We assessed physical risk at the Oaza level by integrating building polygon data with tsunami inundation depth data using GIS. Each building was assigned the largest inundation depth polygon it overlapped with. We then calculated the proportion of buildings in each Oaza likely to be affected by tsunamis, defining this as

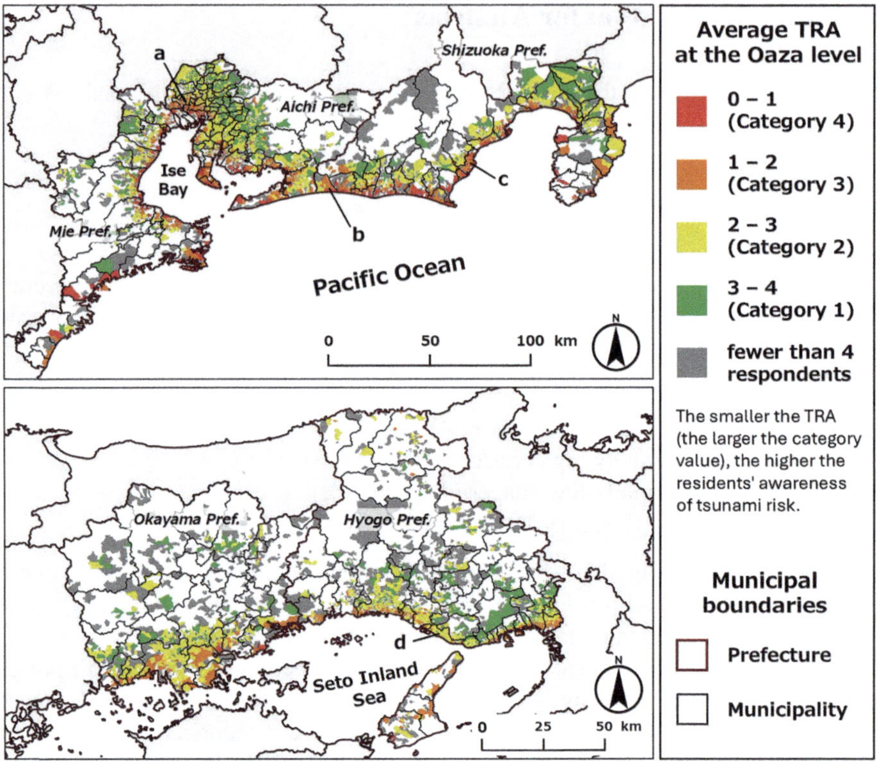

Fig. 4 TRA by Oaza across the entire study area

"physical risk." Based on evaluations from the Great East Japan Earthquake, fragility curves have been proposed, indicating that wooden residential buildings face severe damage or collapse at depths exceeding 1 m, while older buildings are likely to sustain significant damage at depths over 50 cm (Rossetto et al., 2014; Suppasri et al., 2014). Additionally, depths of 30 cm or more pose evacuation challenges and debris-related injury risks (NOAA, 2023). Therefore, in this study, buildings with an estimated inundation depth of 0.3 m or more were considered at risk.

We classified physical risk into four categories using Jenks natural classification:

- ≥50%: High physical risk (Category 4)
- ≥25% and <50%: Moderately high physical risk (Category 3)
- >0% and <25%: Low physical risk (Category 2)
- 0%: No physical risk (Category 1)

Figure 5 shows the distribution of physical risk by Oaza across prefectures. Along the Pacific coast, Oaza in Categories 3 and 4 are continuously distributed, extending inland in the Nōbi Plain, where the Kiso Three Rivers converge, heightening the risk (Fig. 5a). In contrast, along the Seto Inland Sea, there are fewer

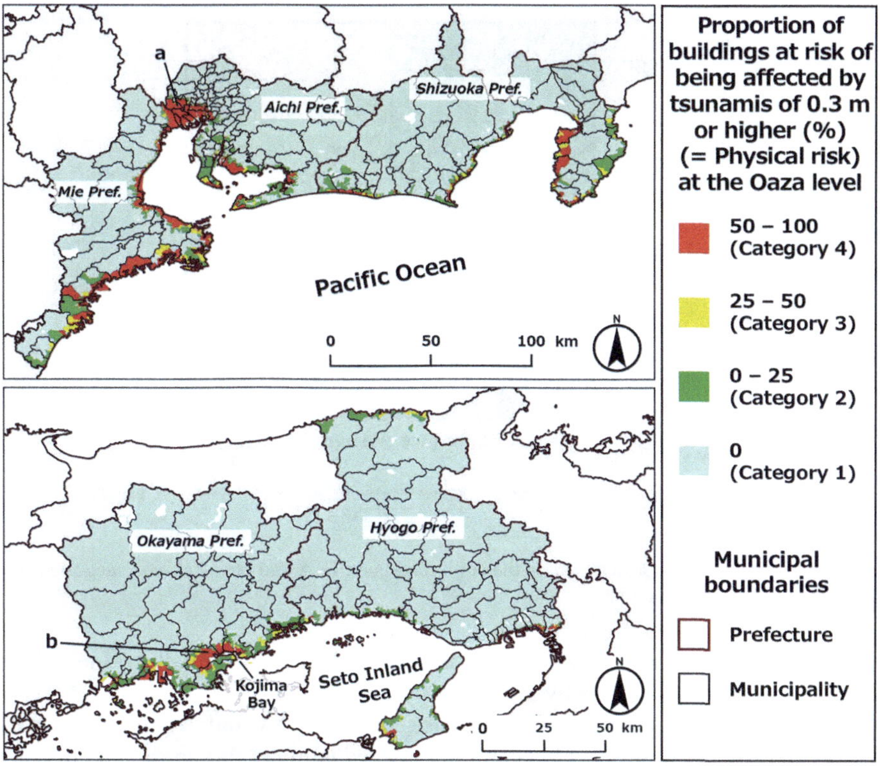

Fig. 5 Physical risk by Oaza across the entire study area

high-risk Oaza. However, areas like southern Okayama City show continuous high-risk distributions, especially near Kojima Bay, where tsunami waves from a Nankai Trough Megaquake could reach inland areas (Fig. 5b).

In summary, while physical risk is higher along the Pacific coast, localized areas of significant risk are also present along the Seto Inland Sea.

5 Result

Finally, using the Oaza-level TRA and physical risk obtained in Sects. 4.1 and 4.2, we investigate the spatial distribution of the gap between TRA and physical risk, as shown in Fig. 1. This proposes a new method for identifying regions that present weaknesses in disaster prevention policies. As shown in Fig. 6, we classified Oaza into eight classes based on the combination of TRA and physical risk categories:

- Class 1, 2: Areas with high physical risk and high TRA.
- Class 3, 4: Areas with low physical risk but high TRA, exhibiting a gap.

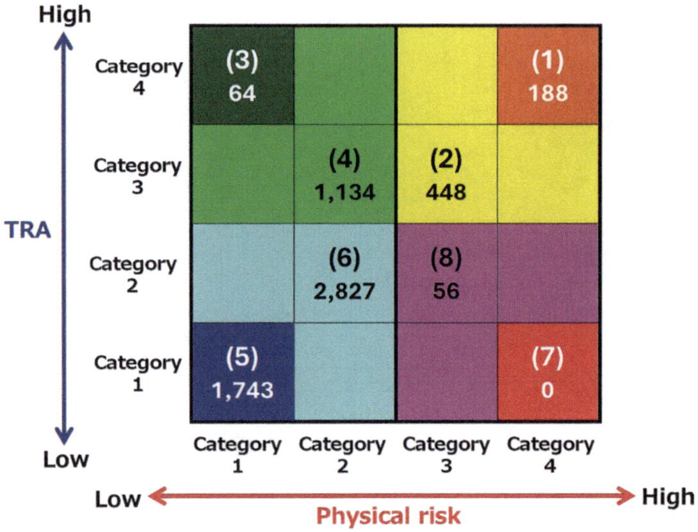

Fig. 6 Classification of Oaza based on the combination of TRA and physical risk categories, and the number of Oaza in each class

- Class 5, 6: Areas with low physical risk and low TRA, potentially vulnerable in the event of an unexpectedly large tsunami.
- Class 7, 8: Areas with high physical risk but low TRA, indicating a significant gap and a high likelihood of becoming a critical vulnerability in tsunami disaster prevention.

Figure 7 shows the classification of Oaza across the study area. Class 1 and 2 Oaza, representing high TRA and physical risk, are continuously distributed along the Pacific coast, indicating that residents have heightened awareness due to the high tsunami risk. Class 3 and 4 Oaza, where physical risk is low, but TRA is high, are also found along the Pacific coast and adjacent inland areas, likely influenced by proximity to high-risk zones and the possibility of an unexpectedly large tsunami. These areas are also widely distributed along the Seto Inland Sea.

Class 5 and 6 Oaza, with low TRA and low physical risk, are generally found inland near Class 3 and 4 areas. In regions like Ise Bay, they are adjacent to higher-risk Oaza, raising concerns about delayed evacuations and inadequate responses in the event of a larger-than-expected tsunami.

Class 7 and 8 represent the most critical vulnerabilities in disaster prevention policies. No Class 7 areas were found, but Class 8 Oaza, with high physical risk and low TRA, were identified in the Nōbi Plain and around Kojima Bay. The Nōbi Plain, located 15–20 km inland from Ise Bay, includes areas near Nagoya Station, where residents might find it hard to imagine a tsunami reaching them. Similarly, Kojima Bay's deep inland location likely fosters complacency among residents. These areas pose a high risk of casualties, making it urgent to not only strengthen

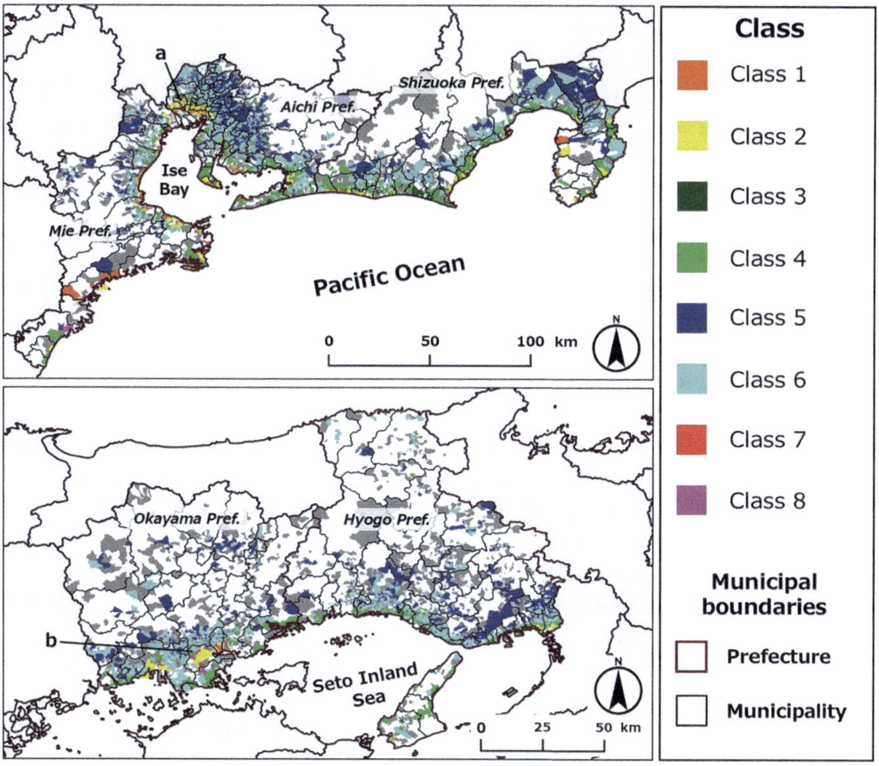

Fig. 7 Results of the classification of Oaza across the entire study area by class

tsunami-resistant infrastructure but also raise public awareness and implement regular evacuation drills.

Furthermore, the results underscore the importance of investigating the socio-economic characteristics and population density of exposed populations. High-density urban areas, such as those along the Pacific coast, face unique challenges in evacuation planning due to limited infrastructure and concentrated populations (Makinoshima et al., 2023). In addition, socio-economic factors, including income disparities and resource accessibility, may further exacerbate vulnerabilities by limiting timely evacuations or preparedness (Deng et al., 2021). Incorporating such socio-economic data into future analyses can provide policymakers with actionable insights to design equitable and effective disaster risk reduction strategies, addressing the specific needs of at-risk populations. To make our study more socially impactful, it is essential to address the challenge of incorporating and examining these factors in future studies.

6 Conclusion

This study provides a novel approach to integrating Tsunami Risk Awareness (TRA) and physical risk assessment at the Oaza level, highlighting critical vulnerabilities in regional disaster prevention strategies. By combining large-scale TRA data and tsunami inundation data, we identified areas where high physical risk coincides with low TRA, particularly in inland regions and along the Seto Inland Sea. These "gap areas" present a significant challenge for disaster management, as insufficient awareness could lead to delayed evacuations and increased casualties during tsunami events. Notably, regions such as the Nōbi Plain and Kojima Bay, where high-risk physical conditions were met with low TRA, underline the importance of targeted interventions.

The results contribute to disaster-resilient urban planning by emphasizing the need for both structural measures, such as tsunami-resistant infrastructure, and non-structural measures, like public education and regular evacuation drills. This approach can provide urban planners and policymakers with actionable insights for closing the awareness gap and enhancing overall community resilience. The integration of spatial risk assessments into urban planning ensures that both awareness and physical risk are addressed comprehensively, paving the way for safer, more resilient cities in tsunami-prone areas.

Future disaster risk reduction policies should incorporate detailed analyses of population density and socio-economic characteristics in high-risk areas. Understanding these factors can provide critical insights into designing equitable evacuation plans and targeted awareness campaigns. By addressing disparities in disaster preparedness and resource access, such measures can enhance the overall resilience of communities exposed to tsunami risks.

In addition, key takeaways for emergency managers include prioritizing vulnerable populations, such as the elderly and disabled, in evacuation planning and implementing awareness education, and training programs. For planners and engineers, relocation and population density management in high-risk zones, alongside structural resilience, should be considered to reduce tsunami risks.

6.1 Future Works

This study identifies several challenges and opportunities for future advancement. Acknowledging the limitations of the Tsunami Risk Awareness (TRA) data is crucial. TRA for Oaza with fewer than four respondents remains uncertain, and even in Oaza with five or more respondents, small sample sizes may introduce bias. Supplementing TRA data with alternative sources and utilizing social media for data collection infrastructure may help overcome these limitations.

Further categorization of building damage risk is also essential. Assessing tsunami-resistant structures beyond wooden buildings will allow for more precise risk evaluations. Technologies for estimating building structures, such as those developed by Furuya and Akiyama (2021), offer significant potential.

To further enhance the precision and equity of disaster risk assessments, future studies should integrate detailed population and socio-economic data into the analysis. We recognize the importance of considering population density and income disparities in high-risk areas to better understand variations in disaster preparedness and vulnerability. As part of our ongoing research, we are developing building-level estimated population datasets (Akiyama et al., 2013) and small-area income datasets (Maneepong et al., 2024). These datasets will enable us to conduct more granular and comprehensive analyses, identifying populations that may require targeted interventions. Incorporating these data into future studies could provide actionable insights for developing equitable disaster risk reduction strategies. Additionally, these datasets hold the potential to evaluate protective actions for mitigating risk by providing detailed insights into population density and socio-economic disparities. Building-level population data can clarify risk distribution, enabling precise evaluation of evacuation plans and protective infrastructure. Similarly, small-area income data can identify socio-economic vulnerabilities, allowing targeted interventions for at-risk populations. By integrating these datasets, future analyses can assess the effectiveness of risk reduction measures more comprehensively and equitably, supporting the development of adaptive and data-driven disaster resilience strategies.

Future studies should also explore the impact of time of day on evacuation dynamics, as human activities and population distribution vary significantly between daytime and nighttime. For instance, the effectiveness of evacuation plans may depend on whether a tsunami occurs during working hours, school hours, or at night when most residents are at home. Additionally, advancements in detection and early warning systems, as well as their integration with evacuation modeling, could further enhance disaster response. In recent years, the movements and distribution patterns of people at different times of the day can now be understood through big data collected from the movement history of mobile devices such as smartphones. Therefore, in future research, incorporating such big data into analyses will enable more precise and dynamic evaluations of damages and evacuation strategies, ultimately contributing to improved disaster preparedness and risk management.

Addressing the gap between physical risk and TRA through disaster risk reduction (DRR) policies is critical. In areas with low TRA, enhancing disaster education and communication strategies is urgent. Tailored approaches that consider cultural and socioeconomic contexts, along with community-based drills, could improve disaster preparedness. Addressing the gap between physical risk and TRA through disaster risk reduction (DRR) policies is critical. In areas with low TRA, enhancing disaster education and communication strategies is urgent. Tailored approaches that consider cultural and socioeconomic contexts, along with community-based drills, could improve disaster preparedness.

This study presents a valuable decision-support tool for urban planning and DRR. The proposed risk visualization method offers planners vital insights into risk distribution and preparedness. To maximize its utility, further technological development is needed to enable more region-specific risk assessments and disaster mitigation strategies.

Acknowledgments This study was supported by JSPS KAKENHI Grant Number 24 K00243 and Prioritized Studies of Advanced Research Laboratories, Tokyo City University. In addition, the source data for the TRA data, the livability data, was kindly provided by Prof. So of Reitaku University, the developer of the dataset. We express our sincere appreciation for their contribution to this study.

References

Akiyama, Y., Takada, T., & Shibasaki, R. (2013). Development of micropopulation census through disaggregation of national population census. In *Proceedings of CUPUM2013, Utrecht, The Netherlands, July 2013*. CUPUM. Paper #110.

Amri, I., Giyarsih, S. R., & Ruslanjari, D. (2024). Tsunami risk awareness, hazard warning knowledge, and intended evacuation behavior among beach users in Bantul, Indonesia. *International Journal of Disaster Risk Reduction, 109*, 104594. https://doi.org/10.1016/j.ijdrr.2024.104594

Cerase, A., Crescimbene, M., La Longa, F., & Amato, A. (2019). Tsunami risk perception in southern Italy: First evidence from a sample survey. *Natural Hazards and Earth System Sciences, 19*(12), 2887–2904. https://doi.org/10.5194/nhess-19-2887-2019

Daito Trust Construction Co., Ltd. (2023). *Livability of cities and ranking of desirable cities to live in 2023: Survey overview*. Retrieved January 30, 2025, from https://www.kentaku.co.jp/sumicoco/ (in Japanese).

Deng, H., Aldrich, D. P., Danziger, M. M., Gao, J., Phillips, N. E., Cornelius, S. P., & Wang, Q. R. (2021). High-resolution human mobility data reveal race and wealth disparities in disaster evacuation patterns. *Humanities and Social Sciences Communications, 8*(1), 1–8. https://doi.org/10.1057/s41599-021-00824-8

Dengler, L. A. (1998). Strategic implementation plan for tsunami mitigation projects. NOAA Technical Memorandum ERL PMEL-113. National Oceanic and Atmospheric Administration, Pacific Marine Environmental Laboratory.

Fire and Disaster Management Agency of Japan. (2021). *Damage situation and police countermeasures associated with the 2011 Great East Japan Earthquake*. Retrieved January 30, 2025, from https://www.fdma.go.jp

Furuya, T., & Akiyama, Y. (2021). Damage assessment of floods for broad urban areas based on insurance payment information using geo big data. In *Proceedings of CUPUM2021, Virtual, June 2021* (p. 15-1). CUPUM.

Goto, K., Chague-Goff, C., Fujino, S., Goff, J., Jaffe, B., Nishimura, Y., Richmond, B. M., Sugawara, D., Szczucinski, W., Tappin, D. R., & Witter, R. C. (2011). New insights of tsunami hazard from the 2011 Tohoku-oki event. *Marine Geology, 290*(1), 46–59. https://doi.org/10.1016/j.margeo.2011.10.004

Harada, K., Imamura, F., & Suppasri, A. (2014). Survey on tsunami risk perception and preparedness in the Tokai Region. *Journal of Disaster Research, 9*(6), 1035–1042.

Hirano, K., Fukushima, Y., Maruya, H., Kido, M., & Sugiura, M. (2023). The anticipated Nankai Trough earthquake and tsunami in Japan: Determinant factors of residents' pre-event evacuation intentions. *Journal of Disaster Research, 18*(3), 233–245. https://doi.org/10.20965/jdr.2023.p0233

Kim, D., Yoshida, A., Okubo, K., & Hayashi, T. (2015). The workshop program of community disaster mitigation planning for mutual support in Sasayama-Fukusumi Preservation Districts, Hyogo. In *Proceedings of Urban Cultural Heritage Disaster Mitigation* (pp. 175–182).

Kitagawa, K. (2019). Co-constructing a narrative of 'never give up' in preparing for a mega-tsunami: An exemplar of 'all-of-society engagement'? *Geosciences, 9*(12), 486. https://doi.org/10.3390/geosciences9120486

Koshimura, S. (2016). Establishing the advanced disaster reduction management system by fusion of real-time disaster simulation and big data assimilation. *Journal of Disaster Research, 11*(2), 164–174. https://doi.org/10.20965/jdr.2016.p0164

Koshimura, S., & Shuto, N. (2014). Tsunami disaster and the role of good spatial planning. *Progress in Disaster Science, 1*, 1–9. https://doi.org/10.1016/j.pdisas.2014.01.001

Koshimura, S., Hayashi, S., & Gokon, H. (2014). The impact of the 2011 Tohoku earthquake tsunami disaster and implications to the reconstruction. *Soils and Foundations, 54*(4), 560–572. https://doi.org/10.1016/j.sandf.2014.06.002

Lewis, J. (2011). *2011 earthquake and tsunami in Japan: Death toll, geology, and hints it was coming*. Facts and details. Retrieved January 30, 2025, from https://factsanddetails.com/japan/cat26/sub161/item1742.html

Makinoshima, F., Oishi, Y., & Imamura, F. (2023). Urban structure reinforces attitudes towards tsunami evacuation. *Scientific Reports, 13*(1), 7992. https://doi.org/10.1038/s41598-023-33562-9

Maneepong, K., Yamanotera, R., Akiyama, Y., Miyazaki, H., Miyazawa, S., & Akiyama, M. C. (2024). Open data-driven 3D building models for micro-population mapping in a data-limited setting. *Remote Sensing, 16*(21), 3922. https://doi.org/10.3390/rs16213922

Matsumaru, R., & Kawaguchi, R. (2015). Awareness on tsunami disaster of visitors—Case of Enoshima, Katase, and Kugenuma Area in Kanagawa, Japan. In *Proceedings of the Japan Society of Civil Engineers (JSCE) 2015 Annual Meeting, IV-044* (pp. 87–88) (in Japanese).

Ministry of Land, Infrastructure, Transport and Tourism (MLIT). (2024). *National Land Numerical Information: Tsunami inundation data*. Retrieved January 30, 2025, from https://nlftp.mlit.go.jp/ksj/gml/datalist/KsjTmplt-A40-v2_1.html (in Japanese).

Mori, T. (2013). The spatial deep structure of cultural landscape in Gokayama based on Oaza. *Journal of City Planning Institute of Japan, 48*(3), 579–584. (in Japanese).

Mori, N., Satake, K., Cox, D., Goda, K., Catalan, P. A., Ho, T.-C., Imamura, F., Tomiczek, T., Lynett, P., Miyashita, T., Muhari, A., Titov, V., & Wilson, R. (2022). Giant tsunami monitoring, early warning and hazard assessment. *Nature Reviews Earth & Environment, 3*(8), 557–572. https://doi.org/10.1038/s43017-022-00327-3

Nakatsuka, M., Kawaguchi, T., & Hoshino, S. (2009). A study on reforming process in regional autonomous organization in rural areas: Focusing on generation of "Ba". *Journal of Rural Planning Association, 28*(3), 135–140. (in Japanese).

Nanjo, K. Z., & Yoshida, A. (2018). A b map implying the first eastern rupture of the Nankai Trough earthquakes. *Nature Communications, 9*, 1117. https://doi.org/10.1038/s41467-018-03514-3

National Oceanic and Atmospheric Administration (NOAA). (2023). *Tsunami dangers*. Retrieved January 30, 2025, from https://www.noaa.gov/jetstream/tsunamis/tsunami-dangers

Rossetto, T., Ioannou, I., & Grant, D. N. (2014). Guidelines for empirical vulnerability assessment. GEM Technical Report 2014-13 V1.0.0, 204 pages. https://doi.org/10.1007/s11069-020-04098-2

Salah, P., & Sasaki, J. (2021). Knowledge, awareness, and attitudes toward tsunamis: A local survey in the southern coast of Iran. *Sustainability, 13*(2), 449. https://doi.org/10.3390/su13020449

Santos, A., & Isidro, M. L. (2020). Tsunami science and future mitigation strategies. *Geosciences, 10*(9), 389. https://doi.org/10.3390/geosciences10090389

So, T. (2021). The relationship between local livability and rent. *Journal of Architecture and Planning (AIJ), 86*(781), 971–979. https://doi.org/10.3130/aija.86.971

Sufri, S., Dwirahmadi, F., Phung, D., & Rutherford, S. (2020). Enhancing community engagement in disaster early warning systems in Aceh, Indonesia: Opportunities and challenges. *Natural Hazards, 103*(3), 2691–2709. https://doi.org/10.1007/s11069-020-04098-2

Suppasri, A., Koshimura, S., & Gokon, H. (2014). Building damage characteristics based on surveyed data and fragility curves of the 2011 Great East Japan tsunami. *Natural Hazards, 66*(2), 319–341. https://doi.org/10.1007/s11069-012-0487-8

Takemoto, S., Shibuya, M., & Sakoda, K. (2021). *Learning from megadisasters: A decade of lessons from the Great East Japan Earthquake*. https://www.worldbank.org/en/news/feature/2021/03/11/learning-from-megadisasters-a-decade-of-lessons-from-the-great-east-japan-earthquake-drmhubtokyo

Takemura, S., Hamada, Y., Okuda, H., Okada, Y., Okubo, K., Akuhara, T., Noda, A., & Tonegawa, T. (2023). A review of shallow slow earthquakes along the Nankai Trough. *Earth, Planets and Space, 75*(1), 164. https://doi.org/10.1186/s40623-023-01920-6

Toronto Metropolitan University. (2015). *Guidelines on anonymity and confidentiality in research*. Retrieved January 30, 2025, from https://www.torontomu.ca/content/dam/research/documents/ethics/guidelines-on-anonymity-and-confidentiality-in-research.pdf

UNESCO. (2020). *The Indian Ocean Tsunami: The global response to a global disaster*. Retrieved January 30, 2025, from https://unesdoc.unesco.org

Wood, N. J., & Schmidtlein, M. C. (2016). Community variations in population exposure to near-field tsunami hazards as a function of pedestrian travel time to safety. *Natural Hazards, 77*(1), 791–816. https://doi.org/10.1007/s11069-012-0434-8

Yamashita, S., & Fujita, T. (2012). Post-tsunami damage and rebuilding in East Japan: Lessons from the 2011 tsunami disaster. *Urban Planning Review, 40*(3), 1–15.

Virtual Reality as a Collaborative Tool in Road Planning: Insights from a Case Study in Perth, Western Australia

Jonathan Pillai, Parisa Izadpanahi, and Shohreh Nematollahi

Abstract The research aims to investigate the effectiveness of virtual reality as a collaborative tool among experts in a road planning project within the context of Perth, Western Australia. Over recent years, virtual reality applications have seen widespread adoption across various fields, assisting stakeholders to evaluate project efficiency prior to implementation, thereby minimising potential errors and risks.

There are studies evaluating the effectiveness of virtual reality as a participatory urban planning tool in the community engagement processes (Christodoulou et al., 2018; van Leeuwen et al., 2018; Schrom-Feiertag et al., 2020), its impact on student learning (Sanchez-Sepulveda et al., 2019a, 2019b), and its utility for initiating informed discussions among experts, particularly within urban planning practices (Jiang et al., 2023).

This paper explores experts' perspectives of VR systems as a collaborative tool in the road planning process. The findings suggest that visual virtual reality is an effective tool whose utility can extend to other areas of project construction and management.

Keywords Virtual reality · Road planning · Collaborative tool

1 Introduction

Virtual reality has evolved into a supportive tool in many fields, spanning from medicine, education, design, and visualisation of urban planning (Chun et al., 2008; Christodoulou et al., 2018; van Leeuwen et al., 2018; Schrom-Feiertag et al., 2020; Jiang et al., 2023).

J. Pillai · P. Izadpanahi (✉) · S. Nematollahi
Curtin University, Bentley, WA, Australia
e-mail: jonathan.pillai@curtin.edu.au; parisa.izadpanahi@curtin.edu.au; shohreh.nematollahi@curtin.edu.au

R. Goodspeed et al. (eds.), *Digital-Era Urban Transformations*, The Urban Book Series, https://doi.org/10.1007/978-3-031-98300-9_4

Virtual Reality creates an alternative reality for experimental purposes (LaValle, 2023). It is a computer-generated, three-dimensional virtual environment that users can interact with (Hamad & Jia, 2022), enabling them to experience and perceive the real world through various images and objects with sound, movement, and other features (Sutcliffe & Kaur, 2000; Schrom-Feiertag et al., 2020). Equipped with a headset and two hand controllers, users immerse themselves in the simulation, where they can explore, move around, and interact with sensory features to test different scenarios (Slater & Sanchez-Vives, 2016; Mihelj et al., 2014).

In the context of this study, VR simulations provide valuable complements to the collaboration process in the project, facilitating a more comprehensive shared understanding of future spatial development and the associated changes (Heldal, 2007). VR simulations are superior to reviewing design iterations compared to conventional methods, such as static visuals, including maps, sketches, and photomontages (Heldal, 2007).

2 Virtual Reality in Road Planning

In a road planning process that spans 7–15 years from design to implementation, incorporating Virtual Reality offers significant advantages in engaging stakeholders, ranging from experts to the local community (Heldal, 2007). It provides various alternatives and scenarios along with their potential impact on the surrounding environment, fostering constructive dialogue between all involved parties (Heldal, 2007).

VR models integrate real-time data and can be tailored to demonstrate scenarios concerning traffic, safety requirements, and more (Heldal, 2007), enabling decision-makers to assess road planning projects effectively. Built on game engines, VR technology is adept at processing heavy real-time graphics (Stock & Bishop, 2005), facilitating comprehension of complex environmental changes and providing detailed analysis of users' experience (Aspin & Roberts, 2005; Stock & Bishop, 2005; Jiang et al., 2023).

Heldal (2004, 2007), Wilson (2003), and Al-Kodmany (2002) also highlight the importance of creating VR scenarios, using real-time data to evaluate project design elements against possible risks for different users.

Further, by incorporating real-time data in virtual reality, VR has assisted road planning for smart cities and urban projects in Australia. Esri Australia, for instance, has integrated VR into its Geographic Information System (GIS) platform, enabling communities and authorities to immerse themselves in visualising planning and development outcomes (Esri Australia, 2024). Consequently, incorporating VR as a permanent tool in various stages of road planning projects offers several advantages, including streamlining processes, mitigating risks, and enhancing transparency. This paper discusses experts' perspectives on a road planning project and proposes enhancements for the VR models to be incorporated in subsequent project stages.

3 Virtual Reality Elements in Case Studies

There are four essential elements in the realm of virtual reality: the virtual environment, virtual presence, sensory feedback (as a response to the user's actions), and interactivity (Sherman & Craig, 2003, cited in Mihelj et al., 2014). The testing process for a project requires the harmonious integration of these elements, alongside a top-tier visual display and a detailed physical model.

The research team aimed to create an engaging virtual reality experience demonstrating its superior potential over passive alternatives.

To create an immersive and engaging virtual environment, the team opted to leverage a game-design methodology as the foundation for their project. This decision was adopted to meticulously evaluate the levels of engagement provided by the experience rather than simply replicating the real-world scenarios in a digital format. According to Schell (2019) (Fig. 1), the thoughtful integration of mechanics, narrative, aesthetics, and technology is crucial in creating an engaging experience, whether it is a game, educational simulation, or virtual experience.

In addition, throughout the development process, it was important to recognise the nuanced technical aspects of the production of a seamless virtual experience. In this case, the team adopted the PBR (Physical Based Rendering) workflow (Kumar, 2020), widely acknowledged as a standard workflow for crafting 3D art assets in game production, specifically for working with game engines. Adopting a game-based methodology in the design process enabled the team to successfully plan out the various stages of production and ensure adequate space for experts' feedback and thorough testing (Gonçalves et al., 2023; Aleem et al., 2016) (Fig. 2).

3.1 Virtual Environment in the Case Study

This study focuses on the Smart Freeway Mitchell Southbound (Fig. 3).

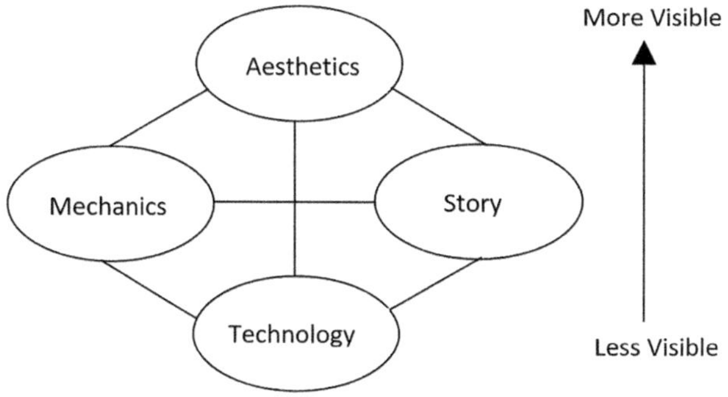

Fig. 1 Jesse Schelle's design tetrad (Schell, 2019)

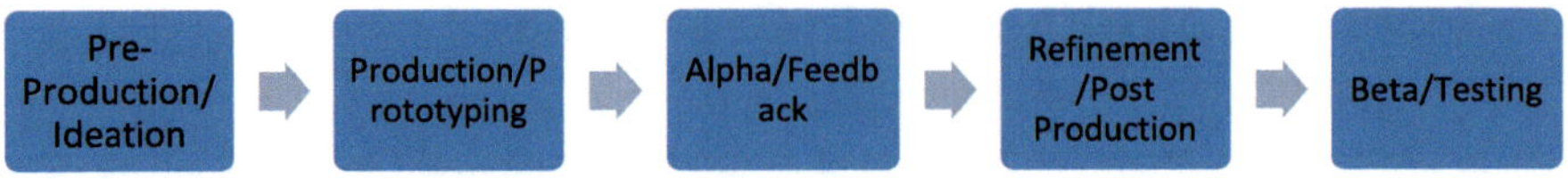

Fig. 2 The stages of production

Fig. 3 The case study location in the Perth Metropolitan Area

During the initial briefing session led by WA main roads experts, it was emphasised that accurate representation of the sites was paramount rather than artistic impressions. Aligning with the experts' requirements and striving for fidelity to the real-world representation of the urban sites, the team chose to construct the site environments using schematic drawings, videos, and reference materials provided by the WA main roads.

Considering various 3D production methodologies, such as photogrammetry and high-resolution modelling, the team ultimately opted for the PBR workflow. Additionally, they embraced a modular approach to ensure accurate texture mapping and fidelity to real-world counterparts while also optimising the assets for optimal performance within the game engine. By implementing the PBR methodology, the team ensured that textures and models were finely tuned, thus enhancing the in-engine experience. Adopting the modular approach alleviates the computational burden of rendering entire scenes, leading to a smoother and more efficient program. The environmental design aimed to achieve a delicate equilibrium between optimising the user experience for factors like lag and VR stuttering while ensuring fidelity and aesthetics of the real world. These considerations remained of utmost importance throughout the design process.

Furthermore, the team used AI technologies like Unity's AI engine and AI Speed Tree as part of the environment design. These technologies played a crucial role in automating the creation of texturing elements, such as roads and plants within the environment. This optimisation allowed for a significantly enhanced experience in virtual reality (Fig. 4).

3.2 *Virtual Presence in the Case Study*

To craft a truly immersive experience, it is crucial to prioritise the design of a virtual presence, which effectively transports the user into the space. When considering one's presence and immersion in a virtual environment, designers often focus on engaging as many senses as feasible (Slater et al., 1994; Kors et al., 2017; Gonçalves et al., 2023). Aspects such as mechanics, aesthetics (both auditory and visual), narrative, and technology-driven elements like haptic feedback are all intertwined in this component.

When conducting this study, the research team carefully considered the existence of diverse user groups, including both drivers and pedestrians. Their primary emphasis was on establishing the fundamental elements necessary to anchor the user's presence. The first aspect the team leveraged was the aesthetic draw of a real-world environment. The experience features a thoughtfully designed introductory segment set in a primary digital environment, providing opportunities for basic interactions. The integration of real-world aesthetics throughout the immersive

Fig. 4 Schematic and initial models. Top left: Initial Spatial Design of Smart Freeway Mitchell Southbound. Top right: Top-down real-time views from Google Earth. Bottom left: Real World image of Greenwood station for Smart Freeway Mitchell Southbound. Bottom right: Schematic model of Greenwood Station

experience was done with the purpose of enhancing user engagement. When initiating the VR experience, the users were seamlessly shifted from the initial non-realistic aesthetic environment to a more realistic, real-world virtual representation of the designated urban environment.

To enhance the feeling of presence within a virtual space, the utilisation of sound and lighting effects played a significant role in replicating real-world sensations. Users were immersed in the sounds of traffic and could experience the proximity of passing vehicles. Additionally, the night and day visibility were programmatically integrated into the system, allowing users to encounter diverse lighting conditions based on the time of day.

The project was designed to evoke a strong sense of place, mirroring the real world. To replicate the real-world settings, the team utilised consistent colour and visual cues, such as signage and traffic indicators. Users were given the option to choose different perspectives or points of view (POV) within the varying experiences, enabling them to embark on a distinct narrative experience. Whether experiencing the VR environment as an observer, a pedestrian, from a top-down perspective, or through a fly-through camera experience (Fig. 5), users could explore the scene from multiple angles, enriching their immersion.

A final aspect that was considered in the design process was the overall flow of the experience. In order to build content that works well on the VR platform, it would be crucial to consider the pipeline. In the context of a game design-based process, a pipeline can be defined as the way in which the content creation task and tools are utilised, considering efficiency and the target content outcomes (Dunlop, 2014; Marklund et al., 2019). Optimisation of the pipeline is crucial for VR-based

Fig. 5 Sample of POVs (a pedestrian view from the Greenwood station bridge)

experience; this can avoid fundamental issues such as screen tearing, lag or user discomfort, such as nausea caused by movement and visual juddering within the VR environment. One optimisation technique employed was the use of HDRI imaging to enhance the skybox and the surrounding background of the sites. This meant the team did not have to model every viewable object.

3.3 *Sensory Experience and Interactivity in the Case Study*

It is widely acknowledged that the thoughtful design of interactions profoundly influences overall engagement with a digital product. There is a symbiotic relationship when considering the interactive and sensory components of the VR experience. It is argued that an embodied interactive and sensory VR experience must carefully integrate different senses in relation to aesthetics and functionality to effectively engage users. In their approach, the team adopted Xiaobo and Yuelin's framework, which is succinctly depicted in the diagram below (Fig. 6).

The following features were designed (Table 1) with the above consideration and their relationship with the senses or extension of our bodily experience in VR.

By incorporating the aforementioned features, users could experience a heightened sense of embodiment and control while immersed in the virtual reality environment (Figs. 7, 8, 9, 10, 11, 12, 13, 14, and 15). This technology replicated real-world experiences in certain interactions while simultaneously pushing the boundaries of what was only achievable in a virtual setting. The intention was for users to have access to both the non-realistic advantages of the virtual environment and the potential for an embodied experience unique to this particular medium.

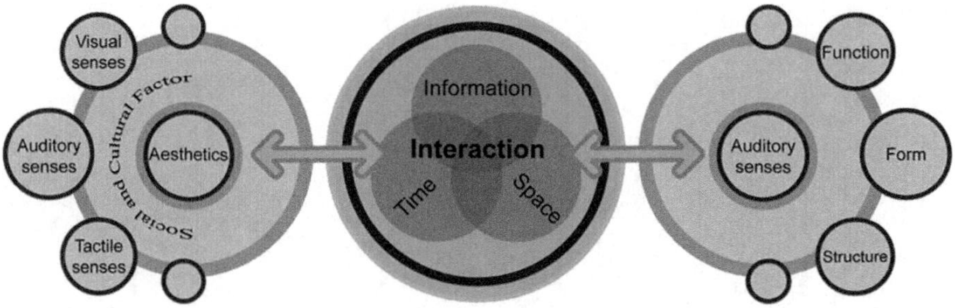

Fig. 6 Xiaobo and Yuelin's design approach (2014, p. 168)

Table 1 Features and sensory interactions

Feature	Description	Sensory/interaction
Point and click for the menu and interactions within the experience	Users can use point and click for selection and deselection as an overall interactive cue for navigation and use of different interactive features	Visual sense: Extension of hands Interactive feedback: To grab, select, and point. See Fig. 7
Teleport-location	Allows users to move across to different locations without physical movement	Visual sense: Extension of movement and feet Auditory sense: Ability to experience sound in different locations Interactive feedback: Move to various locations without physical movement. See Fig. 8
Point of views (varying camera views)	Allows viewers multiple types of views from different 3D camera angles	Visual sense: Different views/different player experiences through views Auditory sense: Sound proximity and distance Interactive feedback: Change in weather correspondence to movement of sliding bar. See Fig. 9
Place traffic collision- AI rerouting	Allows viewers to place a traffic collision on the road	Visual sense: Observable dynamic environment changes based on collision Extension of hands-to-place collision Auditory sense: Sound proximity and distance Interactive feedback: Placement/ planting of collision with clear demarcation of the incident. See Fig. 10
Time	Allows players to change the weather and the time of day, impacting the environment	Visual sense: Observable impact on environment and traffic Auditory: Time of the day and weather triggers traffic density, which corresponds with sound Interactive feedback: Change in weather corresponds to movement of the sliding bar. See Fig. 11

Table 1 (continued)

Feature	Description	Sensory/interaction
Speed	Allows players to manipulate the speed of traffic	Visual sense: Observable impact on environment and traffic Auditory: Traffic density corresponds with sound Interactive feedback: Change in speed corresponds to movement of the sliding bar. See Fig. 12
Traffic	Allows players to manipulate the speed of traffic density	Visual sense: Observable impact on environment and traffic Auditory: Traffic density corresponds with sound Interactive feedback: Change in density of traffic corresponds to movement of sliding bar. See Fig. 13
Physical features of roads and pavement	Allows players to change specific physical environment objects, such as the texture of street and pavement	Visual sense: Observable impact on the environment Interactive feedback: Change in the physical environment, textures correspondence to a selection of material through point and click in the menu system. See Fig. 14
Planting	Allows users to plant foliage and remove foliage along the sidewalk	Visual sense: Observable impact on the environment Interactive feedback: Change in physical environment correspondence to selection and planting of trees. See Fig. 15

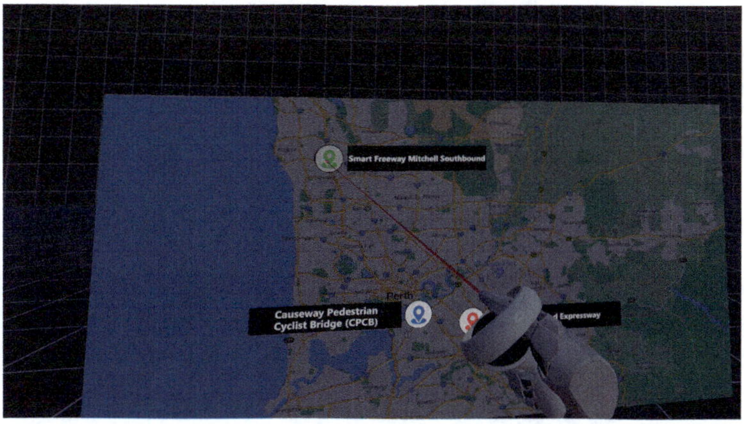

Fig. 7 Point and click for the menu and interactions within the experience

Fig. 8 Teleport-location

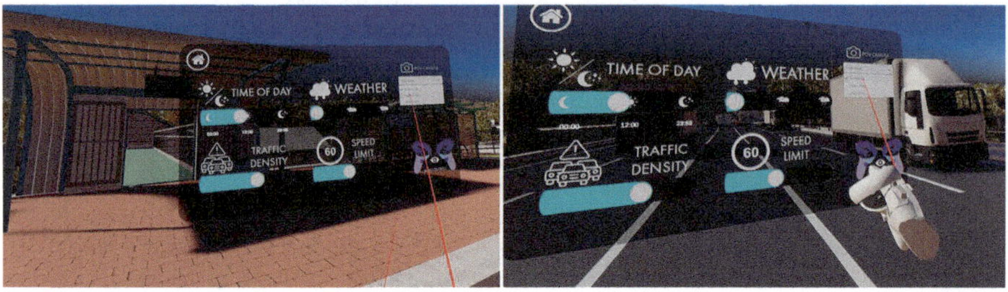

Fig. 9 Point of views (varying camera views)

Fig. 10 Place traffic collision-AI rerouting

Fig. 11 Time

Fig. 12 Speed

Fig. 13 Traffic

Fig. 14 Physical features of roads and pavement

Fig. 15 Planting

Fig. 16 Experts' experience of VR and giving feedback to the research/Curtin team

To further enhance the interactive components of the experience, the team followed a design methodology characterised by continuous iteration. This process involved refining the features based on valuable feedback from experts. Experts' feedback is an indispensable component of interaction design in a virtual environment (Gibbons, 2016). By incorporating and analysing this feedback and closely observing experts' interactions with the VR experience, the team made continuous improvements. This allowed the Curtin University team to refine the design of the interactive components in a manner that ensured the contextual relevance of the case study to stakeholders, particularly experts in WA main roads (Fig. 16).

4 Data Collection

Qualitative research methods were used to gain insights from experts' views. Over the course of 1 week, the research team conducted 30 face-to-face interviews at WA's Main Roads department. Curtin University ethics approval was obtained prior to data collection.[1] Participation in the study was voluntary, with invitations circulated internally.

Upon signing the consent form, research assistants instructed each participant on using the VR tool and navigating through each case study. Participants donned headsets and hand controllers. Following a review of each case study design context, participants engaged in individual interviews. The interviews were

[1] Ethic application number: HRE2023-0145.

audio-recorded, transcribed, and thematically analysed. The sample size of 30 interviewees falls within the acceptable range for qualitative research.

While all participants recommended such a tool for external or winder community consultation, 13% of the participants suggested that the current format, which allows only one person at a time, needs to be made less time-consuming and more convenient, especially for certain age cohorts, such as the elderly.

Furthermore, participants rated VR as the most effective for stakeholder engagement and design stages of a project, followed closely by project management, road planning, and asset management (Figs. 17, 18 and 19). Those participants involved in asset management stated that if VR could be integrated from early design and road planning stages, through to implementation, it could potentially be utilised for other areas involving project management and maintenance of projects by incorporating updated data.

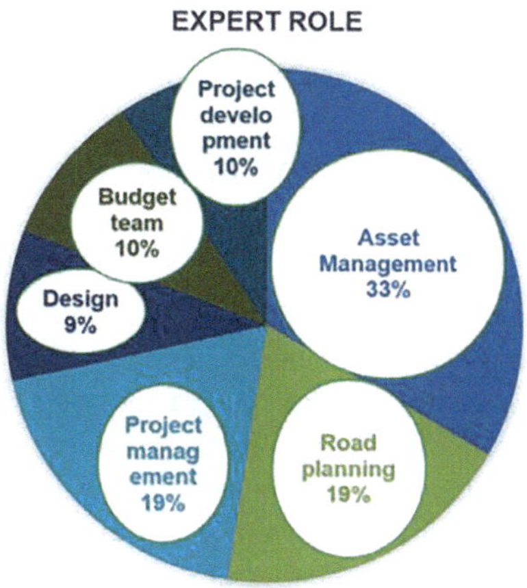

Fig. 17 Interviewees' role in WA main roads

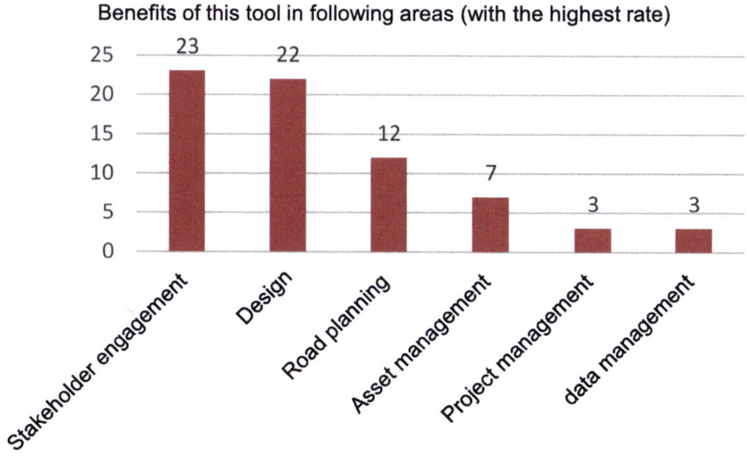

Fig. 18 The effectiveness of VR tool in various departments (the highest)

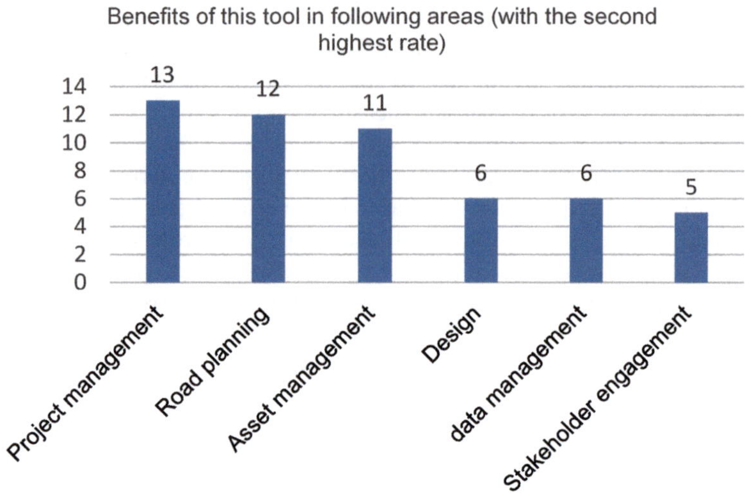

Fig. 19 The effectiveness of VR tool in various departments (the second highest)

5 Qualitative analysis of the Case Study

This section provides qualitative feedback on Smart Freeway Mitchell Southbound case study simulation to highlight areas of improvement. Participants experienced it exclusively from the perspective of a pedestrian due to previous discomfort experienced by experts in a trail driving simulation using VR resulted in feelings of headache and nausea. Each participant was selected to be teleported to critical design points to assess the project design against design standards and safety measures.

Participants were given the option to adjust various parameters, such as the time of the day (peak and off-peak traffic time, nighttime), weather conditions (rainy, etc.), and simulated crash points to evaluate the traffic flow patterns, degree of visibility, and the effectiveness of mounted design notification function. The overall feedback from experts suggested a desire for real-time data on crashes and crash points, particularly during low visibility conditions such as nighttime and rainy weather, to enable testing and modification of corresponding elements within the scenario. Participants highlighted the importance of accurate simulation and real-time design adjustment during collaborative Design Review meetings, which in their belief results in process efficiency.

6 Conclusions

The study was initiated by collaborating teams at Curtin University and project experts at WA Main Roads, Perth, Western Australia. It aimed to evaluate the effectiveness of virtual reality as a collaborative tool between experts when road planning projects are at the early stages of design and review. The visual possibilities that VR provides experts to feel the project at its early stages and test features of design against critical conditions, praised by the interviewees (experts who participated in VR testing).

The interviewees emphasised the further elaboration of the projects in a virtual environment by including ongoing real-world data. It is an important aspect for incorporating this visual tool for other stages of the project, such as external community engagement and managing their maintenance after the project construction is completed. Such thorough elaboration of project models, if it occurs, provides ongoing efficiency for the institution's internal communication and management of projects.

The outcome of this study confirms that designing virtual presence is a dynamic and interdisciplinary endeavour that converges technology, psychology, and creativity. The collaborative efforts of expert experience designers, spatial designers, storytellers, and ethical practitioners shape the future of immersive digital experiences. As road planners navigate this ever-evolving landscape, the focus on user-centred design, spatial awareness, embodiment, and ethical considerations is integral to creating virtual environments that captivate, engage, and truly transport experts to new realms within the digitals domain.

Acknowledgements This research project was co-funded equally by Curtin University and Main Roads Western Australia. The authors gratefully acknowledge their financial support, which made this work possible.

References

Aleem, S., Capretz, L. F., & Ahmed, F. (2016). Critical success factors to improve the game development process from a developer's perspective. *Journal of Computer Science and Technology, 31*(5), 925–950. https://doi.org/10.1007/s11390-016-1673-z

Al-Kodmany, K. (2002). Visualization tools and methods in community planning: From freehand sketches to virtual reality. *Journal of Planning Literature, 17*, 189–211. https://doi.org/10.1177/088541202762475946

Aspin, R. & Roberts, D. (2005). A model for distributed, co-located interaction in urban design/review visualisation (pp. 70–77). https://doi.org/10.1109/DISTRA.2005.3

Christodoulou, N., Papallas, A., Kostic, Z., & Nacke, L. E. (2018). Information visualisation, gamification and immersive technologies in participatory planning. In Extended Abstracts of the 2018 CHI Conference on Human Factors in Computing Systems (CHI EA '18). ACM.

Chun, W., Ge, C., Liu, Y., & Horne, M. (2008). Virtual-reality based integrated traffic simulation for urban planning. In 2008 International Conference on Computer Science and Software Engineering (Vol. 2, pp. 1137–1140). IEEE. https://doi.org/10.1109/CSSE.2008.1074

Dunlop, R. (Ed.). (2014). *Production pipeline fundamentals for film and game*. Focal Press, Taylor Francis Group.

Esri Australia. (2024). ArcGIS 360 VR / CityEngine and immersive planning resources. Esri Australia.

Gibbons, S. (2016). *Design thinking 101*. Nielsen Norman Group. https://www.nngroup.com/articles/design-thinking/

Gonçalves, G., Coelho, H., Monteiro, P., Melo, M., & Bessa, M. (2023). Systematic review of comparative studies of the impact of realism in immersive virtual experiences. *ACM Computing Surveys, 55*(6), 1–36. https://doi.org/10.1145/3533377

Hamad, A., & Jia, B. (2022). How virtual reality technology has changed our lives: An overview of the current and potential applications and limitations. *International Journal of Environmental Research and Public Health, 19*(18), 11278. https://doi.org/10.3390/ijerph191811278

Heldal, I. (2007). Supporting participation in planning new roads by using virtual reality systems. *Virtual Reality, 11*(2), 145–159.

Heldal, I. (2004). *The usability of collaborative virtual environments: Towards an evaluation framework* [Ph.D. thesis, Department of Technology and Society, Chalmers].

Jiang, H., Geertman, S., Zhang, H., & Zhou, S. (2023). Factors influencing the performance of virtual reality in urban planning: Evidence from a view corridor virtual reality project, Beijing. *Environment and Planning B: Urban Analytics and City Science, 50*(3), 814–830. https://doi.org/10.1177/23998083221130709

Kors, M. J. L., Millenaar, K., van der Spek, E. D., Ferri, G., Schouten, B., Sandovar, A., Ketel, C., & Dormans, J. (2017). Arguing on the holodeck: Designing immersive interactive entertainment with persuasive intent. In *Extended Abstracts Publication of the Annual Symposium on Computer-Human Interaction in Play* (pp. 667–671). ACM. https://doi.org/10.1145/3130859.3131444

Kumar, A. (2020). Texturing workflow. In A. Kumar (Ed.), *Beginning PBR texturing: Learn physically based rendering with Allegorithmic's substance painter* (pp. 31–38). Apress. https://doi.org/10.1007/978-1-4842-5899-6_3

LaValle, S. M. (2023). Virtual Reality. Cambridge University Press. https://doi.org/10.1017/9781108182874

Marklund, B. B., Engström, H., Hellkvist, M., & Backlund, P. (2019). What empirically based research tells us about game development. *The Computer Games Journal, 8*(3–4), 179–198. https://doi.org/10.1007/s40869-019-00085-1

Mihelj, M., Novak, D., & Beguš, S. (2014). Virtual reality: Technology and applications. Springer. https://doi.org/10.1007/978-94-007-6910-6

Sánchez-Sepulveda, M. V., Torres-Kompen, R., Fonseca, D., & Franquesa-Sánchez, J. (2019a). Methodologies of learning served by virtual reality: A case study in urban interventions. *Applied Sciences, 9*(23), 5161. https://doi.org/10.3390/app9235161

Sánchez-Sepulveda, M. V., et al. (2019b). [Paper on VR engagement with citizens and students]. In Proceedings of TEEM 2019.

Schell, J. (2019). The art of game design: A book of lenses (3rd ed.). CRC Press.

Schrom-Feiertag, H., Stubenschrott, M., Regal, G., Matyus, T., & Seer, S. (2020). An interactive and responsive virtual reality environment for participatory urban planning. In Proceedings of the Symposium on Simulation for Architecture and Urban Design (SimAUD 2020) (pp. 119–125).

Sherman, W. R., & Craig, A. B. (2003). Understanding virtual reality: Interface, application, and design. Morgan Kaufmann.

Slater, M., & Sanchez-Vives, M. V. (2016). Enhancing our lives with immersive virtual reality. *Frontiers in Robotics and AI, 3*, 74. https://doi.org/10.3389/frobt.2016.00074

Slater, M., Usoh, M., & Steed, A. (1994). Depth of presence in virtual environments. *Presence: Teleoperators and Virtual Environments, 3*(2), 130–144. https://doi.org/10.1162/pres.1994.3.2.130

Stock, C., & Bishop, I. (2005). Helping rural communities envision their future. In *Visualization in landscape and environmental planning* (pp. 145–151). Taylor & Francis.

Sutcliffe, A., & Kaur, K. (2000). Evaluating the usability of virtual reality interfaces. *Interacting with Computers, 13*(2), 327–341.

van Leeuwen, J. P., Hermans, K., Jylhä, A., Quanjer, A.-J., & Nijman, H. (2018). Effectiveness of virtual reality in participatory urban planning: A case study. In Proceedings of the 4th Media Architecture Biennale Conference (MAB18) (pp. 128–136). ACM.

Wilson, J. R. (2003). If VR has changed, then have its human factors? In D. de Waard et al. (Eds.), *Human factors in the age of virtual reality* (pp. 9–30). Shaker Publishing.

Rapid Development of a Cloud-Based Chatbot for Housing Policy Enquiries

Wayne Williamson

Abstract The use of artificial intelligence (AI) chatbots for urban planning in the literature is exploratory, fragmented, and lacking collaboration with practicing planners. Moreover, the potential for chatbots to complement or replace daily planning processes is not well understood. This study does not involve evaluating the performance of chatbots or selecting the most suitable tool; instead, the study sought to engage planners in the development of a chatbot to leverage existing question-answer data and socialise planners with the potential usefulness of AI for repetitive planning tasks. In-depth workshops were conducted with managers and planning officers in a state government planning agency in Australia. Insights were gained through discussion about planners subconsciously adapting to different jargon used in questions, how planners review and reuse previous answers, how recording questions and answers will aid the planning team, and low-risk pathways for planning officers to start using chatbots as an internal resource before making them public.

Keywords Chatbot · Artificial intelligence · Cloud computing · Planning practice · Housing policy

1 Introduction

Industry 4.0 transforms how organisations generate and capture value through advanced technologies. Key technologies driving this change include simulation technology, system integration, the Internet of Things (IoT), cybersecurity, cloud computing, and big data (Adel, 2022). Cloud-based chatbots, the focus of this chapter, use the internet to store and manage data on remote servers that can then deliver on-demand computing services with significant processing power (Adel, 2022; Hassan et al., 2021). Cloud-based chatbots are widely available today, particularly

W. Williamson (✉)
Department of Geography and Planning, Macquarie University, North Ryde, NSW, Australia
e-mail: wayne.williamson@mq.edu.au

R. Goodspeed et al. (eds.), *Digital-Era Urban Transformations*, The Urban Book Series, https://doi.org/10.1007/978-3-031-98300-9_5

in the form of personal assistants like Google Assistant, Siri, and Alexa, on smartphones and in people's homes (Harms et al., 2019).

For some practicing planners, part of their daily tasks includes providing information and advice on their planning system to various users. The customers, clients, and consumers of a planning system include, but are not limited to, applicants, residents, the wider public, small to large businesses, community groups, government agencies, other planning authorities, and elected officials (Kitchen, 2007: 105). Planners also deal with applicants both in the pre- and post-lodgement phases of a planning application (Schoneboom et al., 2024). Daily planning practice involves the clear provision of planning information, which is a skill to be mastered when communicating with those not versed in planning policy and processes (Schoneboom et al., 2024). Given that the customer is central to the planning service for most planning systems (Harris & Thomas, 2011), the use of chatbots to provide at least an initial level of customer service could be advantageous for planning agencies.

The goal of this chapter is to contribute to current literature on AI and urban planning by developing a proof-of-concept chatbot in collaboration with the housing policy team at the New South Wales (NSW) Department of Planning, Housing and Infrastructure (Department). Drawing on previous question-answer data and workshops with the housing policy team, Google's *Dialogflow* was employed to demonstrate how planners could use a no-code Software as a Service (SaaS) platform to create a chatbot. This follows on from observations by Sanchez et al. (2023: 191) that urban planners expect AI to "have a significant impact on planning practice in the future", while having low levels of knowledge about AI. Moreover, Fu (2024) notes there is an urgent need to collaborate with urban planners to understand how Natural Language Processing (NLP) could be adopted in planning practice. Hence, this chapter does not evaluate the performance of the chatbot beyond basic testing in workshop demonstrations but seeks to improve planners' knowledge of NLP concepts and the discussions this may generate. Dialogflow was chosen because it offers a simple user interface, and the question-answer data provided by the housing policy team was easily transformed into the required data format. The project used the Dialogflow Trial Edition, which is available at no charge.

A chatbot, also referred to as a dialogue system or conversational agent, is software designed to mimic human conversation with a machine (Adamopoulou & Moussiades, 2020). Typically embedded into online interfaces like web pages, chatbots aim to provide a personalised user experience. They use NLP, a subset of machine learning (ML), to understand, generate, and interpret natural language (Adamopoulou & Moussiades, 2020; Senadheera et al., 2024). The purposeful nature of NLP allows it to conduct conversations by extracting information from utterances and utilising it as a digital service. According to Lin (2022), chatbots draw on data from various sources, including websites, databases, social media, transcriptions of customer support conversations, and question-answer datasets.

This chapter begins by providing an overview of the literature on chatbots, chatbots used in urban planning, and the case study. To present this case study, the paper is then organised into two sections. The first section outlines the main concepts for developing a chatbot using Dialogflow. The second section discusses the data and training phases. The concluding sections provide a discussion regarding the role of planning jargon when dealing with public enquiries, ongoing training of a chatbot, and low-risk pathways for planning officers to start using chatbots.

2 Chatbots Overview

Chatbots are defined as sophisticated computer systems designed to simulate human conversation for automated online advice and support. The first notable chatbot, ELIZA, was developed by Joseph Weizenbaum in 1966. ELIZA employed pattern matching and pre-written scripts to generate responses based on keywords provided by users (Weizenbaum, 1966). ELIZA served as an initial trial for the Turing Test proposed by computer scientist Alan Turing to determine whether artificial intelligence could mimic human behaviour (Turing, 1950). Drawbacks for ELIZA were its limited knowledge of any particular topic, inability to engage in longer conversations, and iinability to learn from discussions. The late 1990s saw advancements in chatbot technology with the introduction of Artificial Intelligence Markup Language (AIML) (Wallace, 2009). Wallace's creation—A.L.I.C.E. (Artificial Linguistic Internet Computer Entity) enhanced the capability of conversational programs by employing conversational rules, heuristic pattern matching, and an expandable database of topics and categories, thereby enabling AIML chatbots to generate more human-like responses (Wallace, 2009).

Intelligent Personal Assistants (IPAs) such as Apple's Siri (Apple, 2025), Google Assistant (Google, n.d.), Microsoft Cortana (Microsoft, 2025), and Amazon's Alexa (Amazon, 2025) have become increasingly integral to daily life, reflecting substantial investments in AI chatbots equipped with NLP capabilities (Suta et al., 2020). IPAs utilise extensive datasets, web resources, and machine learning techniques to handle natural language queries across a broad spectrum of topics (de Barcelos Silva et al., 2020). Furthermore, there have been significant advancements in NLP that have revolutionised how people interact with businesses. Social media platforms have provided a platform for developers to deploy chatbots tailored to their brands or services, enabling customers to conduct routine tasks directly within messaging apps. This era witnessed the creation of numerous text-based chatbots with specialised functions across popular messaging platforms, industries, and research applications (Dale, 2016).

The next phase in the evolution of NLP emerged in late 2022 with the launch of ChatGPT (Fu et al., 2023). Large Language Models (LLM) such as ChatGPT, developed by OpenAI (openai.com), are Generative Pre-Trained Transformer (GPT) models, which can generate fluent and comprehensive responses to human questions and text summarisation (Zhong et al., 2023). The distinguishing factor for LLMs compared to the previous iterations of NLP technology is that they require little to no directions or examples for task recognition (Fu et al., 2023), due to the scale of the datasets used for training (Floridi & Chiriatti, 2020). However, researchers have found general-purpose LLMs underperform on domain-specific knowledge

(Shen et al., 2023). Fu (2024) suggests a domain-specific urban planning LLM could be built or fine-tuning an existing LLM could increase performance but would require significant amounts of resources, expertise, and time.

2.1 Chatbot Categories and Approaches

The primary objectives of a chatbot can be categorised into Informative, Conversational, and Task-based types. Depending on their knowledge domain, chatbots can access various levels of information. Generic chatbots can respond to user queries across different domains. Interpersonal chatbots focus on specific services like booking reservations or answering frequently asked questions (FAQ) without engaging in friendly conversations. Intrapersonal chatbots serve as close companions within a user's environment, often integrated into messaging platforms like Slack or WhatsApp. Lastly, Inter-agent chatbots facilitate communication between other chatbots, exemplified by the interaction between Alexa and Cortana (Nimavat & Champaneria, 2017).

The methods used by chatbots to generate responses can be split into three types: Rule-based, Retrieval-based, and GPT-based. Rule-based chatbots employ pattern-matching algorithms to match user input with predefined rules, pioneered by early chatbots like ELIZA (Weizenbaum, 1966). Retrieval-based models are more adaptable, selecting responses by analysing available resources, often using APIs to find the most appropriate answer (Hien et al., 2018). GPT models produce responses that mimic human-like natural language, drawing from training data and previous interactions (Boonstra, 2021). It is acknowledged that training GPT models is challenging due to the need for extensive data to facilitate meaningful conversations (Adamopoulou & Moussiades, 2020).

2.2 Natural Language Understanding and Generation

Chatbots extract the content from user utterances using NLP. Therefore, they do not need a specific response for every potential user input, but do require a training dataset, which can be a significant challenge when existing datasets are insufficient (Adamopoulou & Moussiades, 2020). Chatbots that rely on NLP techniques rely on Natural Language Understanding (NLU) to retrieve context from user utterances, and Natural Language Generation (NLG) to generate and return text to the user (Adamopoulou & Moussiades, 2020). NLU involves several steps of data preprocessing to clean text, transform it into a workable format, and highlight any features in the text that an algorithm can work with. At a basic level, the preprocessing includes removing common words, grouping different versions of the same word, and tagging nouns, verbs, or adjectives (Gillis, 2024). Chatbots using NLU to retrieve context from unstructured human language based on the user's intention use *intent classification* and *entity extraction* (Jung, 2019). The intent classification model can be a statistical algorithm or a pretrained model using manual classification of text phrases into topics, also known as *intents* (Intents, 2020). After training a chatbot, they can automatically classify the user's new inputs to the *intents* they were trained on (Hien et al., 2018).

2.3 The Dialogflow Approach

Dialogflow uses NLU to extract values from text. This approach seeks to recognise the *intent*, *entities*, and *context* of a user's utterances (Dialogflow, 2019; Boonstra, 2021). Dialogflow allows creators to train their chatbot agents to become smarter over time. Since Dialogflow is available through Google Cloud, it adheres to the Google Cloud Terms of Service. Being part of cloud services ensures reliability, low latency, and easy integration with other Google Cloud services, such as data analytics tools (Boonstra, 2021).

2.4 Chatbots and Urban Planning

There has been exponential growth in literature interested in government use of chatbots in recent years (Senadheera et al., 2024). Moreover, there is a need for guidance on harnessing NLP techniques in planning practice, which should also discuss the challenges, risks, and future research (Fu et al., 2024). The existing literature focuses on how all levels of government are using chatbots for citizen complaints and enquiries for a wide range of government services (Bright et al., 2019). For many government agencies, a chatbot is deployed as the first level of assistance (Makasi et al., 2020). While chatbots have been adopted for numerous customer-oriented services, there is limited literature focused on urban planning services. However, Bright et al. (2019) do provide an example from the UK where a chatbot was deployed to facilitate the process of applying for planning permission for certain housing development types. Moreover, Hasan et al. (2023) used planning guideline documents sourced from local government websites in Sri Lanka to create a vectorised database and utilised OpenAI's ChatGPT model to integrate with their database using Application Programming Interface (API) services. This approach allowed Hasan et al. (2023) to leverage the ChatGPT model in conjunction with specific knowledge contained in the planning documents. Feedback from targeted testing was generally positive and found that specialised urban planning chatbots may be beneficial for both citizens and city planners (Hasan et al., 2023).

There are also public-facing urban planning and urban management chatbots emerging. For example, New York's *MyCity* Chatbot (chat.nyc.gov), the *OpenCityAI* chatbot (opencityai.com) developed by Urban Information Lab at University of Texas and an AI planning assistant called *MyLot* (mylot.ai) developed by an

Australian start-up. Othengrafen et al. (2024) suggest future AI systems, like chatbots, will be highly specialised for processing specific planning tasks but will ultimately only make suggestions. Hence, AI will be regarded as a planning tool rather than a planning solution (Othengrafen et al., 2024). The existing literature is exploratory and fragmented, and suggests future research should seek to benchmark different NLP techniques, particularly given the latest developments with ChatGPT (Fu, 2024). Acknowledging the fast-paced nature of NLP research, Lin et al. (2025) have recently published the results of using a fine-tuned LLM for the classification of planning text sourced from planning applications.

3 The Case Study: Housing Policy Hotline

Housing policy in NSW is legislated by *State Environmental Planning Policies* and *Local Environmental Plans* (Williams, 2012). State Environmental Planning Policies (SEPP) deal with issues of significance to the state and people of NSW and are overseen by the NSW State Government. Local Environmental Plans (LEP) are statutory instruments developed and implemented by local governments (Williams, 2012). The *State Environmental Planning Policy (Housing) 2021 (DPHI, 2021)*, known as the Housing SEPP, facilitates the development of affordable and diverse housing. The Housing SEPP includes nine housing types and planning provisions or frameworks for short-term rental uses, apartment design, and collecting affordable housing contributions (Table 1).

Given the breadth of housing provisions in the Housing SEPP and the statewide application of the provisions, the Department maintains a phone number and email account as a contact for further information and advice. The housing policy team maintains a weekly roster for planners to respond to any phone calls and emails received (Manager, 2024).

The Housing Policy team maintains a basic contract tracking spreadsheet for phone and email enquiries, which provides the following high-level statistics. From January to December 2023, there were 523 recorded contacts consisting of 369 emails and 154 phone calls. This equates to 2.1 enquiries per workday throughout the year. It is noted that the enquiries can range from a single simple question through to a complex set of questions that may require hours to resolve and answer (Manager, 2024). The most common contact type are individuals, which include property owners and the public (Fig. 1). These enquiries are typically from people that are not familiar with the planning legislation and are looking to validate information provided to them by another source, for example a member of the public or a builder (Manager, 2024). Consultants and property developers are also regular users of the housing policy email account. These enquiries are more detailed and commonly include two or more questions from people who are familiar with the planning system and building regulations. While consultant planners provide

Table 1 Provisions in the Housing SEPP

Housing type	Summary of provisions
In-fill affordable housing	Provides floor space incentives to deliver both market and affordable housing
Boarding houses	Purpose built rental accommodation that must be managed by a community housing provider
Co-living	Purpose built rental accommodation that can be built and managed by a private entity
Build-to-rent housing	Purpose-built rental housing that is held in single ownership and management
Seniors housing	Housing designed to meet the needs of seniors and people with a disability
Caravan parks and manufactured home estates	Planning provisions for caravan parks, camping grounds, manufactured home estates and moveable dwellings
Group homes	Temporary or permanent accommodation for people with a disability or people that are socially disadvantaged
Retention of existing affordable rental housing	Provisions to mitigate the loss of existing affordable housing
Secondary dwellings	Planning provisions for a self-contained dwelling located within, attached to, or separate from another dwelling on the same lot
Social housing	Self-assessment powers for state agencies delivering housing
Short-term rental accommodation	Planning framework for management of dwellings used for short-term rental purposes
Design quality of residential apartment development	Planning and design standards for apartments
Affordable housing contribution schemes	Enabling provisions to develop an affordable housing contribution scheme

planning advice to their clients (Schoneboom et al., 2024), they often collect information from public service planners during the process.

The most common enquiries are in relation to secondary dwellings (Fig. 2). This reflects the high number of enquiries from individuals. Secondary dwellings, locally known as "granny flats" but also referred to as in-law units or accessory dwelling units (ADU) are small self-contained dwellings that may be built on the same lot as an existing single-family home. Hence, individuals considering building a secondary dwelling on their lot often start by conducting their own research, which can involve seeking initial advice from the housing policy team.

Affordable housing, which includes provisions for floor space incentives, retention of existing affordable housing, and contribution schemes, is the second most common enquiry. These enquiries are mostly from industry actors such as consultants, property developers, architects, and council planning officers, and are detailed questions regarding requirements and process (Manager, 2024). Similarly, seniors housing and group homes provisions are unique housing types that have detailed planning provisions contained in the Housing SEPP and generate detailed questions

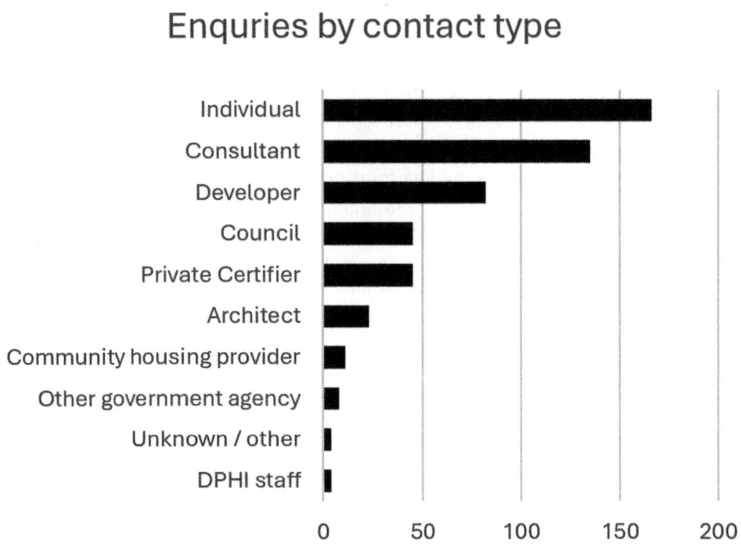

Fig. 1 Number of enquiries by contact type

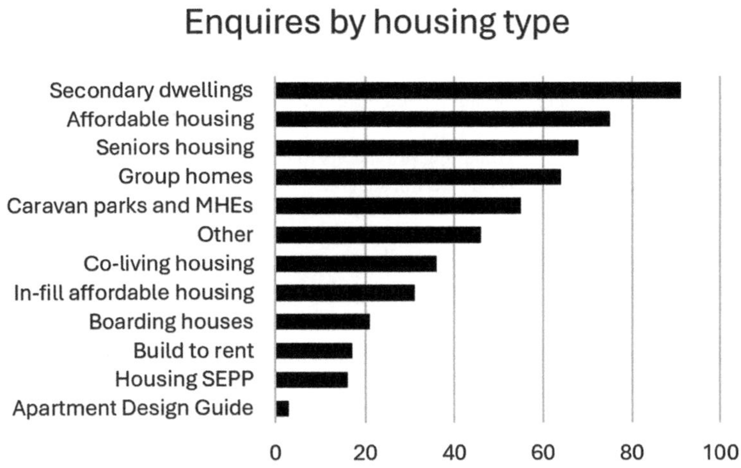

Fig. 2 Number of enquiries by housing type

from the industry. The Other category in Fig. 2 represents enquiries that are not applicable to the Housing Policy team. In most cases, they are people looking for advice and/or information on a different SEPP or function undertaken by the Department.

4 Development and Training

The chatbot was developed with Google's *Dialogflow*, a cloud-based SaaS platform that allows users to build chatbots with minimal coding skills. The reasons for choosing Dialogflow were twofold. First, it offers a simple user interface, and secondly, the question-answer data provided by the housing policy team was easily transformed into the *intents, training phrases,* and *responses* format required for Dialogflow's data architecture. It was useful to be able to show the planning staff how their data was formatted and loaded into the chatbots back-end through the content manager. We also considered Microsoft's Azure AI Language Services, which utilises the more powerful OpenAI GPT model. However, Azure's AI Services require some technical background to set up computational resources and training data. For this proof-of-concept chatbot, Dialogflow was deemed suitable.

In Dialogflow, the chatbot creator defines user enquiries and specifies how the chatbot may respond to enquiries based on keywords and concepts. Once an account has been created on the Google platform, the first step is to create an agent, in this case, the *housing policy chatbot*. The main elements needed to build the chatbot are *intents, training phrases,* and *responses* (Boonstra, 2021; Intents, 2020, Contexts, 2020).

Intents: are used to categorise a user's utterances. When a user asks the chatbot a question, it triggers an intent. Each intent contains training phrases, contexts, and responses.

Training phrases: are the questions, phrases, and words that users may add to the text field of the chatbot during their enquiries. A user's utterance will be matched against the training phrases. Training phrases are developed for each individual intent (Boonstra, 2021).

Responses: example text the chatbot uses to respond to training phrases.

Chatbots can also use *actions* and *parameters* to execute tasks. For example, the chatbot might log a reservation based on a customer request. The housing policy chatbot is designed to handle standard text-based enquiries, so these options were not necessary. More complex chatbots can also include *context* as part of *intents*. Chatbots use *context* to control a conversation's flow when turn-taking occurs and is executed with follow-up questions (Contexts, 2020). Dialogflow keeps track of previous user utterances, which can help to gain additional information from the user (Boonstra, 2021; Intents, 2020).

During back-end content creation and editing, Dialogflow provides the creator with the ability to add a list of *intents* based on potential enquiries. For each *intent*, training phrases and valid responses for those phrases are added (Fig. 3). Boonstra (2021) advises that it is good practice to include up to 15 different training phrases.

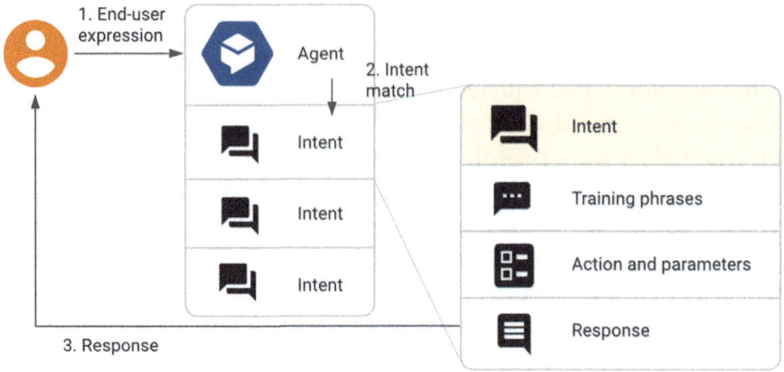

Fig. 3 Architecture of a Dialogflow agent and intents. (Source: Intents, 2020)

4.1 Content Creation

To populate the chatbot with *training phrases* and *responses*, the FAQ page on the Housing SEPP website (DPHI, 2021) was reviewed. The FAQ document and associated webpages do not contain enough information for chatbot training purposes. The housing policy team provided examples of question–answer data that have been submitted through the housing policy email account. Once the initial *intents* were written, workshops with the housing policy team (3 managers and 11 planning officers) were conducted in August 2024 to test and discuss responses. During these workshops, a conversational style like the responses currently provided for email enquiries was established.

Given that secondary dwellings are the most common enquiry received through the existing email account, a proof-of-concept chatbot was developed for that housing type. It is also one of the simplest housing types in the Housing SEPP. Notwithstanding, 32 *intents* were created for secondary dwellings, and it was trained on 105 training phrases and 52 unique responses. The tasks of reviewing and formatting frequently asked questions, grouping them into *intents*, loading the training phrases and responses into Dialogflow, and conducting the workshops took approx. 40 h. The project used the Dialogflow Trial Edition, which is available at no charge.

To provide a more conversational experience, default welcome and fallback *intents* were used, and each *intent* includes 'yes', 'no' and 'more' follow-up *intents* that either close off the conversation, direct the user to the housing policy email account or direct the user to information on the Department's website. Enquiries are not always related to the Housing SEPP. Thus, an *intent* was implemented for questions that may be about other planning laws as a way of helping people to find the right person or team within the Department to manage their enquiry.

4.2 Training and Learning

One advantage of using an AI chatbot is its capacity to learn and enhance its ability to respond to enquiries. Dialogflow facilitates this by allowing the addition of user enquiries to the *intents*, thereby refining the backend algorithm. Additionally, Dialogflow enables backend administrators to review past chatbot interactions and align them with more accurate *intents* and *responses*. These alignments then guide future responses based on familiar enquiries. It is recommended to regularly review chatbot transactions to improve content and fill in any gaps based on user questions and interactions (Boonstra, 2021). However, this aspect is more than just performance improvement, as Sanchez et al. (2024) note, human oversight and continuous monitoring are essential to ensure ethical practices in AI-driven urban planning.

Dialogflow offers features for reviewing transcripts of chatbot interactions. The training pages display a list of conversations between users and the chatbot. By selecting a conversation, administrators can see the date, the user's enquiry, and the chatbot's responses. Additionally, Dialogflow provides history and analytics options. The history feature identifies enquiries that did not match any *intents*. Building chatbots is an ongoing process. A chatbot will capture insights about what enquiries it is being asked. If conversations continue to be used to train an existing agent, the chatbot will become more complex over time (Boonstra, 2021).

5 Key Topics Discussed at Workshops

The workshops with the housing policy team were conducted in August 2024 and were split into two parts. First, testing the chatbot with recent questions that had been received through emails. This is important as testing must be undertaken with questions that are different from the training data. Secondly, a walkthrough of the training phrases and responses. The testing component of the workshops used eight new questions to test the chatbot's ability to respond to questions about secondary dwellings. The chat example in Fig. 4 demonstrates the chatbot's ability to correctly answer a common question received through the email account, asking if a secondary dwelling can be built on top of an existing garage.

Fallback responses were also tested by using questions like; "I've had a look at the planning legislation and I'm still a bit confused". Given the lack of keywords in the users' utterance, the chatbot prompts the user to rephrase the question. If the user cannot provide more details, the chatbot will direct them to the housing policy email account. The workshops generated enthusiastic discussions on several topics. Three key topics planning jargon, ongoing training, and a low-risk pathway to start using the chatbot are discussed below.

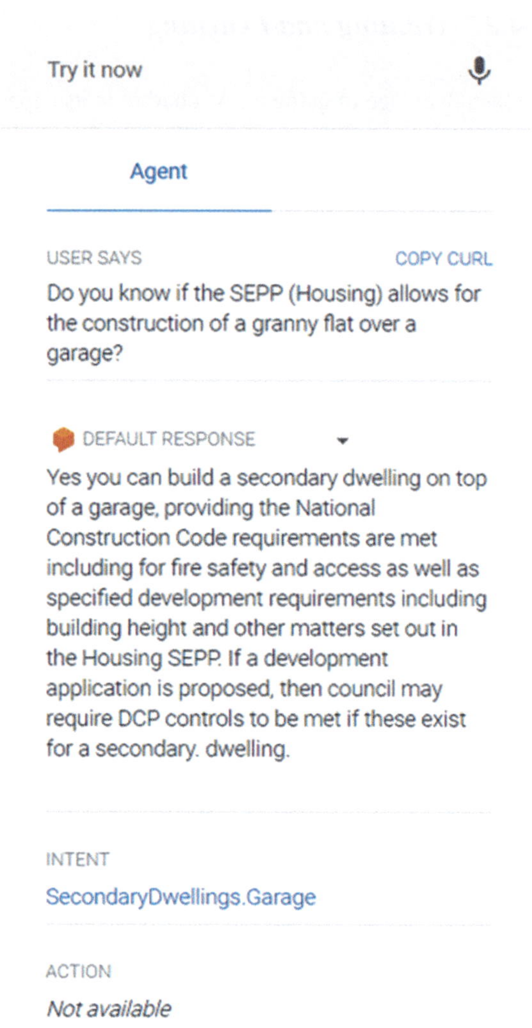

Fig. 4 Testing the *Housing Policy Chatbot*

5.1 Planning Jargon

The workshop attendees noted the need to recognise the phrases and words that the public uses. These are often different from what a planner would use. For example, a planner will commonly refer to a piece of land as a "site" or a "lot". However, the public is more likely to refer to their land as a "block". Moreover, a planner is also likely to use the term "dwelling", while the public is more likely to use the terms "house" or "home", but planners subconsciously adapt to the wording used.

> Well, and you kind of do that anyway when you're on the hotline. We know when somebody says block, you adapt your understanding. (Planning officer, 2024)

> When people say house versus dwelling, you know, they are interchangeable in our heads. (Planning officer, 2024)

It is only through training and testing that the chatbot creator gets an appreciation of the *Intents* architecture used by Dialogflow to extract values out of text (Dialogflow, 2019). The training data must include as many keywords as possible that the public may use. However, the sequence of words and the presence of *stop words*—words that do not add value to the sentence such as "hi", "my", "me" in the question(s) are removed as part of NLP process that breaks down text into a machine-readable format (Gillis, 2024). Dialogflow performs well in this regard. The workshop attendees were shown the various forms of planning jargon included in the *training phrases*. A manager remarked:

> Good training for us anyway, isn't it? Like, understanding the way different people express things outside planning jargon (Manager, 2024)

5.2 *Ongoing Training*

The use of planning jargon was a segue to a discussion on how to analyse the chatbot's performance if the public were given access to it. A demonstration of the analytics and history functions in Dialogflow was shown to the workshop members. This discussion prompted a comment from a planning officer that highlights the complexity of dealing with questions from the public:

> Like, I got a question the other day, which was can I convert a non habitable building to a secondary dwelling? I was able to tweak an answer that we have previously used relating to garages. But we didn't get enough information from that person to clarify. When you say non habitable, is it a shed? (Planning officer, 2024)

This comment highlights the countless ways in which the public may be considering the use of planning provisions. Moreover, it shows that planners tasked with answering questions from the public often undertake a similar process to a chatbot by reviewing previous answers and devising a bespoke answer to the current question. Essentially, the housing policy team routinely records these questions-answers and often reuses them. The step to training a chatbot with this data has been a relatively quick process over 5 days, not weeks or months.

5.3 *Low-Risk Pathway*

As the focus for creating the chatbot was an interface for questions from the public, it was unexpected when the workshop discussion turned to using the chatbot as an internal team resource:

> Would it be useful for us to just use it as an internal thing while we're still getting the courage to put it out. Would it make it easier for people to find answers to things rather than going through the spreadsheets and everything? Or would it not make much difference? (Manager, 2024)

These comments from a manager were well received, with others suggesting the chatbot will "train us, while we're training it" (Planning officer, 2024). This discourse raises questions about a chatbot's potential role in internal planning authority processes as opposed to the more well-known public-facing question-answer service.

6 Conclusions

This study provides insights from workshopping a specialised proof-of-concept chatbot with managers and planning officers. The first finding suggests that chatbots have the potential to influence planning agency processes. As chatbots are increasingly used as a communication channel to distribute information, they may also significantly influence how planners address the public's information needs. Secondly, the introduction of chatbots may allow planners to focus more on unique cases and utilise data analytics tools to learn from the public's questions while enhancing their own communication strategies. Finally, the use of chatbots may not produce short-term efficiencies. Although it is suspected that chatbots could be mobilised to manage repetitive information requests, it is envisaged that staff workloads would not decrease in the short term, as planners would need to dedicate time to assessing the performance of the chatbot, refreshing training data, and conducting further testing. Amendments to planning legislation would also trigger additional work. However, legislation amendments often include updates to web pages and a media release. Adding chatbot training phrases and responses to the communications package would be a complimentary task. The Housing Policy chatbot is not a set-and-forget type of project. It is envisaged that there would be ongoing training and maintenance.

Given that chatbots are at an early stage of adoption in urban planning, initial chatbot implementations may fail to cater to the public's needs. To this end, the current rate of emails and phone calls would persist. However, the managers who attended the workshop saw benefits in pursuing an internal implementation in the first instance. The benefits of this approach are twofold. Firstly, the housing policy team would have the opportunity to understand how AI might improve productivity and gain an understanding of the risks and limitations of using AI in the workplace. This would be done in parallel with existing work processes until stable results are achieved. Secondly, the internal implementation would function like a socialisation of AI with a view to becoming proficient in the use of chatbot technologies. Success would create AI champions in their planning agency.

Chatbots can offer both benefits and challenges for planning agencies, particularly when used to assist and inform the public. The real-time conversational data that would be recorded by chatbots could be invaluable for planning agencies and allow them to better respond to policy issues more effectively. However, this would require planning agencies to implement and manage chatbots and incorporate associated analytics into decision-making processes. Essentially, a chatbot would perform as a repository of knowledge. This knowledge is currently held by the more experienced planning officers and inconsistently recorded in various documents and spreadsheets.

Building a proof-of-concept chatbot using existing question-answer data in a cloud-based platform such as Dialogflow was an illuminating experience for all involved. The usable interface, simple concepts, and clear data architecture of the Dialogflow platform present planners with a genuine opportunity to create a public-facing chatbot with no coding skills. The cloud-based approach puts planners in the driving seat to create a chatbot, instead of trying to explain planning concepts and jargon to IT professionals to build a chatbot in an environment that may require coding skills. Further research could use longitudinal studies to assess workloads prior to and post chatbot implementation and monitor if cloud-based chatbots learn and improve responses over time.

References

Adamopoulou, E., & Moussiades, L. (2020). Chatbots: History, technology, and applications. *Machine Learning with Applications, 2*, 100006.

Adel, A. (2022). Future of industry 5.0 in society: Human-centric solutions, challenges and prospective research areas. *Journal of Cloud Computing, 11*(40), 1–15.

Amazon (2025). What exactly is alexa? Retrieved September 24, 2025 from https://developer.amazon.com/en-GB/alexa

Apple (2025). Siri Apple. Retrieved September 24, 2025 from https://www.apple.com/siri/

Boonstra, L. (2021). *The definitive guide to conversational AI with Dialogflow and Google Cloud*. APress.

Bright, J., Ganesh, B., Seidelin, C., et al. (2019). *Data science for local government*. University of Oxford.

Contexts. (2020). Retreived July 15, 2024, from https://cloud.google.com/dialogflow/es/docs/contexts-overviews

Dale, R. (2016). The return of the chatbots. *Natural Language Engineering, 22*(5), 811–817.

de Barcelos Silva, A., Miguel Gomes, M., Andre da Costa, C., et al. (2020). Intelligent personal assistants: A systematic literature review. *Expert Systems with Applications, 147*, 113193.

Department of Planning, Housing and Infrastructure (DPHI). (2021). *Housing SEPP*. Retrieved July 15, 2024, from https://www.planning.nsw.gov.au/policy-and-legislation/housing/housing-sepp

Dialogflow. (2019). Retrieved July 15, 2024, from https://cloud.google.com/dialogflow

Floridi, L., & Chiriatti, M. (2020). GPT-3: Its nature, scope, limits, and consequences. *Minds and Machines, 30*(4), 681–694.

Fu, X. (2024). Natural language processing in urban planning: A research agenda. *Journal of Planning Literature, 39*(3), 395–407.

Fu, X., Wang, R., & Li, C. (2023). Can ChatGPT evaluate plans? *Journal of the American Planning Association, 90*(3), 525–536.

Fu, X., Brinkley, C., Sanchez, T., et al. (2024). Text mining public feedback on urban densification plan change in Hamilton, New Zealand. *Environment and Planning B Urban Analytics and City Science*. https://doi.org/10.1177/23998083241272097

Gillis, A. (2024). *Natural language processing*. Retrieved August 2, 2024, from https://www.techtarget.com/searchenterpriseai/definition/natural-language-processing-NLP

Google (n.d). Meet your Google Assistant. Retrieved September 24, 2025 from https://assistant.google.com/

Harms, J. G., Kucherbaev, P., Bozzon, A., et al. (2019). Approaches for dialog management in conversational agents. *IEEE Internet Computing, 23*(2), 13–22.

Harris, N., & Thomas, H. (2011). Clients, customers and consumers: A framework for exploring the user-experience of the planning service. *Planning Theory and Practice, 12*(2), 249–268.

Hasan, F., Jayasinghe, A., & Sathsarana, S. (2023). Specialized urban planning chatbot: A participatory approach for evaluating efficiency. *Bhúmi, The Planning Research Journal, 10*(2), 3–16.

Hassan, H., Barakat, S., Qusay, I., et al. (2021). Survey on serverless computing. *Journal of Cloud Computing: Advances, Systems and Applications, 10*(39), 1–29.

Hien, H., Cuong, P., Nam, L., et al. (2018). *Intelligent assistants in higher-education environments: The FIT-EBot, a chatbot for administrative and learning support*. Paper presented at Ninth International Symposium on Information and Communication Technology, New York, NY, 6-7 December.

Intents. (2020). Retrieved July 15, 2024, from https://cloud.google.com/dialogflow/es/docs/intents-overview

Jung, S. (2019). Semantic vector learning for natural language understanding. *Computer Speech & Language, 56*, 130–145.

Kitchen, T. (2007). *Skills for planning practice*. Palgrave Macmillan.

Lin, P. (2022). *Training a chatbot: How to decide which data goes to your AI*. Retrieved July 9, 2024, from https://landbot.io/blog/chatbot-training-data

Lin, Y., Thackway, W., Soundararaj, B., et al. (2025). Transforming urban planning through machine learning: A study on planning application classification using natural language processing. *Environment and Planning B: Urban Analytics and City Science*. https://doi.org/10.1177/23998083251369142

Makasi, T., Nili, A., Desouza, K., et al. (2020). Chatbot-mediated public service delivery. *First Monday, 25*(12).

Microsoft (2025). Cortana voice assistance. Retrieved September 24, 2025 from https://www.microsoft.com/enus/cortana

Nimavat, K., & Champaneria, T. (2017). Chatbots: An overview types, architecture, tools and future possibilities. *International Journal for Scientific Research and Development, 5*(7), 1019–1024.

Othengrafen, F., Sievers, L., & Reinecke, E. (2024). From vision to reality: The use of artificial intelligence in different urban planning phases. *Urban Planning, 9*(8576), 1–18.

Sanchez, T., Shumway, H., Gordner, T., et al. (2023). The prospects of artificial intelligence in urban planning. *International Journal of Urban Sciences, 27*(2), 179–194.

Sanchez, T., Brenman, M., & Ye, X. (2024). The ethical concerns of artificial intelligence in urban planning. *Journal of the American Planning Association, 91*, 294–307. https://doi.org/10.1080/01944363.2024.2355305

Schoneboom, A., Slade, J., Tait, M., et al. (2024). *What town planners do: Exploring planning practices and the public interest through workplace ethnographies*. Bristol University, Policy Press.

Senadheera, S., Yigitcanlar, T., Desouza, K., et al. (2024). Understanding chatbot adoption in local governments: A review and framework. *Journal of Urban Technology, 32*, 35–69. https://doi.org/10.1080/10630732.2023.2297665

Shen, X., Chen, Z., Backes, M., et al. (2023). ChatGPT we trust? Measuring and characterizing the reliability of ChatGPT. arXiv preprint arXiv:2304.08979.

Suta, P., Lan, X., Biting, W., et al. (2020). An overview of machine learning in chatbots. *International Journal of Mechanical Engineering and Robotics Research, 9*(4), 502–510.

Turing, A. M. (1950). Computing machinery and intelligence. *Mind, LIX*(236), 433–460.

Wallace, R. S. (2009). The anatomy of A.L.I.C.E. In R. Epstein, G. Roberts, & G. Beber (Eds.), *Parsing the Turing Test* (pp. 181–210). Springer.

Weizenbaum, J. (1966). ELIZA—A computer program for the study of natural language communication between man and machine. *Communications of the ACM, 9*(1), 36–45.

Williams, P. (2012). Statutory planning. In S. Thompson & P. Maginn (Eds.), *Planning Australia: An overview of urban and regional planning* (2nd ed., pp. 98–123). Cambridge University Press.

Zhong, Q., Ding, L., Liu, J., et al. (2023). Can ChatGPT understand too? A comparative study on ChatGPT and fine-tuned BERT. arXiv preprint, arXiv:2302.10198.

Urban Streets and Mobility Systems

A City's Response to Platformized Mobility: Emerging Interactions of Ride-Hailing Beyond Automobile Travel

Si Qiao, Fang Bian, and Anthony Gar-On Yeh

Abstract The rapid emergence of ride-hailing services has introduced a significant shift in urban mobility, challenging conventional transportation paradigms and raising critical questions about interactions with urban functional zones. This chapter explores whether ride-hailing replicates general road automobile travel dynamics or embodies a distinct mobility mode requiring new urban planning frameworks by utilizing point-of-interests, mobile phone trajectories, and ride-hailing orders. We analyze unique interactions between ride-hailing and urban functional zones, finding significant differences compared to general road automobile travel. Proximity to shopping centers, airports, and metro stations negatively correlates with ride-hailing usage, contrasting with positive correlations for road automobile travel. These findings underscore the need for tailored urban planning to support evolving platform-based mobility systems.

Keywords Smart mobility · Built environment · Automobile travel · Urban planning · Land use

1 Introduction

Ride-hailing typically refers to a digital platform service that enables users to book a vehicle ride via a mobile application, connecting passengers with drivers directly. Traditionally, ride-hailing is often compared with private car usage, taxis, and public transportation modes (such as buses). These are well-accepted in the transport

S. Qiao (✉) · A. G.-O. Yeh
Department of Urban Planning and Design, The University of Hong Kong, Hong Kong, SAR, China
e-mail: siqiao@connect.hku.hk; hdxugoy@hku.hk

F. Bian
School of Architecture, Harbin Institute of Technology (Shenzhen), Shenzhen, China
e-mail: bianfang@hit.edu.cn

R. Goodspeed et al. (eds.), *Digital-Era Urban Transformations*, The Urban Book Series, https://doi.org/10.1007/978-3-031-98300-9_6

literature as common alternatives or complements to ride-hailing (e.g., Huang et al., 2021; Wang & Noland, 2021; Zhang et al., 2020). However, how platform-based services like ride-hailing differ in their interaction with urban functions (e.g., commercial, residential, parking zones) compared to more established, non-platformized forms of road automobile travel remains a persistent gap. This gap highlights a deeper theoretical challenge: whether ride-hailing merely replicates the established dynamics of road automobile travel or embodies a fundamentally distinct mode of smart mobility that requires reevaluating urban planning frameworks.

Two potential scenarios emerge in the examination of these debates. The first scenario posits that the travel purposes facilitated by ride-hailing services may mirror those of general automobile travel, indicating a convergence in how urban functional zones impact both forms of transportation, even though some argue that technology-based smart mobility subverts traditional travel mobility, including its organization, operation, and governance (Docherty et al., 2018). As urban mobility systems are deeply entrenched in historical patterns of car dependency and related planning policies and governance structures, ride-hailing services, operating within a socio-technical niche, may not be able to transform urban functions depending on the extent to which they can gain traction and disrupt established mobility regimes, or whether they remain a complement to existing road automobile travel (Schwanen, 2016).

Many cases examined the substitute effect of ride-hailing and other road automobile modes (e.g., Bian et al., 2023; Qiao & Yeh, 2021, 2023a), suggesting ride-hailing services act as *substitutes for traditional road automobile modes*—especially taxis and private cars. Thus, ride-hailing could be considered more of a travel alternative, simply the response to the explosive growth in data production brought about by the global penetration of digital technologies (Kong & Woods, 2018; Luque-Ayala & Marvin, 2015). In this scenario, the mechanism of ride-hailing would not undergo profound changes despite the broader discourse surrounding smart mobility and the advent of a new mobility era. The interaction between urban functions and ride-hailing would likely follow the car-dependency patterns.

The second scenario suggests that the interactions between urban functional zones and ride-hailing services may diverge significantly from those associated with general road automobile travel. This divergence could indicate that *ride-hailing operates according to its own inherent travel logic*, one that is distinct from traditional forms of mobility. As digital platforms become increasingly central to urban mobility, platform urbanism offers a lens suggesting cities are navigating a complex platform-based ecosystem that includes private companies, public organizations, and citizens (Van der Graaf & Ballon, 2019). Ride-hailing platforms, thus, are empowered to reconfigure the urban landscape by shifting mobility practices and influencing both individual travel behavior and broader spatial dynamics. These platforms do not simply optimize transportation; they introduce new socio-spatial constructs that challenge traditional urban planning paradigms, which would necessitate the development of a unique planning framework specifically tailored to guide the future of smart mobility, such as urban air mobility, autonomous vehicle fleets, and Mobility-as-a-Service (MaaS) systems (Qiao et al., 2022).

So, in this study, we compare the interactions between *ride-hailing* and urban functions with *other road automobile travel*, which encompasses traditional forms of vehicle use such as private cars and taxis. This comparison is supported by transportation research that groups these modes under road-based travel (Shaheen & Chan, 2016; Clewlow & Mishra, 2017). To accomplish this objective, this study initially generates the distribution of urban functions by reclassifying the point-of-interest (POI) dataset through the semantic context transformation based on natural language processing (NLP). After that, this study analyses the effects of identified functional zones on ride-hailing usage and general road automobile travel. Ride-hailing usage is estimated by one-month user orders with pick-up and drop-off locations, and it collects one-month mobile phone trajectories to identify general road automobile travel.

The remainder of this paper is organized as follows. Section 2 reviews the theoretical underpinnings of urban functional zones and road automobile travel behavior, particularly focusing on ride-hailing. Section 3 describes the study area and data and proposes to extract urban functional zones. Section 4 reports the results and discusses the empirical findings. Lastly, Section 5 provides the conclusions and future research directions.

2 Literature Review

2.1 Effects of Urban Functions on Ride-Hailing Usage

Recent research has explored the relationship between urban built environments and smart mobility, particularly focusing on how factors like density, diversity, and design influence ride-hailing usage (Tirachini & del Río, 2019; Ghaffar et al., 2020). Studies in cities like Austin, Chicago, and Chengdu show that high employment density often correlates with increased ride-hailing for commuting (Yu & Peng, 2020; Ghaffar et al., 2020). However, the effects of commercial and shopping centers are more consistent, with most studies indicating that these zones generally attract higher ride-hailing activity (Bian et al., 2023; Wang & Noland, 2021).

For infrastructural amenities, such as transport hubs and hospitals, ride-hailing services can act as feeders for intercity transit (Wang & Noland, 2021). However, in some cases, competition with metro systems has been observed (Ghaffar et al., 2020). In areas with sufficient medical facilities, ride-hailing demand for such services tends to decrease, as residents prefer other transportation options (Li et al., 2023).

Recreational facilities, such as parks and leisure areas, are also linked to ride-hailing demand, but findings vary based on accessibility. High concentrations of entertainment facilities increase ride-hailing usage for leisure trips, but spatial proximity often leads people to walk instead (Bian et al., 2023; Zheng et al., 2022).

2.2 Effects of Urban Functions on General Road Automobile Travel

Research on the effects of urban functions on road automobile travel has shown that accessibility to employment and retail centers generally reduces vehicle miles traveled (VMT) (Ewing et al., 2015). In cities like Mexico City and Jinan, high employment accessibility has been linked to fewer road automobile trips (Guerra, 2014; Jiang et al., 2017). However, evidence on the influence of grocery stores and other commercial outlets is mixed, with some studies reporting increased car use for shopping trips (Chatman, 2013; Li & Zhao, 2017).

The interaction between transport infrastructure and road automobile travel has been extensively studied, particularly the impact of proximity to public transit hubs. In the USA, public transit availability tends to reduce car use, while findings in cities like Shanghai suggest varying effects based on local conditions (Cervero & Murakami, 2010; Li & Zhao, 2017). For healthcare trips, shorter distances to medical facilities encourage walking over car travel, especially among elderly populations (Du et al., 2020). Recreational amenities pose one of the most complex links to road automobile travel, with car ownership and usage often increasing in neighborhoods close to leisure facilities (Chatman, 2013). However, the availability of comprehensive empirical studies on this subject remains limited.

In summary, while studies have explored the interactions between urban functions and both ride-hailing and road automobile travel, gaps remain. Specifically, a comparative understanding of how different urban functions affect ride-hailing and road automobile travel in the same geographic and social context is still unclear. This research aims to address these gaps, providing insights for planners on optimizing urban functional zones for both private cars and ride-hailing services.

3 Methodology

3.1 Study Area and Spatial Unit

Chengdu was selected as the case study due to two key factors. First, Chengdu is the second-largest ride-hailing market for DiDi in China,[1] with over 8.5 million active users (CBN Data, 2016). The city's well-established ride-hailing market provides an ideal context for analyzing the relationship between ride-hailing demand and neighborhood characteristics. Second, Chengdu's monocentric urban structure simplifies the study of travel behaviors.

This study uses 614 neighborhoods as the primary spatial unit, each representing a residential block or quarter. In the Chinese context, neighborhoods serve as key

[1] DiDi refers to providers of on-demand and application-based transport services similar to those provided by various transportation network companies, such as Uber, Lyft and Grab.

administrative and geographical units, with residents typically sharing similar sociodemographic attributes, making them suitable for this analysis.

3.2 *Data Descriptions*

This study utilizes multiple data sources, summarized in Table 1, with descriptive statistics provided in Table 2. The primary data on ride-hailing usage was derived from DiDi, one of China's largest transport network companies. The dataset includes 6,618,725 trips made by DiDi users in Chengdu during the period from November 1 to November 30, 2016.

To compare ride-hailing with general road automobile travel, we supplemented the data with mobile phone signaling trajectories from China Unicom. This dataset captures residents' travel behavior in Chengdu over the same month, comprising 29,988,792 home-based travel records. The definition of general automobile travel here refers to including all forms of road-based travel, encompassing private cars, taxis, and ride-hailing services. Thus, from this dataset, 17,704,041 road automobile trips were identified based on travel speeds between 20 km/h and 120 km/h. Speed thresholds were determined based on established studies on transportation modes (Aarts & Van Schagen, 2006; Keller et al., 1996; Qiao & Yeh, 2023b). The distribution of automobile and non-automobile travel was found to be consistent with the findings presented in the report on transportation characteristics in Chengdu (Chengdu, 2021).

A key objective of this study is to characterize urban functional zones using open-source point-of-interest (POI) data. The POI dataset, obtained from the Gaode Map API, consists of 1,037,630 data points, categorized across three hierarchical levels: 23 top-level categories, 232 mid-level categories, and 1571 bottom-level categories. To enhance classification clarity and reduce redundancy, we applied a word embedding process based on situational semantics to refine the classification of POIs.

Proximity to various functional zones was calculated as a key variable, representing accessibility to urban services. Additionally, a mixed land-use indicator, reflecting the diversity of neighborhood services, was included to capture land-use heterogeneity.

We also accounted for transportation infrastructure by including variables such as proximity to metro stations, bus stop density, and road network characteristics (e.g., road types and intersections).

Finally, we incorporated sociodemographic factors—such as population size, age distribution, gender, and income levels—that are well-documented in influencing ride-hailing use (e.g., Qiao & Yeh, 2023a; Rayle et al., 2016; Tirachini & del Río, 2019). Education was excluded to avoid multicollinearity with income, as these variables are often closely correlated.

Table 1 Variables used in the model and their measures and data sources

Variables	Variable measures	Data source	Time of data
Dependent variable			
Ride-hailing orders	Recorded orders of DiDi within a neighborhood for 30 days	DiDi company	November 2016
Automobile travel	Trips of automobile trips with a speed between 20-120 km/h for 30 days	Mobile phone	November 2018
Independent variable			
Urban characteristics			
Urban functions			
Proximity to different types of functional zones	Distance to closest functional zones by corresponding types	Calculated by authors	2018
Mixed land use	Entropy index calculated from POI dataset with categories of medical service, educational service, industrial land use, leisure facilities, living facilities, shopping and caterings, firms, and public sectors	Based on POIs	2018
Transport facilities			
Bus stop density	Log of the number of bus stops in a neighborhood/area of a neighborhood	Gaode map	2018
Distance to metro station	Distance to closest metro station	Gaode map	2018
Road network design			
Expressway density	Log of the length of the expressway in a neighborhood/area of a neighborhood	Gaode map	2018
Arterial density	Log of the length of arterial road in a neighborhood/area of a neighborhood	Gaode map	2018
Secondary road density	Log of the length of secondary road in a neighborhood/area of a neighborhood	Gaode map	2018
Branch road density	Log of the length of branch road in a neighborhood/area of a neighborhood	Gaode map	2018
Sidewalk density	Log of the length of sidewalk in a neighborhood/area of a neighborhood	Gaode map	2018
Intersection density	Log of number of three-leg and four-leg intersections in a neighborhood/area of a neighborhood	Gaode map	2018
Sociodemographic characteristic			
Resident density	Resident population living in the neighborhood /the area of a neighborhood	Lands department	2018
Age level	Sorted by average age of neighborhood, the top 1/3 is marked as 1, the middle 1/3 is marked as 2, and the last 1/3 is marked as 3	MBR	2018
Gender	If the proportion of the male population exceeds 0.5, it is recorded as 1; otherwise, 0	MBR	2018
Income level	Sorted by the average income of the neighborhood, the top 1/3 is marked as 1, the middle 1/3 is marked as 2, and the last 1/3 is marked as 3	MBR	2018

Table 2 Descriptive statistics of variables

Variables	Unit	Mean	Std. Dev.
Ride-hailing orders	Number	10779.93	16159.67
Automobile travel	Number	15585.40	14648.40
Urban characteristics			
Functional zones			
Distance to different types of functional zones			
Shopping center	Kilometer	5.92	5.16
Business zone	Kilometer	1.02	1.81
Working zone	Kilometer	0.16	0.77
Park and green space	Kilometer	0.56	0.80
University	Kilometer	2.03	3.10
Hospital	Kilometer	2.93	3.77
Airport	Kilometer	6.87	3.99
Mixed land use	Entropy	0.57	0.24
Transport facilities			
Bus stop density	Number/km^2	12.86	9.85
Distance to metro station	Km	0.97	1.26
Road network design			
Expressway density	Meter/km^2	779.90	1483.80
Arterial density	Meter/km^2	2377.57	2197.64
Secondary road density	Meter/km^2	6117.27	3171.40
Branch road density	Meter/km^2	5341.78	3498.8
Sidewalk density	Meter/km^2	448.50	1129.34
Intersection density	Number/km^2	79.25	53.13
Sociodemographic characteristics			
Population density	1000 residents/km^2	22.57	18.04
Gender: Male percentage	Percentage	0.50	0.01
Age	Year	42.37	4.27
Income	1 RMB	47157.90	6447.22

3.3 *Identification of Urban Functional Zones*

A key challenge in identifying urban functional zones lies in the limitations of POI data, which provides only two-dimensional information: coordinates and categories. To address this, we employed a semantic analysis approach using natural language processing (NLP) to estimate zoning types, improving accuracy in classification (e.g., Gao et al., 2017; Yao et al., 2017; Zhai et al., 2019).

We divided the study area into 4140 functional analysis zones (FAZs), defined by the main road network, including highways and arterial roads. This finer spatial granularity allowed us to better infer urban functions compared to broader neighborhood units (Zhai et al., 2019).

We then used a word embedding model to reclassify POIs based on their semantic context. The dataset, containing 1751 POI types, was trained using over eight million Chinese words from Tencent's AI Lab.[2] Using the skip-gram model, we generated a 768-dimensional vector space, mapping similar POIs together. Dimensionality was reduced to three using the T-SNE method, confirming that POI types were well-clustered based on the cosine similarity of their word embeddings (Liu et al., 2020). Finally, POIs unrelated to human activity (e.g., road signs, indoor facilities) were eliminated, and functional zones were identified via clustering based on the reclassified POIs.

3.4 Elasticity Log–Log Model Regression

This study aims to examine the unique interactions between urban functions and ride-hailing usage/automobile travel by using the elasticity log-log model regression. This model is particularly advantageous for determining the nonlinear effects of proximity to different types of functional zones on travel trips. The reason is that the model enables the estimation of percentage change in trips for a given percentage change in distance to various functional zones by transforming the dependent and independent variables into natural logarithms. The corresponding formula is as follows:

$$\ln(Y) = \beta_0 + \sum_{n=1}^{N} \beta_n \ln(X_n) + \varepsilon$$

where $ln(Y)$ is the natural logarithm of the dependent variables ride-hailing and automobile trips; $ln(X_n)$ are the natural logarithms of the independent variable distance to different types of functional zones; β_n are the coefficients for the independent variables, representing the elasticities of the percentage change in Y for a 1% change in the respective independent factors, and ε is the error term, which represents the unexplained variation in $ln(Y)$.

4 Results

4.1 Spatial Patterns of Urban Functions, Ride-Hailing Usage, and Road Automobile Travel

Figure 1 shows the clustering results of Chengdu's urban functional zones. The undeveloped zone has a minimal correlation with any POI types, while the residential zone is dominated by housing and limited services. Large shopping centers are strongly associated with retail services, particularly shopping malls and hypermarkets, alongside elevated entertainment and catering services. Business zones feature

[2] https://ai.tencent.com/ailab/nlp/data/Tencent_AILab_ChineseEmbedding.tar.gz.

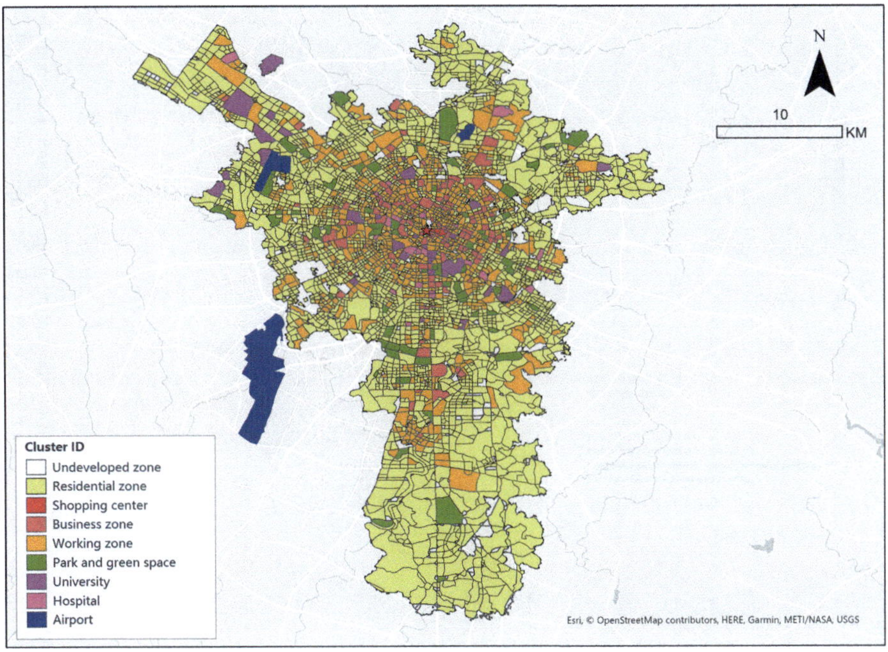

Fig. 1 Illustration of urban functional zones derived from the two-step semantic method

a high concentration of firms and general daily services, while working zones primarily consist of companies and commercial buildings with few daily life amenities. Parks, universities, hospitals, and the airport show strong correlations with their respective POI types. The clustering results were validated using high-resolution remote sensing images obtained from Sentinel-2 satellites and local functional data provided by the Chengdu Urban Planning Bureau. The validation process involved cross-referencing the identified clusters with land-use classifications from the functional data and visual inspections of satellite imagery to confirm the presence and characteristics of urban functions.

Figure 2 illustrates the spatial and temporal distributions of ride-hailing usage and road automobile travel. Ride-hailing usage is evenly distributed throughout the day, while road automobile travel peaks during rush hours. Spatially, ride-hailing is concentrated around the city center, whereas automobile trips are more dispersed across the urban area.

4.2 *Effects of Urban Functions on Ride-Hailing Usage*

Table 3 lists the results of the elasticity log–log model regression for estimating the effects of urban functions on ride-hailing usage. VIFs for all variables are under 5, showing that the model has no evident multicollinearity. Additionally, the 0.848 of R-squared shows the model has a good fit.

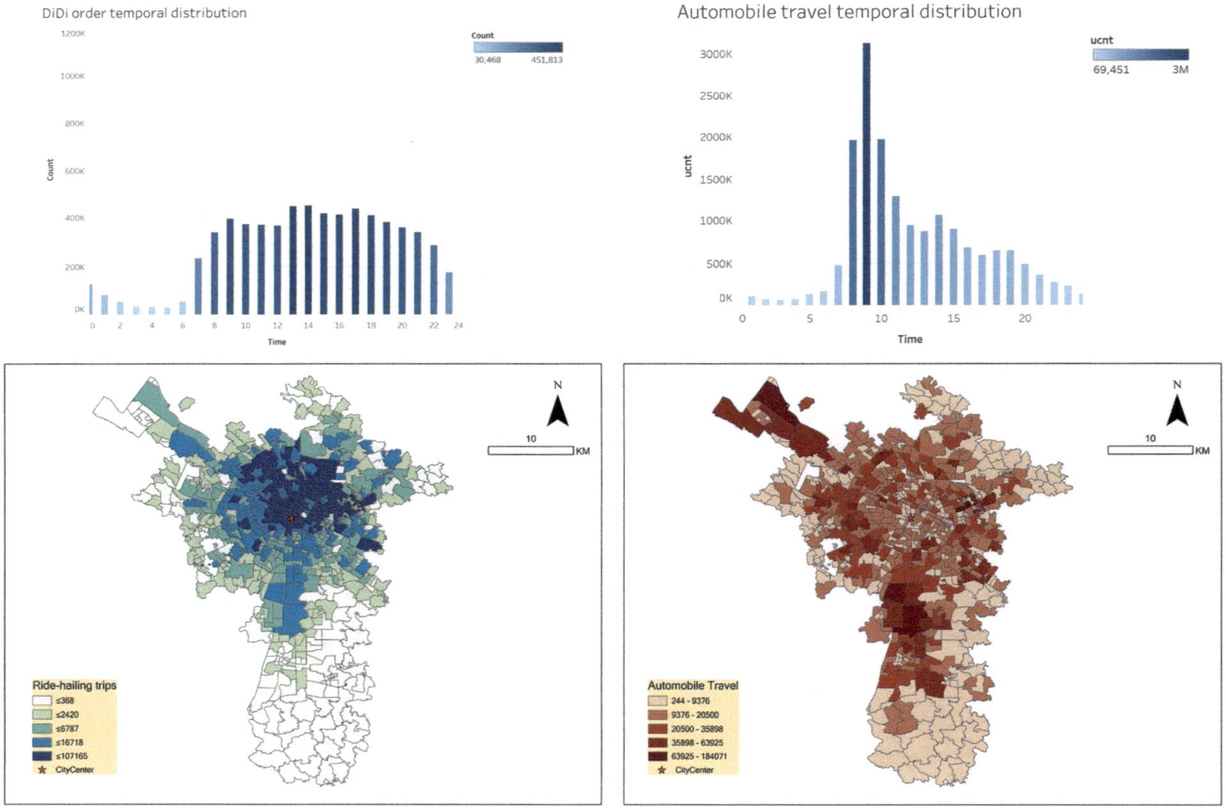

Fig. 2 Comparison of ride-hailing usage and general road automobile travel. (Left top: temporal distribution of ride-hailing usage by hours; Left bottom: spatial distribution of ride-hailing usage by neighborhoods; Right top: temporal distribution of road automobile travel by hours; Right bottom: spatial distribution of road automobile travel by neighborhoods)

Table 3 Regression results of elasticity log-log model for ride-hailing usage

Ride-hailing orders (log)	Coefficient	Robust S.E.	p-value	Sig	VIF
Independent variables					
Urban characteristics					
Urban functions					
Distance to different functional zones (log, km)					
Shopping center	−0.423	0.055	0.000	***	2.39
Business zone	−0.277	0.041	0.000	***	2.39
Working zone	−0.649	0.088	0.000	***	1.66
Park and green space	−0.140	0.039	0.000	***	1.18
University	−0.176	0.034	0.000	***	1.60
Hospital	−0.226	0.039	0.000	***	1.99
Airport	−0.324	0.046	0.000	***	1.21
Mixed land use	−0.184	0.164	0.263		1.10
Transportation infrastructure					
Bus stop density (log, 1/km^2)	0.378	0.140	0.007	***	2.05
Distance to metro (log, km)	0.017	0.041	0.676		1.54
Road network design (log, 1/km^2)					
Expressway density	−0.012	0.027	0.650		1.12
Arterial density	0.180	0.034	0.000	***	1.24
Secondary road density	0.814	0.141	0.000	***	1.91
Branch road density	0.190	0.119	0.112		1.54
Sidewalk density	0.149	0.031	0.000	***	1.37
Intersection density	0.262	0.094	0.005	***	3.90
Sociodemographic characteristics					
Population density (1000/km^2)	0.027	0.009	0.002	***	2.51
Gender (male percentage)	0.833	0.153	0.000	***	3.31
Age (3 levels)	0.674	0.070	0.000	***	2.05
Income (3 levels)	0.013	0.087	0.877		3.49
Cons.	−0.642	0.803	0.424		
Observations	614				
R-squared	0.848				
Akaike crit. (AIC)	1710.799				
Bayesian crit. (BIC)	1803.619				

Note: *** $p < 0.01$, ** $p < 0.05$, * $p < 0.1$

The regression results show that the seven types of functional zones significantly impact ride-hailing usage. The negative effect coefficients of functions indicate that the closer to the type of functional zone, the more ride-hailing orders were generated. Sorted in order of coefficients, working zones (−0.649***), shopping centers (−0.423***), airports (−0.324***), business zones (−0.277***), hospitals (−0.226***), universities (−0.176***) and parks and green space (−0.140***), are reported separately. Considering that the average distances to different functional zones vary, the generated trips converted according to elastic effects for the entire month are 4372 trips for every meter closer to the working zone, 292 trips for the business zone, 269 trips for parks and green space, 93 trips for university, 83 trips for hospital, 77 trips for shopping center and 50 trips for the airport. However, the land use mix did not show any significant signs.

Apart from the effects of urban functions, bus stop density (0.378***) shows a positive link to ride-hailing usage, indicating that a neighborhood with denser bus stops relates to higher ride-hailing usage. This correlation is consistent with the results of Circella et al. (2018), suggesting that the substitution effect may exist between public transit and ride-hailing demands.

In terms of road design factors, the road network density of arterial (0.180***), secondary road (0.814***), sidewalk (0.149***), and intersections (0.262***) are statistically significant and positive to neighborhood ride-hailing usage, consistent with the findings from previous studies (Wang & Mu, 2018; Yu & Peng, 2019, 2020; Qiao & Yeh, 2023a, 2023b). This result indicates that higher road density could attract more ride-hailing orders and reduce the response time of ride-hailing vehicle drivers through better road infrastructure accessibility.

Results of sociodemographic factors reveal that ride-hailing usage is higher in neighborhoods with a high proportion of males (0.833***) and senior groups (0.674***), consistent with a previous individual-level investigation (Rayle et al., 2016). The average age of all neighborhoods is about 42 years old, which indicates that neighborhoods with high shares of middle-aged groups have more ride-hailing needs. Nevertheless, considering that the middle-aged group may have a higher income or higher travel demands, this indicator does not mean that aging neighborhoods will generate more ride-hailing trips than younger neighborhoods. By contrast, previous studies have shown that elderly residents are a vulnerable group subject to smart-induced restrictions owing to the digital divide affecting access to smart mobility for the aging population (Zhang et al., 2024).

4.3 Comparison Between Ride-Hailing and Road Automobile Travel

Table 4 lists the results of the elasticity log-log model regression for estimating the effects of urban functions on road automobile travel. The regression result shows different effects of the same influencing factors on road automobiles compared with ride-hailing usage. The five types of functional zones (out of seven) have a significant impact on road automobile travel. The negative effect 1coefficients of the business zone (−0.198***), working zone (−0.373***), and park and green space (−0.104***) indicate that the 1 m closer to these types of functional zones, the more ride-hailing orders were generated, converting separately into 302 trips for business zone, 3633 trips for working zone, 289 trips for park and green space. Meanwhile, distance to shopping centers and the airport shows positive correlations (0.199*** and 0.082***) with road automobile travel, indicating that being closer to shopping

Table 4 Regression results of the elasticity log-log model for other automobile travel

Automobile travel (log)	Coefficient	Robust S.E.	p-value	Sig	VIF
Independent variables					
Urban characteristics					
Urban functions					
Distance to different functional zones (log, km)					
Shopping center	0.199	0.042	0.000	***	2.39
Business zone	−0.198	0.036	0.000	***	2.39
Working zone	−0.373	0.059	0.000	***	1.66
Park and green space	−0.104	0.032	0.001	***	1.18
University	0.025	0.029	0.382		1.60
Hospital	−0.015	0.029	0.615		1.99
Airport	0.082	0.034	0.016	**	1.21
Mixed land use	0.038	0.133	0.777		1.10
Transportation infrastructure					
Bus stop density (log, 1/km^2)	0.427	0.109	0.000	***	2.05
Distance to metro (log, km)	−0.211	0.034	0.000	***	1.54
Road network design (log, 1/km^2)					
Expressway density	0.047	0.021	0.029	**	1.12
Arterial density	0.073	0.027	0.008	***	1.24
Secondary road density	0.128	0.131	0.329		1.91
Branch road density	0.040	0.096	0.678		1.54
Sidewalk density	0.100	0.026	0.000	***	1.37
Intersection density	−0.274	0.087	0.002	***	3.90
Sociodemographic characteristics					
Population density (1000/km2)	0.013	0.009	0.121		2.51
Gender (male percentage)	−0.329	0.142	0.021	**	3.31
Age (3 levels)	−0.228	0.063	0.000	***	2.05
Income (3 levels)	−0.239	0.082	0.004	***	3.49
Cons.	8.417	0.628	0.000	***	
Observations	614				
R-squared	0.437				
Akaike crit. (AIC)	1454.246				
Bayesian crit. (BIC)	1547.066				

Note: *** $p < 0.01$, ** $p < 0.05$, * $p < 0.1$

centers and the airport relates to less road automobile travel (52 and 18 trips less, respectively, for each meter closer).

Apart from the effects of functions, bus stop density (0.427***) and distance to the metro (−0.211***) show significant links to road automobile travel. This result indicates that a neighborhood with denser bus stops and closer to the metro relates to higher road automobile travel. The impact of the road networks on both automobile and ride-hailing travel is similar, suggesting that a markedly concentrated road infrastructure would enhance their utilization.

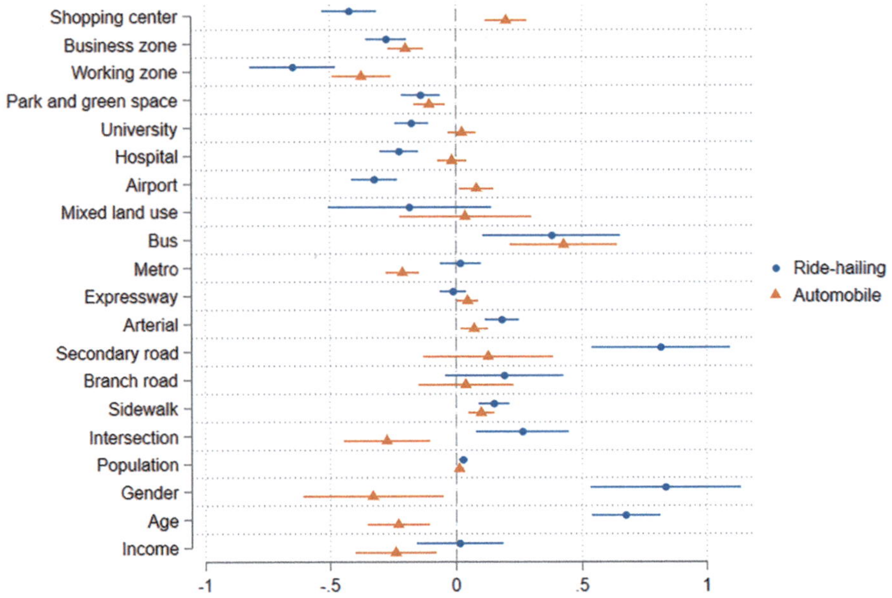

Fig. 3 Comparison of coefficients between ride-hailing and other road automobile travel

Figure 3 illustrates the comparison between ride-hailing usage and road automobile travel. The varied effects of shopping centers, universities, and airports elaborate on the mechanism that using ride-hailing is different from other road automobile travel. Combining the temporal distribution of their utilization, ride-hailing could be considered a supplementary travel mode used for irregular travel purposes that require the most flexibility. For road networks, secondary road networks and intersections are important designs for ride-hailing calls. Moreover, the demographic characteristics of ride-hailing users may significantly differ from those who opt for other automobile modes, as evidenced by socioeconomic indicators, especially age and gender.

5 Discussion and Conclusions

This study estimates the unique effects of urban function zones on ride-hailing usage compared with general road automobile travel. Firstly, this study determined the heterogeneous effects of proximity to different types of functional zones on ride-hailing usage. The current research observed that proximity to urban functional zones tends to induce a high degree of ride-hailing usage after controlling for other influencing factors. The number of ride-hailing trips increases throughout the month by 4372 for every meter closer to the working zone, 292 for the business zone, 269 for parks and green space, 93 for the university, 83 for the hospital, 77 for the shopping center, and 50 for the airport. The positive effect of functions on ride-hailing usage is consistent with most studies conducted in the same city (Qiao & Yeh, 2021; Wang & Noland, 2021; Zheng et al., 2022).

However, when comparing ride-hailing usage with road automobile travel, the effects of functions exhibit variations. For general road automobile travel, each meter closer to the working zone, business zone, and park and green space corresponds to the generation of 3633, 302, and 289 trips, respectively. Conversely, each meter closer to shopping centers and the airport relates to reducing 52 and 18 trips during the month. The finding is comprehensible because travel demands to shopping centers or the airport are characterized by greater irregularity and flexibility, as well as the necessity for on-demand and long-distance movement capability compared with demands for work. This finding is consistent with the conclusion that ride-hailing could provide necessary access to the airport and nighttime entertainment areas when public transit cannot fill the needs (Qiao & Yeh, 2021).

Therefore, this study's findings substantiate the unique effects of urban functions on ride-hailing compared with other road automobile travel, implying that ride-hailing services possess a distinct travel logic for certain purposes. Proximity to shopping centers, airports, and metro stations exhibits an increase in ride-hailing compared with general road automobile travel. This finding suggests that these urban facilities attract a flexible, on-demand, and, importantly, asset-free/leasing mode of travel, which lies in the dynamic interplay between bottom-up supply and demand mechanisms and top-down platform regulations. Ride-hailing platforms enable drivers to strategically position themselves in areas of high demand, often guided by real-time data and incentives provided by the platform to maximize their profits. This flexibility contributes to the observed differences in spatial and temporal patterns compared to traditional automobile travel. Additionally, the platform-imposed constraints, such as dynamic pricing and algorithmic dispatching, introduce a layer of complexity that influences both driver behavior and user demand. These unique dynamics underscore the need for urban planning frameworks that account for the adaptive and regulated nature of platform-based mobility systems.

This observed uniqueness necessitates the development of a tailored functional planning framework to guide the evolution of future smart mobility solutions with similar features, such as autonomous vehicles and urban air mobility. The expected benefit would be a reduction in parking spaces in high-density areas if additional flexible asset-free/leasing mobility services are planned according to behavioral habits. By comprehensively understanding the relationship between ride-hailing services, urban functions, and the needs of diverse communities, we can ensure the responsible and equitable development of smart mobility solutions, paving the way for a significantly sustainable and inclusive urban landscape.

This research has some limitations that warrant further investigation. First, thereis a two-year time gap between the datasets (Didi ride-hailing data from November 2016 and mobile phone data from November 2018). During this period, travel behavior and urban land use patterns may have changed, potentially introducing bias into the findings. However, we believe the relatively stable pace of urban development and functional zone reclassification in densely populated areas like Chengdu minimizes the impacts of this limitation. Future studies could benefit from synchronized datasets to reduce such discrepancies. Second, the study's reliance on aggregated neighborhood data may not capture individual travel behaviors, suggesting future research could benefit from more detailed, disaggregated surveys. Finally, big data and models offer simplified representations of reality. Identifying urban functional zones involves numerous factors, including population growth, planning, and social dynamics, meaning the zones presented here are indicative rather than comprehensive.

Acknowledgements We would like to thank the financial support from National Natural Foundation of China (NSFC), Grant Number: 42301224; The RGC Postdoctoral Fellowship Scheme (PDFS), Ref. No.: PDFS2425-7H01; Chan To-Haan Endowed Professorship Fund of the University of Hong Kong.

References

Aarts, L., & Van Schagen, I. (2006). Driving speed and the risk of road crashes: A review. *Accident Analysis & Prevention, 38*(2), 215–224.

Bian, F., Qiao, S., & Yeh, A. G.-O. (2023). Mobility resilience: Transit-oriented development, ride-hailing, and car ownership. *Transportation Research Part D: Transport and Environment, 123*, 103924.

Cervero, R., & Murakami, J. (2010). Effects of built environments on vehicle miles traveled: Evidence from 370 US urbanized areas. *Environment and Planning A, 42*(2), 400–418.

Chatman, D. G. (2013). Does TOD need the T? On the importance of factors other than rail access. *Journal of the American Planning Association, 79*(1), 17–31.

Chengdu Report on the Characteristics of Green Transportation (2021). Retrieved September 5, 2024, from https://www.cdipd.org.cn/ueditor/php/upload/file/20221111/1668159271355997.pdf

Circella, G., Alemi, F., Tiedeman, K., Handy, S., & Mokhtarian, P. (2018). The Adoption of Shared Mobility in California and Its Relationship with Other Components of Travel Behavior (No. NCST-RR-201802).

Clewlow, R. R., & Mishra, G. S. (2017). Disruptive transportation: The adoption, utilization, and impacts of ridehailing in the United States.

Docherty, I., Marsden, G., & Anable, J. (2018). The governance of smart mobility. *Transportation Research Part A: Policy and Practice, 115*, 114–125.

Du, M., Cheng, L., Li, X., & Yang, J. (2020). Factors affecting the travel mode choice of the urban elderly in healthcare activity: Comparison between core area and suburban area. *Sustainable Cities and Society, 52*, 101868.

Ewing, R., Tian, G., Goates, J., Zhang, M., Greenwald, M. J., Joyce, A., Kircher, J., & Greene, W. (2015). Varying influences of the built environment on household travel in 15 diverse regions of the United States. *Urban Studies, 52*(13), 2330–2348.

Gao, S., Janowicz, K., & Couclelis, H. (2017). Extracting urban functional regions from points of interest and human activities on location-based social networks. *Transactions in GIS, 21*(3), 446–467.

Ghaffar, A., Mitra, S., & Hyland, M. (2020). Modeling determinants of ridesourcing usage: A census tract-level analysis of Chicago. *Transportation Research Part C: Emerging Technologies, 119*, 102769.

Guerra, E. (2014). The built environment and car use in Mexico City: Is the relationship changing over time? *Journal of Planning Education and Research, 34*(4), 394–408.

Huang, G., Qiao, S., & Yeh, A. G.-O. (2021). Spatiotemporally heterogeneous willingness to ridesplitting and its relationship with the built environment: A case study in Chengdu, China. *Transportation Research Part C: Emerging Technologies, 133*, 103425.

Jiang, Y., Gu, P., Chen, Y., He, D., & Mao, Q. (2017). Influence of land use and street characteristics on car ownership and use: Evidence from Jinan, China. *Transportation Research Part D: Transport and Environment, 52*, 518–534.

Keller, T. S., Weisberger, A., Ray, J., Hasan, S., Shiavi, R., & Spengler, D. (1996). Relationship between vertical ground reaction force and speed during walking, slow jogging, and running. *Clinical biomechanics, 11*(5), 253–259.

Kong, L., & Woods, O. (2018). The ideological alignment of smart urbanism in Singapore: Critical reflections on a political paradox. *Urban Studies, 55*(4), 679–701.

Li, S., & Zhao, P. (2017). Exploring car ownership and car use in neighborhoods near metro stations in Beijing: Does the neighborhood built environment matter? *Transportation Research Part D: Transport and Environment, 56*, 1–17.

Li, X., Xu, J., Du, M., Liu, D., & Kwan, M.-P. (2023). Understanding the spatiotemporal variation of ride-hailing orders under different travel distances. *Travel Behaviour and Society, 32*, 100581.

Liu, K., Yin, L., Lu, F., & Mou, N. (2020). Visualizing and exploring POI configurations of urban regions on POI-type semantic space. *Cities, 99*, 102610.

Luque-Ayala, A., & Marvin, S. (2015). Developing a critical understanding of smart urbanism? *Urban Studies, 52*(12), 2105–2116.

Qiao, S., & Yeh, A. G.-O. (2021). Is ride-hailing a valuable means of transport in newly developed areas under TOD-oriented urbanization in China? Evidence from Chengdu City. *Journal of Transport Geography, 96*, 103183.

Qiao, S., & Yeh, A. G.-O. (2023a). Is ride-hailing competing or complementing public transport? A perspective from affordability. *Transportation Research Part D: Transport and Environment, 114*, 103533.

Qiao, S., & Yeh, A. G.-O. (2023b). Understanding the effects of environmental perceptions on walking behavior by integrating big data with small data. *Landscape and Urban Planning, 240*, 104879.

Qiao, S., Huang, G., & Yeh, A. G.-O. (2022). Mobility as a service and urban infrastructure: From concept to practice. *Transactions in Urban Data, Science, and Technology, 1*(1–2), 16–36.

Rayle, L., Dai, D., Chan, N., Cervero, R., & Shaheen, S. (2016). Just a better taxi? A survey-based comparison of taxis, transit, and ridesourcing services in San Francisco. *Transport Policy, 45*, 168–178.

Schwanen, T. (2016). Innovations to transform personal mobility. In *Low carbon mobility transitions* (pp. 206–212). Goodfellow Publishers Ltd.

Shaheen, S., & Chan, N. (2016). Mobility and the sharing economy: Potential to facilitate the first- and last-mile public transit connections. *Built Environment, 42*(4), 573–588.

Tirachini, A., & del Río, M. (2019). Ride-hailing in Santiago de Chile: Users' characterisation and effects on travel behaviour. *Transport Policy, 82*, 46–57.

Van der Graaf, S., & Ballon, P. (2019). Navigating platform urbanism. *Technological Forecasting and Social Change, 142*, 364–372.

Wang, M., & Mu, L. (2018). Spatial disparities of Uber accessibility: An exploratory analysis in Atlanta, USA. *Computers, Environment and Urban Systems, 67*, 169–175.

Wang, S., & Noland, R. B. (2021). Variation in ride-hailing trips in Chengdu, China. *Transportation Research Part D: Transport and Environment, 90*, 102596.

Yao, Y., Li, X., Liu, X., Liu, P., Liang, Z., Zhang, J., & Mai, K. (2017). Sensing spatial distribution of urban land use by integrating points-of-interest and Google Word2Vec model. *International Journal of Geographical Information Science, 31*(4), 825–848.

Yu, H., & Peng, Z.-R. (2019). Exploring the spatial variation of ridesourcing demand and its relationship to built environment and socioeconomic factors with the geographically weighted Poisson regression. *Journal of Transport Geography, 75*, 147–163.

Yu, H., & Peng, Z.-R. (2020). The impacts of built environment on ridesourcing demand: A neighbourhood level analysis in Austin, Texas. *Urban Studies, 57*(1), 152–175.

Zhai, W., Bai, X., Shi, Y., Han, Y., Peng, Z.-R., & Gu, C. (2019). Beyond Word2vec: An approach for urban functional region extraction and identification by combining Place2vec and POIs. *Computers, Environment and Urban Systems, 74*, 1–12.

Zhang, B., Chen, S., Ma, Y., Li, T., & Tang, K. (2020). Analysis on spatiotemporal urban mobility based on online car-hailing data. *Journal of Transport Geography, 82*, 102568.

Zhang, X., Zhou, Z., Xu, Y., & Zhao, X. (2024). Analyzing spatial heterogeneity of ridesourcing usage determinants using explainable machine learning. *Journal of Transport Geography, 114*, 103782.

Zheng, Z., Zhang, J., Zhang, L., Li, M., Rong, P., & Qin, Y. (2022). Understanding the impact of the built environment on ride-hailing from a spatio-temporal perspective: A fine-scale empirical study from China. *Cities, 126*, 103706.

Towards Net Zero: Converting Car Commutes to Bicycling in Australia

Parisa Izadpanahi, Scott N. Lieske, and Simone Z. Leao

Abstract Achieving net zero transportation emissions requires energy efficiency improvements and shifts to less carbon-intensive travel. This article addresses the gap in Australian transportation research regarding the impact of shifting short commuting trips from cars to bicycles on greenhouse gas (GHG) emissions. We model potential emissions reduction at the suburb scale across Australia centering on scenarios of increasing mode shift towards bicycling. Results indicate if all short commute car trips (≤10 km) shift to bicycles, 3.4 million cars per day will no longer be used for commuting, leading to a reduction of up to 5.4% of car commuting GHG emissions. This study also identifies where such mode shifts would result in the highest emissions reductions. The magnitude and location of potential emissions reduction brings a new perspective to emissions reduction, transportation, and bicycle planning, suggesting mode shift to bicycling in areas with a high number of short car trips could reduce GHG emissions.

Keywords GHG emissions · Short car trips · Dashboard · Decarbonisation

P. Izadpanahi · S. Z. Leao
Curtin University, Bentley, WA, Australia
e-mail: Parisa.izadpanahi@curtin.edu.au

S. N. Lieske (✉)
St. Lucia, QLD, Australia
e-mail: scott.lieske@uq.edu.au

R. Goodspeed et al. (eds.), *Digital-Era Urban Transformations*, The Urban Book Series, https://doi.org/10.1007/978-3-031-98300-9_7

1 Introduction

Reducing greenhouse gas (GHG) emissions in all sectors, including transportation, is a critical component to meeting net zero targets (Kuramochi et al., 2018; Lamb et al., 2021; Levin et al., 2020). Globally, road transport accounts for 73% of transportation emissions (Lamb et al., 2021). In 2019, transport accounted for 19% of total GHG emissions in Australia, (Australian Government, 2021a) with transport emission growing since 2005, counteracting emissions reductions in Australia's electricity sector (Infrastructure Australia, 2019).

Historically, bicycle transport mode share in Australia is very low. According to the census, more than two-thirds of people drive to work, and nationally only 1.1% commute by bicycle (ABS, 2016). However, some recent trends are pointing in a more promising direction. The number of people cycling at least once a week in Australia increased by over 4% from 2019 to 2021 and around one-third of riding is for transportation while two-thirds of riding is for recreation (Munro, 2021 p. 21). The number of riders on key bike paths in Melbourne city centre increased 2.7 times between November 2019 and April 2020 based on a survey by Bicycle Network. In Brisbane, bicycle traffic increased substantially on city bikeways in 2020. In Sydney's west, Parramatta City Council saw an increase of 50 to 60% on bikeways, while Sydney City Council saw a 25 to 50% increase in cycle traffic despite a drop in overall people movement (Mark, 2020).

The total amount of GHG emissions from commuting by car in Australia is estimated at approximately 15 million tonnes of CO2e per year. Among the possibilities for reducing transportation emissions is replacing short commuting trips made by car with bicycling. Short commuting trips made by car are an opportunity for GHG emissions reduction as short trips up to 5 km on a bicycle are feasible for a great number of commuters (Castro et al., 2019). This distance does not require a high level of fitness, nor expose the cyclist to an unsafe traffic environment for a long time or cause excessive sweating. Prior research indicates that around 40% of commuters driving cars to work up to 5 km, are able to shift to walking or bicycling (Neves & Brand, 2019).

Shifting short car trips to bicycling is a low-cost opportunity that can be implemented in the short term, but this option is overlooked in current emissions reduction national plans and strategies for Australia. For example, 'Australia's Long-Term Emission Reduction Plan—A whole-of-economy plan to achieve net zero emissions by 2050' (Australian Government, 2021b) states that enabling the deployment of Electric Vehicles (EVs) can help reduce transport emissions by 58% by 2050 but opportunities for mode shift are unexplored.

This chapter addresses the gap in Australian transportation research on the impact of GHG emissions from the mode shift from short commuting trips made by car to bicycling. The analysis considers a set of scenarios based on a combination of conversion rates (percent of commuters using short car trips to work who would be willing to convert to bicycling), and frequency of conversion (number of working days per week for the conversion from car to bicycling). All estimations are presented on an online interactive dashboard, designed to disseminate results visually to urban and transport planners and other interested stakeholders. The results here bring a new perspective for the planning of bicycling infrastructure based on the goal of achieving the highest GHG emissions reduction.

The chapter structure is as follows: Section 2 overviews prior research on replacing short car trips with bicycling. Section 3 describes the data used in the study, while Sect. 4 presents the method used to estimate GHG emissions reduction. Section 5 introduces the project dashboard. Section 6 shows the results for GHG emissions reduction potentials for Australia for four different scenarios. Section 7 discusses the main findings of the research then concludes the paper.

2 Background

Regardless of the purpose of the journey, most active travel occurs within relatively short distances: 2, 5, and 10 km distance for walking, cycling, and e-biking respectively (Castro et al., 2019). Research shows that the majority of the trips in this range are made by private cars (Beckx et al., 2013; Keall et al., 2018; Neves & Brand, 2019). These short car trips contribute disproportionately to GHG emissions because of vehicles' 'cold starts', especially in colder climates (Beckx et al., 2013; de Nazelle et al., 2010). However, these short trips may be amenable to a partial mode shift towards active travel (Beckx et al., 2013; Brand et al., 2021a; Brand et al., 2021b; de Nazelle et al., 2010; Keall et al., 2018; Neves & Brand, 2019).

Research has addressed the topic of mode shift from short car trips to cycling in the last two decades. In the early 2010s, studies began to estimate potential mode shift (Godefroy & Morency, 2012; Halldórsdóttir et al., 2011) and to assess its limits (Beckx et al., 2013), particularly through modelling approaches. Studies began to establish associations between the mode shift to bicycling with reduction in fuel consumption and GHG emissions. Lovelace et al. (2011) assessed the potential for mode shift from short car trips to bicycling and estimated the amount of fuel that could be saved in different scenarios. Lindsay et al. (2011) developed a similar study but went further and associated reduced fuel consumption with a reduction in air pollution and GHG emissions. They concluded that transport-related GHG emissions could be reduced by 0.4% if 5% of vehicle travels shift to cycling. The European Cyclists' Federation released a (2011) report estimating typical GHG emissions produced by different modes of transport in grams of CO2e by passenger and unit of distance travelled: 271 g CO2e /passenger-km for cars; 101 g for public transport; and 21 g for conventional bicycles. Tainio et al. (2017) looked at the effects of replacing short car trips with bicycling in England using the National Travel Survey. They found this mode shift offers benefits for health, reduction in GHG emissions, and in consumers' fuel costs.

Using a mixed-methods approach with GPS tracking and in-depth interviews in the UK, Neves and Brand (2019) found that 41% of car trips of up to 5 km could be replaced by walking or cycling and this shift leads to a 5% reduction in GHG emissions from car travel. Brand et al. (2021b) estimated the lifecycle carbon emissions effects of changes in active travel in cities, considering the following GHG emissions by mode of transport: 190 g CO2e/passenger-km for cars; 67 g for public transport; and 5 g for conventional bicycles. They found that if a person converts from 1 car trip/day to 1 cycling trip/day for 200 days a year, the transport-related GHG emissions will drop by 0.5 tonnes over a year which constitutes a substantial share of average per capita GHG emissions from transport. The difference between the ECF (2011) and Brand et al. (2021b) estimates of GHG emissions by mode of transport may be due to technology improvements, energy efficiency increases in car engine manufacturing, and new emissions regulations. Whitehead et al. (2018) estimated that a typical Australian petrol vehicle generates 355 g of CO_2e per km in real-world fuel life cycle emissions. This parameter is per vehicle, and not per passenger. When considering the average car patronage in Australia of 1.85 passengers per vehicle, it results in around 192 g of CO_2e per passenger-km, which is very similar to the Brand et al. (2021b) estimate.

Limiting the most severe impacts of global warming will require net zero emissions in the next 30 to 50 years. Opportunities to reduce GHG emissions in the transport sector must go beyond fuel efficiency increases and vehicle electrification. Australia's low bicycling mode share and a high percentage of short distance car trips present an opportunity for GHG emissions reduction through transport mode shift.

This study aims to (1) evaluate the potential for greenhouse gas emissions reduction due to a mode shift from short commuter car trips to bicycling across Australia, and a comparative assessment of the shift to electric vehicles; and (2) identify locations where such transport mode shift would result in the highest impacts of emissions reduction through consideration of four different scenarios.

3 Data

All input data used in this study are sourced from open data platforms. Commuter transport data for Australia are obtained from the census Journey to Work (JTW) 2016. These data include the number of commuting trips by modes of transport at the suburb scale (statistical area level 2, SA2), median travel distance by modes of transport at regional spatial scales (statistical area level 4, SA4), and number of trips by distance and by transport mode at the national level.

Estimated reductions in GHG emissions are calculated and mapped at the SA2 scale. Cycling infrastructure data are from OpenStreetMap (OSM), which offers consistent coverage across Australia. Cycleways road segments were overlaid to SA2 boundaries, and the total length of cycling infrastructure segments within each SA2 was summed. Parameters of GHG emissions by mode of transport associated with vehicle production, maintenance, operation, and fuel (in gCO2e /passenger-km) are based on Brand et al. (2021a).

Contextual factors that could influence bicycling adoption include weather, congestion, and standard of public health (Pearson et al., 2023; Pucher & Buehler, 2012). Weather data were obtained from the Australian Bureau of Meteorology. Congestion data were obtained from the 'Urban Transport Crowding and Congestion—The Australian Infrastructure Audit 2019' developed by Infrastructure Australia. This study produced a modelled characterisation of congestion during the AM peak on major roads of six relevant and most congested urban regions across Australia in 2016. In this study, congestion has been characterised as the sum of the length of road segments with High or Medium 'Volume/Capacity Ratio' by SA2. Public health data are measured as the proportion of the population by LGA (then converted to SA2) who (1) undertake low, very low or no exercise, and (2) are obese or overweight (AHPC, 2020).

4 Methods

Four scenarios that replace short car trips by bicycling or explore shifts to e-bikes or EVs are used to estimate the potential emissions reductions possible through these mode shifts. Modelling and scenarios are developed through a series of seven interrelated steps (Fig. 1). The approach is to (1) estimate existing GHG emissions from commuting, (2) identify the number of short car trips, and (3) estimate emissions from these trips. These estimates are used to calculate impacts from four mode/fuel shift scenarios including 5 new bike share, and 6 new GHG emissions for each scenario as well as map GHG emissions reductions. Finally, we explore opportunities, and barriers to model shift. An overview of these methods is presented below.

Step 1: Estimate total GHG emissions from commuting by travel mode by SA4 in Australia and proportionally distribute them by SA2

Step 1 calculates the total GHG emissions by mode of transport (car, public transport, bicycling) for each SA4 based on the number of trips, median commute travel distances and GHG emissions (in Kg CO2e /passenger-km) for each transport mode. SA4-scale emissions are then disaggregated to the SA2 level. The disaggregation process is performed for each transport mode and is proportional to the number of commute trips by mode in each SA2 in relation to the total for the same mode within the containing SA4.

Step 2: Identify the number of short car trips and the potential for mode shift from cars to bicycling in Australia by SA2

Step 2 identifies the potential mode shift from cars to bicycling in Australia by SA2 based on the number of short car trips. Two ranges are considered: trips with a distance below 5 km are considered as suitable for mode shift to conventional bicycling ('very short trips'); and trips with a distance between 5 and 10 km are considered suitable for mode shift to e-bikes ('short trips').

JTW provides the average distribution of commute trips by mode of transport and by travel distance up to 300 km. Analysis of JTW data indicate that (1) cars are

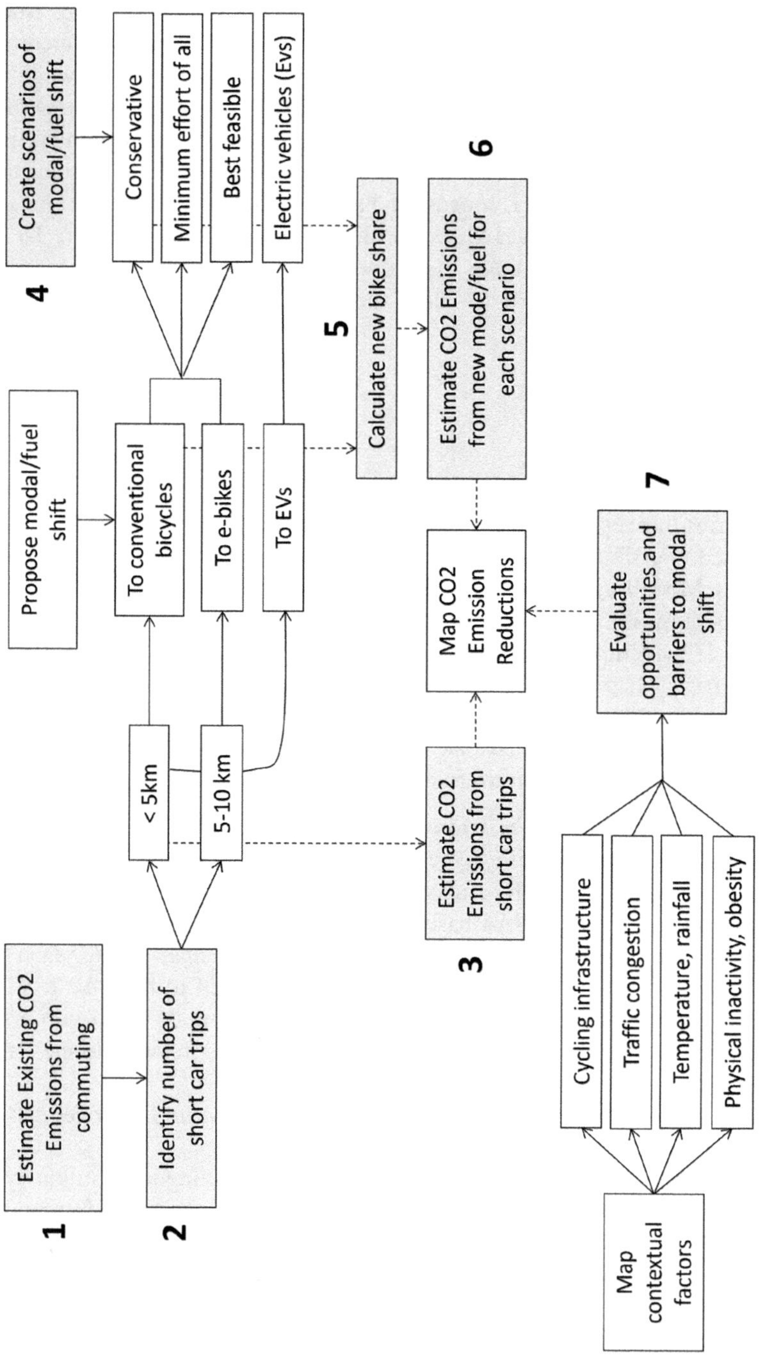

Fig. 1 Overview of research approach (Steps 1 to 7)

the predominant transport mode in Australia for all travel distances, including very short trips; (2) walking mode share is high at very short distances, but decreases quickly with increasing distances, becoming negligible for trips above 5 km; (3) the use of public transport grows with increasing distances; and (4) bicycling is generally low in Australia, with maximum share (2–2.5% of trips by bike) between 1 to 5 km trips, and declining to around 1% mode share around 10 km. Based on JTW data, on average 76% of commute trips shorter than 5 km in Australia are made by car (with an average distance of 3 km), and 82% of commute trips 5–10 km in Australia are made by car (with an average distance of 8 km).

Step 3: Modelling annual GHG emissions produced by short car commuting trips by SA2.

Step 3 estimates annual GHG emissions produced by the 2016 very short/short car commuting trips by SA2. This was based on the total number of commute trips by car, the national proportion and average distances of car trips which are very short or short, and the typical GHG emissions for petrol cars.

Step 4: Create scenarios of mode shift from car to bicycling and fuel shift to EVs by SA2

Three scenarios of progressively increasing mode shift towards bicycling are analyzed. The first scenario, 'Conservative', considers a low engagement in which only 20% of the very short/short car trip commuters would shift their travel mode from car to bicycle (called 'conversion rate', c), and only once a week (called 'frequency', f). The second scenario assumes a 'Collective Effort' towards GHG emissions reduction, in which all (100%) very short/short trip car commuters would shift their mode once a week. The third scenario assumes that 40% of the short car trip commuters would switch to bicycling on a daily basis (5 days/week). As the literature indicates that around 40% mode shift is a possible and realistic goal (Delso et al., 2018; Neves & Brand, 2019), this context has been called the 'Best Feasible' scenario. For the sake of comparison, a fourth scenario assessed the result of the conversion of all very short/short trips made by conventional petrol cars to electric vehicles (EVs), 5 days/week. See Table 1. For each scenario, this study calculates the number of very short/short car trips that can potentially be replaced by bicycling.

Table 1 Parameters for the proposed scenarios

		Scenarios			
Conversion to	Parameters	Conservative	Minimum effort of all	Best feasible	Electric cars
Bicycling	Conversion rate (cr)	20%	100%	40%	0
	Frequency (f), annually	1 day/week	1 day/week	5 days/week	0
EVs	Conversion rate (cr)	0	0	0	100%
	Frequency (f), annually	0	0	0	5 days/week

Step 5: Estimate the annual number of short commute trips and new bike mode share by SA2 for each scenario

Step 5 calculates the new bike mode share (% of commute by bicycle, out of all commute trips) by SA2 for each scenario. This is the result of the sum of the bike share already existing in the area plus the additional proportion from the mode shift.

Step 6: Estimate annual GHG emissions reduction of each scenario by SA2

Step 6 estimates the annual GHG emissions reduction that can be obtained from each scenario. GHG emissions reduction estimates consider the savings of emissions avoided by short car trips not taken, added to the emissions produced by replacing trips by bicycling. Emissions reduction is based on the number of trips, average distance of the trips, and the GHG emissions by mode of transport. Estimates for very short trips (<5 km) consider an average distance of 3 km for the trips and a shift to conventional bicycles. Short trips (5–10 km) are estimated based on an average distance of 8 km and a shift to e-bikes. The fourth scenario estimates GHG emissions produced by EVs. The results are presented in tonnes of CO2e /year for each SA2 and % of the overall commute-related GHG emissions by SA2.

Step 7: Consideration of contextual factors for the promotion of mode shift from cars to bicycles

Step 7 maps contextual factors that can create opportunities or barriers for the promotion of the mode shift from cars to bicycling. These include the amount of bicycling infrastructure (in km), the proportion of the population which are physically inactive or are overweight/obese, presence of congestion, and weather conditions (temperature and rainfall).

5 Data Visualisation Dashboard

While there are several examples of net zero dashboards, only the UK Climate Policy Dashboard has a specific section addressing emissions from the transport sector with supporting data. Policy recommendation number 17 in this dashboard proposes 'putting cycling at the heart of decision-making' and 'empowering and encouraging local authorities' to do so (Policy Connect, 2022). The data in the dashboard, however, is aggregated at the national level, which has limited use for decision-making at the local level.

Despite the recognition that mode shift to bicycling can assist in GHG emissions reduction, existing dashboards associated with net zero goals do not provide data specific to the topic, or at a scale that can support planning and decision-making at a local level. To overcome this limitation, we developed the 'Bicycling Towards Net Zero Dashboard' built using Tableau and ArcGIS Online.

6 Results

6.1 *The 'Bicycling Towards Net Zero Dashboard'*

The 'Bicycling Towards Net Zero Dashboard' organises selected results from this study in a meaningful and user-friendly manner, including an introduction, followed by three infographics/map components: (1) comparing scenarios; (2) the Best Feasible scenario; and (3) opportunities and barriers for bicycling and concluding remarks. Dashboard functionalities allow users to easily query and visualise data by theme or geography. The maps and plots are synchronised. Linking robust databases with an easy-to-use interface promotes the dissemination of the results to a large audience and can stimulate a process of learning and insight development. The dashboard is available online through the link: https://arcg.is/1CvKG9.

The dashboard presents the potential mode shift, GHG emissions reduction, and new bike mode share for the four proposed scenarios. Interacting with the map and plots, a user can analyse specific scenarios or compare multiple scenarios. For example, mode shift and resulting emissions reduction for the Best Feasible Scenario are presented in Dashboard-2 (Fig. 2). Dashboard users are able to analyse differences between states, capital cities, regional areas, and specific local government areas. A user can also identify the largest potentials for emissions reduction, and then check where these can be achieved across the country. Maps and plots provide information on opportunities and barriers for the switch, such as existing cycling infrastructure, weather, and population health.

6.2 *Opportunities for Mode Shift: Number of Short Commute Car Trips*

Out of a total of 8.8 million commute trips per day in Australia, 7.3 million (83%) are made by car. This study found that if all short commute car trips (≤10 km) would switch to bicycling daily, 3.4 million cars per day would no longer be used for commuting (1.8 million cars for trips shorter than 5 km), particularly in capital cities. There is a large potential in Australia for GHG emissions reduction if part of these short car trips is shifted to bicycling. Table 2 summarises the results, indicating the number of short commute car trips by State, by distance (<5 km or 5-10 km), and by location (Capital or Regional areas).

Results indicate opportunities for mode shift from car to bicycling are unevenly distributed across Australia. The states New South Wales (NSW), Victoria (Vic), and Queensland (Qld) account for 78% of all short commute car trips (2.6 million trips/day). Moreover, two-thirds of the opportunities across Australia are concentrated in capital cities (2.3 million trips/day), and one-third are located in regional areas. Figure 3 shows the total number of trips by state, as well as the breakdown of these trips by distance and by location.

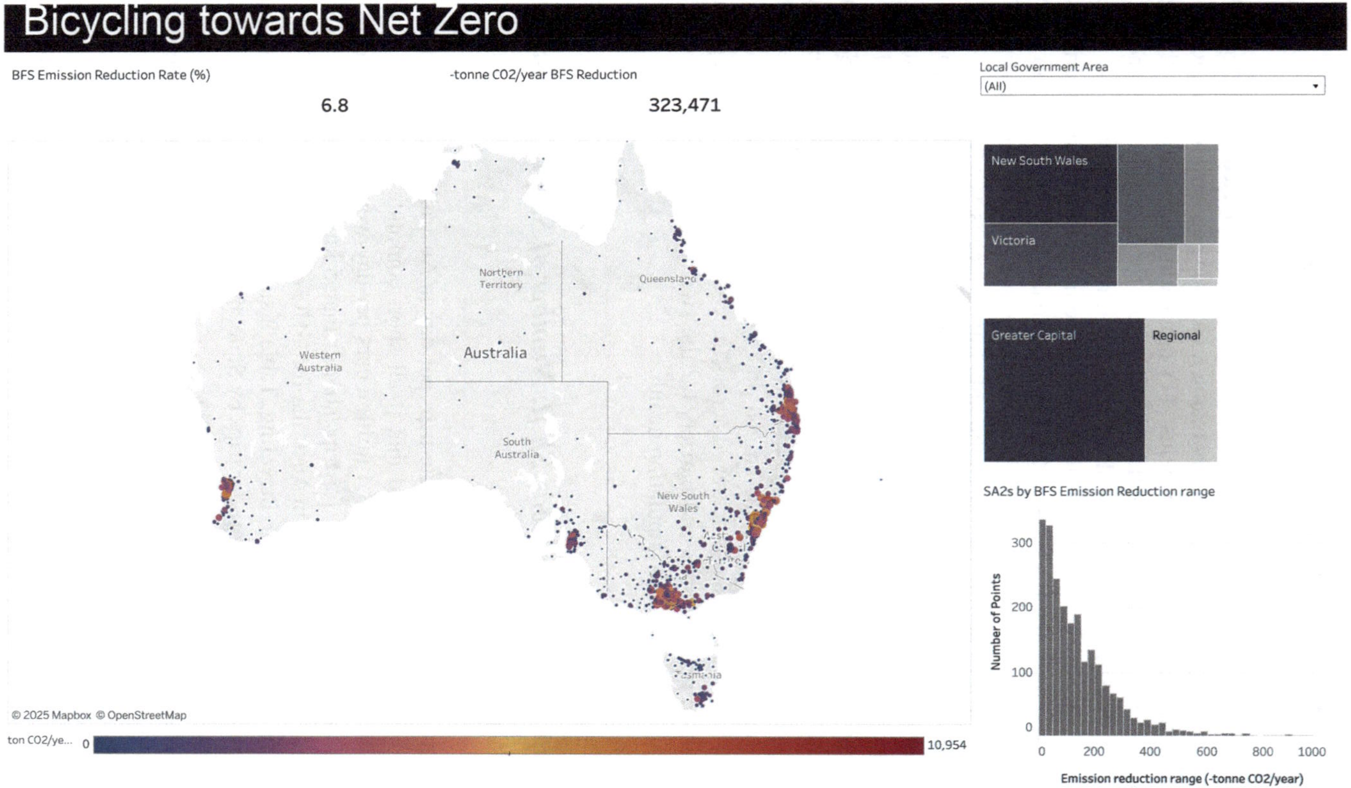

Fig. 2 Bicycling towards net zero dashboard: best feasible scenario

Table 2 Number of short commute car trips in Australia by State, by distance, and by location

State	2016 Total Population	Trips < 5 km			Trips 5-10 km			All short trips		
		Capital	Regional	%	Capital	Regional	%	Sum	%	Trips by pop
NSW	7,725,900	360,314 (57%)	268,322 (43%)	55	361,739 (70%)	156,839 (30%)	45	1,147,214	34	0.15
Vic	6,068,000	304,077 (72%)	118,913 (28%)	52	332,934 (84%)	64,226 (16%)	48	820,150	24	0.14
Qld	4,844,500	150,063 (44%)	191,938 (56%)	51	173,230 (54%)	149,425 (46%)	49	664,656	20	0.14
WA	2,617,200	118,551 (69%)	52,562 (31%)	50	139,557 (83%)	28,409 (17%)	50	339,079	10	0.13
SA	1,708,200	89,681 (69%)	40,877 (31%)	51	114,196 (91%)	10,672 (9%)	49	255,426	8	0.15
Tas	519,100	20,537 (46%)	23,949 (54%)	57	17,996 (54%)	15,069 (46%)	43	77,551	2	0.15
ACT	396,100	30,082 (100%)	–	46	36,015 (100%)	–	54	66,097	2	0.17
NT	244,900	13,618 (59%)	9524 (41%)	66	10,105 (83%)	2083 (17%)	34	35,330	1	0.14
Sum	23,879,000	1,086,923	706,085		1,185,772	426,723		3,405,503	100	
Mean		65% ± 17	35% ± 17	53 ± 6	77% ± 16	23% ± 17	47% ± 6			0.15 ± 0.01

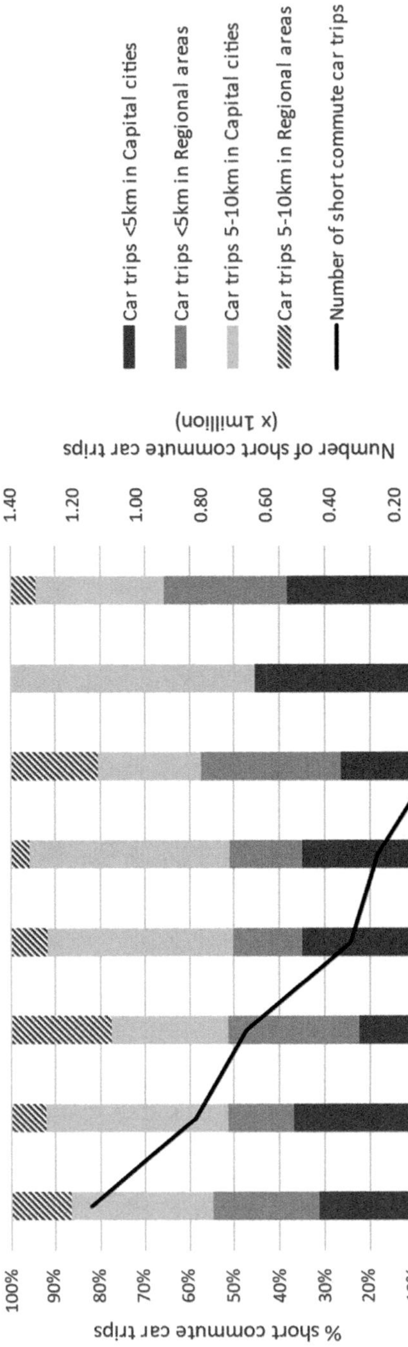

Fig. 3 Number and share of short commute car trips by state, location, and distance

Mapping short car commute trips by SA2 in the 'Bicycling Towards Net Zero Dashboard' provides an additional layer of information to support local planning and action. Strategies to promote mode shift can be different for urban and regional residents, and for trips up to 5 km or 5–10 km. Users of the dashboard can identify which areas have a large number of short car trips, or query the number of short trips for an area of interest. The map of commute car trips <5 km for Victoria by SA2, for example, clearly indicates opportunities for mode shift not only within the boundaries of Greater Melbourne, but also in regional areas such as Geelong, Ballarat, and Echuca, among others (Fig. 4).

6.3 GHG Emissions Reduction from Commute Mode Shift in Australia for Different Scenarios

This study found that if all short commute car trips (≤ 10 km) would switch to bicycling daily, 809 kilo tonnes of GHG emissions could be saved per year, which is equivalent to 5.4% of GHG emissions from all commuting cars in Australia.

This section presents and compares the GHG emissions reduction from shifting short commute car trips to bicycling for the proposed scenarios (Conservative; Collective Effort; Best Feasible), and from replacing petrol cars by EVs (electric vehicles). Table 3 describes the scenarios, summarises the GHG emissions reduction that can be achieved, and assesses the impact of this reduction in relation to the overall national emissions from commuting by car.

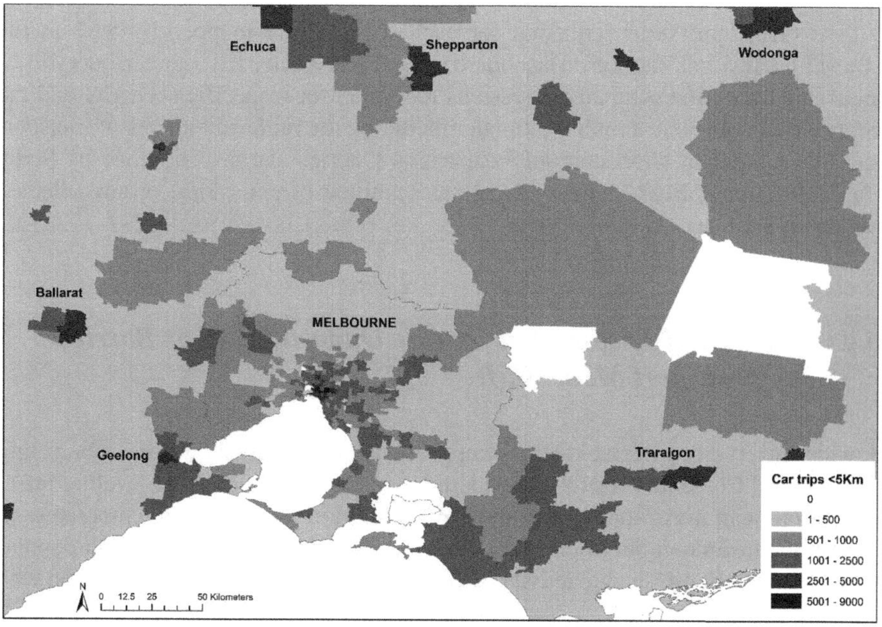

Fig. 4 Number of short commute car trips <5 km by SA2 for the State of Victoria

Table 3 Scenarios of GHG emissions reduction from shifting short commute car trips to bicycling

Scenario	Description	Total GHG emissions reduction (k tonnes CO2e /year) from shifting short car trips to bicycling or e-bicycling			% Of total GHG emissions from car commuting
		trips <5 km	trips 5-10Km	Sum	
Conservative	20% of the short car trips are converted to bicycling once a week	9.6	22.8	32.4	0.2
Collective effort	100% of the short car trips are converted to bicycling once a week	47.8	113.9	161.7	1.1
Best feasible	40% of the short car trips are converted to bicycling 5 days/week	95.6	227.9	323.5	2.1
Electric vehicles	100% of the short car trips are converted to electric vehicles 5 days/week	98.2	235.3	333.5	2.2

If all petrol cars involved in short commute car trips were replaced by electric vehicles, without any mode shift, the reduction of GHG emissions achieved would be equivalent to the Best Feasible scenario for mode shift to bicycling. Note that the Best Feasible scenario also promotes health benefits from more physical activity, and potentially still enables the 60% of the car trips not shifted to bicycling to be replaced by electric cars with additional GHG emissions reduction.

The GHG emissions reduction for each scenario are mapped by SA2 in the 'Bicycling Towards Net Zero Dashboard' (Fig. 5). A dashboard user can query local areas and check the estimated emissions reduction for a specific scenario, and the context that contributes to this estimate (including the preferred modes of transport and the number of short-commute car trips). Queries can also be done by State, Capital/Regional areas, name of local area, amount of emissions, or any selected region on the map.

6.4 Contextual Factors that Can Be Opportunities or Barriers for Transport Mode Shift

A wide variety of contextual factors can positively or negatively influence bicycling mode share. Contextual factors considered here are the extent of the cycling infrastructure; mean maximum temperature in summer; mean minimum temperature in winter; the mean number of rainy days per year; the proportion of the adult population with habit of physical inactivity; and the proportion of the adult population who are overweight or obese.

Figure 6 illustrates the context for Mount Isa, located in the hot and dry Australian Outback. The 'Bicycling Towards Net Zero Dashboard' describes Mount Isa as

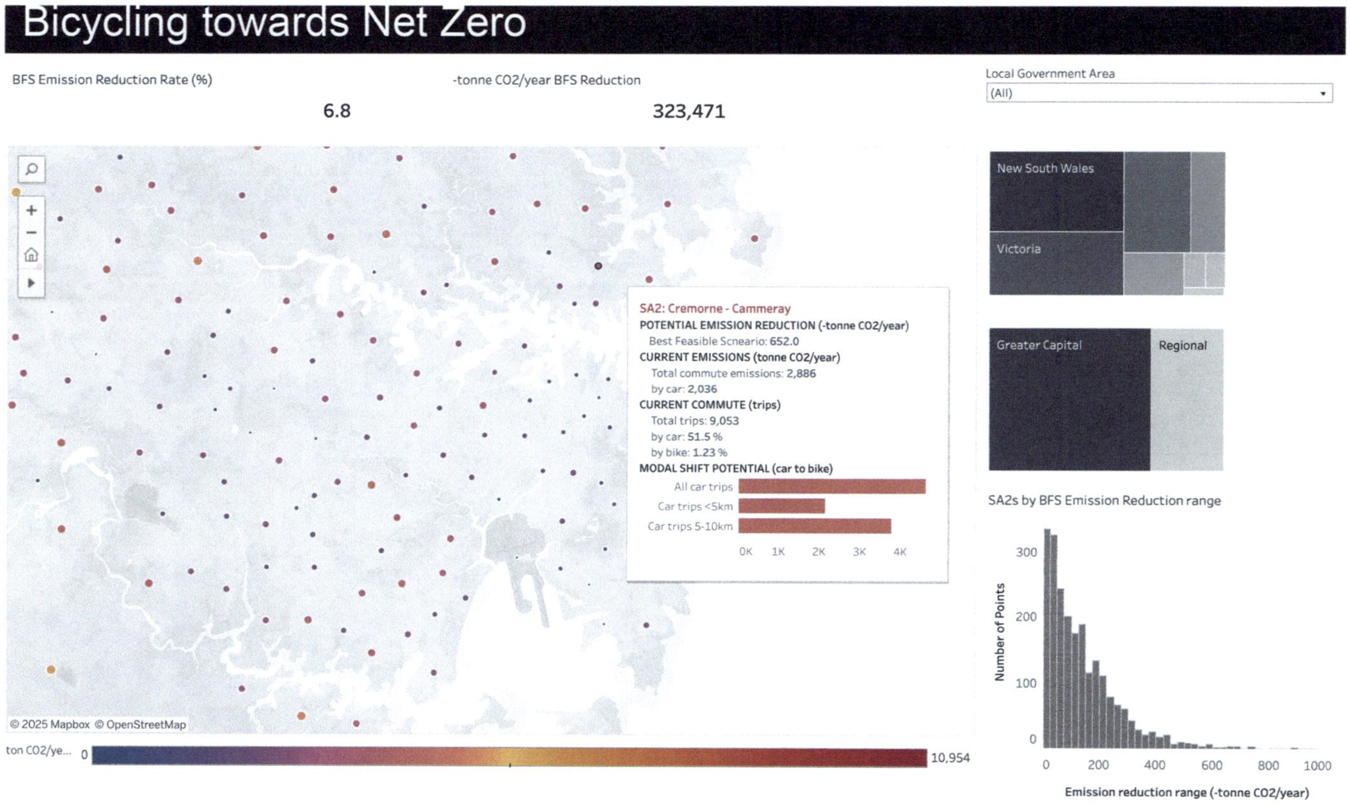

Fig. 5 GHG emissions reduction from shifting short commute car trips to bicycling in Australia for the Best Feasible Scenario: detail of Sydney region in the Bicycling Towards Net Zero Dashboard

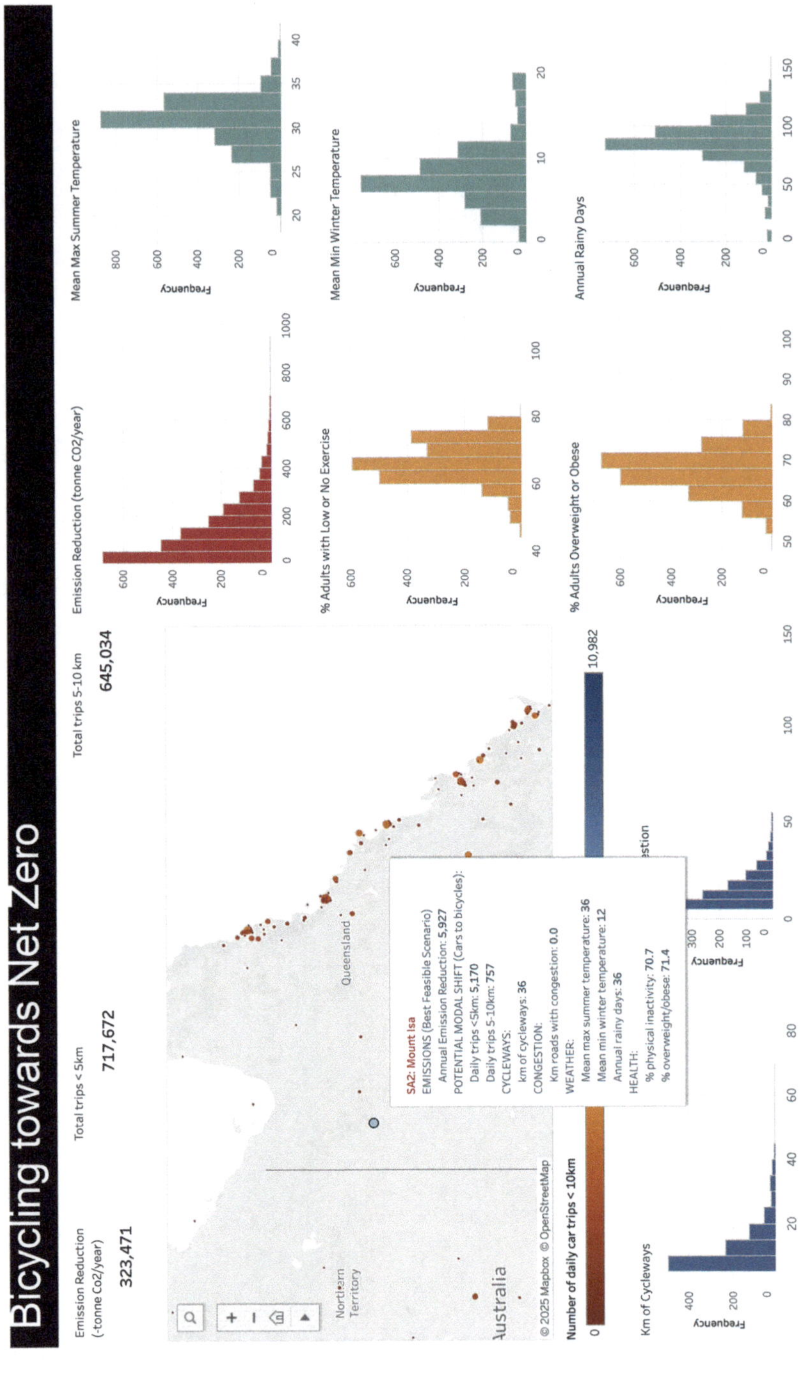

Fig. 6 Contextual factors which may affect mode shift to bicycling: detail of Mount Isa, Queensland, in the Bicycling Towards Net Zero Dashboard

having a very large amount of short commute car trips less than 5 km, and therefore a high potential for emissions reduction from mode shift to bicycling. However, the dashboard also indicates an average high temperature in summer and a very low annual rainfall. These conditions may reduce the attractiveness of bicycling due to low thermal comfort. Moreover, there is a limited bicycling network (only 36 km) which may adversely affect the sense of safety of potential new cyclists. The dashboard also shows that more than 70% of the adult population in Mount Isa is overweight or obese and is physically inactive. On the one hand, this may be a restriction to bicycling, as people are not seen as fit enough. On the other hand, this situation may be a reason to promote a mode shift to bicycling to enhance individual and public health through physical activities. This information can support local planners in finding the best solutions for handling transport, health and net zero goals in an integrated manner.

Each of the contextual factors was correlated with the following variables: (1) the overall proportion of commuter bicycling trips by SA2; (2) the overall proportion of commuter car trips by SA2; and (3) the proportion of commute car trips <5 km and 5–10 km by SA2. No correlation was found between the variables and the contextual factors. Despite these factors not having a statistically strong relation to affect the rate of bicycling as a mode of transport or car use, they may provide relevant information for planners considering the development of policies, plans or programs to increase bicycling in a specific area.

7 Discussion and Conclusion

This chapter addresses the gap in Australian transportation research on the impact of GHG emissions from the mode shift from short commuting trips made by car to bicycling. This study aimed to (1) evaluate the potential for greenhouse gas emissions reduction due to a mode shift from short commuter car trips to bicycling across Australia, and a comparative assessment of the shift to electric vehicles; and (2) identify locations where such transport mode shift would result in the highest impacts of emissions reduction. To address this, four different scenarios are considered.

The large number of short car trips to work in Australia indicates a high potential for GHG emissions reduction if these trips are replaced by walking, cycling, or electric vehicles. It was found that 3.4 million short car trips/day could lead to the reduction of up to 5.4% of all GHG emissions from car commuting in Australia annually. The findings of this study suggest that replacing just part of the short car trips could increase bicycling share in Australia to up to 16%. For reference, in the Netherlands, one of the countries with highest rates of bicycling in the world, 25% of workers commute by bicycle only, and around 40% of the train commuters use bicycles to get to train stations (Oakil et al., 2016). Bike share rates in Australia would increase from 1.2% to approximately 3%, 8%, and 16% for the Conservative, Collective Effort, and Best Feasible scenarios, respectively. Defining the four scenarios is an effective method to break down the overall research problem and address the gap.

The ‘Bicycling Towards Net Zero Dashboard’ dashboard maps short commute car trips and GHG emissions reduction opportunities at local jurisdictions across the country for different scenarios. The dashboard also provides information on contextual factors associated with health conditions, weather, congestion, and existing bicycling infrastructure, which can influence mode shifts. The dashboard provides information to support a better understanding of current context for mode shift and its consequent benefits to net zero goals, and to inform decision-making for a more sustainable, healthier, and equitable transportation in Australia.

The GHG emissions reduction demonstrated in this study and others (Astegiano et al., 2019; Brand et al., 2021a; Brand et al., 2014; Brand et al., 2021b; Bucher et al., 2019; Neves & Brand, 2019) from higher adoption of bicycling as a mode of transport can offer important support for developing public policy interventions. This can be relevant and timely, considering the extent of climate change issues currently faced in cities and the high level of exposure to climate change and net zero debate in the media. As a low-cost solution making use of current technology and infrastructure, campaigns for mode shift can be seen as a strategic investment. The new ‘Bicycle towards Net Zero dashboard’ provides important information to support the planning of such strategies, indicating locations and conditions that can favour highest outcomes of GHG emissions reduction for efforts to motivate the mode shift to bicycling.

Planning bicycling infrastructure is generally focused on areas with higher volumes of cyclists considering accessibility and safety concerns. This study brings a new perspective, suggesting the mode shift to bicycling in areas with a high number of short car trips could reduce GHG emissions. These two perspectives are not exclusive and can be complementary.

It is acknowledged this research has a number of limitations. The authors selected the average distance of the car trips for calculating the emission reductions. The actual distributions may be different. This study utilised census data from 2016 which was the latest available at the time this research was conducted. Thus this research does not capture travel activity patterns influenced by COVID and its aftermath. This study assumed equal lifecycle GHG emissions for EVs and conventional fuel cars. All GHG emission calculations are based on the difference in source of energy for running cars. A more accurate approach would be to consider the manufacturing country’s energy source. Given Australia‘s high carbon content in electricity from coal compared to the average European peer, choosing a manufacturer from a country with renewable energy sources could potentially result in lower Life Cycle Carbon emissions. There may also be forces impacting mode shift that were not considered due to limitations in the research data and scope.

This research demonstrates that Australia has a high potential for a mode shift from short car trips to bicycling, and that the resulting emissions reduction achieved is not negligible. This study also indicates specific opportunities in regional areas that have not been fully explored, as major investments still concentrate on capital cities. Mode shift will occur only with the development of policies, programs, and plans to support bicycling across Australia (IEA, 2021).

Acknowledgement This research project has been funded by the Net Zero Research Program of Curtin University, from July to December 2021.

References

AHPC. (2020). *Australian Health Tracker by Area, 2020*. AHPC. https://www.atlasesaustralia.com.au/ahpc/atlas_double/atlas.html?indicator=i2&date=2014-15&indicator2=i2&date2=2017-18

Astegiano, P., Fermi, F., & Martino, A. (2019). Investigating the impact of e-bikes on modal share and greenhouse emissions: A system dynamic approach. *Transportation Research Procedia, 37*, 163–170.

Australian Bureau of Statistics. (2016). *More than two in three drive to work, Census reveals*. Australian Bureau of Statistics. https://www.abs.gov.au/ausstats/abs@.nsf/mediareleasesbyreleasedate/7DD5DC715B608612CA2581BF001F8404?OpenDocument

Australian Government. (2021a). *National greenhouse gas inventory quarterly update: Department of Industry, Science and Resources*. Australian Government. https://www.industry.gov.au/data-and-publications/national-greenhouse-gas-inventory-quarterly-update-september-2021

Australian Government. (2021b). *Australia's long-term emissions reduction plan, a whole-of-economy plan to achieve net zero emissions by 2050*. Australian Government. https://www.industry.gov.au/sites/default/files/October%202021/document/australias-long-term-emissions-reduction-plan.pdf

Beckx, C., Broekx, S., Degraeuwe, B., Beusen, B., & Panis, L. (2013). Limits to active transport substitution of short car trips. *Transportation Research Part D, Transport and environment, 22*, 10–13.

Brand, C., Goodman, A., & Ogilvie, D. (2014). Evaluating the impacts of new walking and cycling infrastructure on carbon dioxide emissions from motorized travel: A controlled longitudinal study. *Applied Energy, 128*, 284–295.

Brand, C., Dons, E., Anaya-Boig, E., Avila-Palencia, I., Clark, A., de Nazelle, A., Gascon, M., & Panis, L. (2021a). The climate change mitigation effects of daily active travel in cities. *Transportation Research. Part D, Transport and Environment, 93*, 102764.

Brand, C., Dons, E., Anaya-Boig, E., et al. (2021b). The climate change mitigation impacts of active travel: Evidence from a longitudinal panel study in seven European cities. *Global Environmental Change, 67*, 102224.

Bucher, D., Buffat, R., Froemelt, A., & Raubal, M. (2019). Energy and greenhouse gas emission reduction potentials resulting from different commuter electric bicycle adoption scenarios in Switzerland. *Renewable & Sustainable Energy Reviews, 114*, 109298.

Castro, A., Gaupp-Berghausen, M., Dons, E., et al. (2019). Physical activity of electric bicycle users compared to conventional bicycle users and non-cyclists: Insights based on health and transport data from an online survey in seven European cities. *Transportation Research Interdisciplinary Perspectives, 1*, 100017.

de Nazelle, A., Morton, B. J., Jerrett, M., & Crawford-Brown, D. (2010). Short trips: An opportunity for reducing mobile-source emissions? Transportation research. *Part D, Transport and Environment, 15*(8), 451–457.

Delso, J., Martín, B., & Ortega, E. (2018). Potentially replaceable car trips: Assessment of potential modal change towards active transport modes in Vitoria-Gasteiz. *Sustainability (Basel, Switzerland), 10*(10), 3510.

European Cyclist Federation. (2011). *Cycle more often 2 cool down the planet! Quantifying CO2 savings of Cycling, Brussels*. European Cyclist Federation. https://ecf.com/system/files/Cycle_More_Often_2_Cool_Down_the_Planet.pdf

Godefroy, F., & Morency, C. (2012). Estimating latent cycling trips in Montreal, Canada. *Transportation Research Record, 2314*(1), 120–128.

Halldórsdóttir, K., Christensen, L., Jensen, T. C., & Prato, C. G. (2011). *Modelling mode choice in short trips-shifting from car to bicycle*. ETC.

IEA. (2021). *Tracking transport*. IEA. https://www.iea.org/reports/tracking-transport-2021

Infrastructure Australia. (2019). *An assessment of Australia's future infrastructure needs. A. Government*. Infrastructure Australia.

Keall, M., Chapman, R., Shaw, C., Abrahamse, W., & Howden-Chapman, P. (2018). Are people who already cycle and walk more responsive to an active travel intervention? *Journal of Transport & Health, 10*, 84–91.

Kuramochi, T., Höhne, N., Schaeffer, M., et al. (2018). Ten key short-term sectoral benchmarks to limit warming to 1.5°C. *Climate Policy, 18*(3), 287–305.

Lamb, W. F., Wiedmann, T., & Pongratz, J. (2021). A review of trends and drivers of greenhouse gas emissions by sector from 1990 to 2018. *Environmental Research Letters, 16*(7), 73005.

Levin K, Rich D, Ross K, Fransen T, Elliott C (2020) *Designing and communicating net-zero targets*. https://www.wri.org/research/designing-and-communicating-net-zero-targets?utm_source=twitter&utm_medium=promoted&utm_campaign=socialmedia&utm_term=video

Lindsay, G., Macmillan, A., & Woodward, A. (2011). Moving urban trips from cars to bicycles: Impact on health and emissions. *Australian and New Zealand Journal of Public Health, 35*(1), 54–60.

Lovelace, R., Beck, S. B. M., Watson, M., & Wild, A. (2011). Assessing the energy implications of replacing car trips with bicycle trips in Sheffield, UK. *Energy Policy, 39*(4), 2075–2087.

Mark, D. (2020). *Australia is facing a 'once-in-a-lifetime opportunity' as cycling booms, advocates say*. ABC News. https://www.abc.net.au/news/2020-05-17/coronavirus-brings-once-in-a-lifetime-opportunity-for-cycling/12247870

Munro, C. (2021). *National Walking and cycling participation survey*. Cycling and Walking Australia and New Zealand (CWANZ). https://www.cwanz.com.au/wp-content/uploads/2021/08/NWCPS_2021_report_v2.0.pdf

Neves, A., & Brand, C. (2019). Assessing the potential for carbon emissions savings from replacing short car trips with walking and cycling using a mixed GPS-travel diary approach. *Transportation Research Part A, Policy and Practice, 123*, 130–146.

Oakil, A. T. M., Ettema, D., Arentze, T., & Timmermans, H. (2016). Bicycle commuting in The Netherlands: An analysis of modal shift and its dependence on life cycle and mobility events. *International Journal of Sustainable Transportation, 10*(4), 376–384.

Pearson, L., Gabbe, B., Reeder, S., & Beck, B. (2023). Barriers and enablers of bike riding for transport and recreational purposes in Australia. *Journal of Transport & Health, 28*, 101538.

Policy Connect. (2022). *UK climate policy dashboard: Tracking government progress to deliver net zero*. Policy Connect. https://www.policyconnect.org.uk/sustainability/climate-policy-dashboard

Pucher, J., & Buehler, R. (2012). *City cycling*. MIT Press.

Tainio, M., Monsivais, P., Jones, N., Brand, C., & Woodcock, J. (2017). 1976—Mortality, greenhouse gas emissions, and consumer cost impacts of replacing short car trips with cycling: A health impact assessment study. *Journal of Transport & Health, 5*, S71.

Inferring On-Street Parking Occupancy with Smart Meter Data

Daniel Jordon, Robert C. Hampshire, and Tayo Fabusuyi

Abstract The excessive search for parking, known as cruising, causes pollution and congestion. To mitigate these negative effects, cities are exploring new methods. However, accurately measuring the number of vehicles searching for parking is challenging and typically requires sensing technologies. In this paper, we propose an alternative method that eliminates the need for such technology by leveraging parking meter payment transactions to estimate parking occupancy and the number of cars searching for parking. Our estimation scheme is based on Particle Markov Chain Monte Carlo (PMCMC). We validate the PMCMC approach using data simulated from a *GI*/*GI*/*s* queue, demonstrating that it produces asymptotically unbiased Bayesian estimates of parking occupancy and key model parameters such as arrival rates, average parking times, and payment compliance rates. Finally, we apply our method to parking meter data from SF*park*, a large-scale parking experiment, and subsequently compare the Particle Markov Chain Monte Carlo parking occupancy estimates to the actual data from parking sensors. Our approach is easy to replicate and scalable, relying solely on historical parking payment transaction data that cities already have.

Keywords Smart parking · Queueing inference engine (QIE) · Particle filter · Particle Markov Chain Monte Carlo · Metropolis-Hastings

D. Jordon
Independent Scholar, Philadelphia, PA, USA

R. C. Hampshire · T. Fabusuyi (✉)
University of Michigan, Ann Arbor, MI, USA
e-mail: hamp@umich.edu; Fabusuyi@umich.edu

R. Goodspeed et al. (eds.), *Digital-Era Urban Transformations*, The Urban Book Series, https://doi.org/10.1007/978-3-031-98300-9_8

1 Introduction

Finding an available parking space is an annoyance of modern life. The search for parking has external costs as well, given that excess driving increases traffic congestion and pollutes the environment. In response to these negative externalities, city officials are deploying parking information and pricing systems. Parking occupancy information and market-based parking pricing strategies reduce the time to find a parking space and pollution (Chatman & Manville, 2014; Fabusuyi et al., 2014; Millard-Ball, 2014; Pierce & Shoup, 2013). These strategies require cities to install both parking occupancy sensors and automated payment stations. However, on-street parking sensors are expensive to install and maintain, and many cities cannot afford to deploy parking sensors. In contrast to parking sensors, the automated payment stations are more affordable and more widely deployed. In this paper, we attempt to answer the following question: is it possible to estimate parking occupancy and the number of drivers searching for parking using only parking meter payment data?

Two challenges stand in the way of answering this question. First, a large fraction of drivers do not pay at all or do not pay enough to cover their parking time. For example, in San Francisco, the non-payment rate is roughly 30%, driven in part by the perennial problem with the abuse of disabled parking placards for on-street parking (SFpark San Francisco Metropolitan Authority, 2014). Figure 1 plots the probability distribution of the non-payment rates in San Francisco. During metered hours, only 70% of parking time is paid for on the modal block. The second challenge is that drivers almost never pay for the exact amount of parking that they need, as they either pay too much or too little.

It is these challenges that our estimation approach seeks to address. Specific contributions of our approach include:

- The introduction of a modeling framework to infer parking occupancy and the time to find parking given a sequence of parking meter payments using a variant of the Particle Markov Chain Monte Carlo (PMCMC) method. We show that this algorithm generates asymptotically unbiased estimates of parking occupancy of

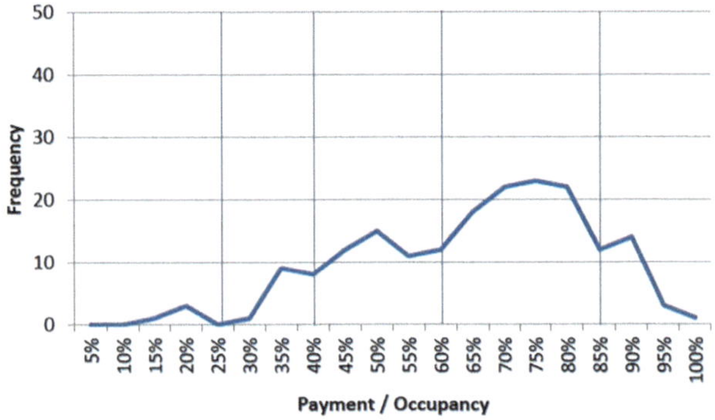

Fig. 1 Probability distribution of ratio of paid parking time to occupied time. We observe that many blocks experience a low payment compliance rate. Seventy-five percent of the occupied time on the modal block is paid for. Source: San Francisco Metropolitan Transportation Authority

the underlying model parameters, such as arrival rates, average parking time, and non-compliance rates (Sect. 4.3).
- The use of a stochastic queueing model to provide constraints for inferring the unobserved parking occupancy and the model parameters. (Sect. 3.1).
- The validation of the performance of the proposed methods with data from a large-scale parking experiment called SF*park* (Sect. 5.2) and with simulated data from *GI*/*GI*/*s* queue (Sect. 5.1).

The sample path construction for the *GI*/*GI*/*s* queue is the foundation of our queueing inference methods. The sample path formulation leads to a flexible modelling framework for inference. Our proposed method extends to time-varying model parameters (arrival rates, service rates, etc.). For the ease of exposition, we focus on the case of constant model parameters. Furthermore, our methods extend to the case of random order of service. There exists an analogous sample path construction of a GI/GI/s queue with random order of service (Sutton et al., 2011). For ease of exposition, we focus on the first-come first first-served service discipline.

The balance of the paper is organized as follows. Section 2 provides a review of the existing literature. The details of the application of a stochastic queuing model to parking occupancy are discussed in Sect. 3. Information on the computational algorithm and how the queuing behavior is inferred from smart parking meter transaction data is provided in Sect. 4. Numerical experiments are carried out in Sects. 5 and 6, the last section, presents a discussion of the findings and concludes.

2 Literature Review

Traffic congestion has a huge negative impact not only on local municipalities but also on the global economy. It increases pollution, wastes energy, and costs billions in lost time and productivity. A significant contributor to traffic congestion is vehicles searching for parking, and it is estimated that in busy areas, an average of 30% of traffic congestion is due to drivers cruising for open parking spaces (Shoup, 2006). Also contributing to this problem is the increase in vehicle ownership (Schaller, 2021), a development which adds to the scarcity of parking spaces, particularly on-street parking. Monitoring on-street parking is a much more difficult task than monitoring garage parking, which is relatively easy to compute through gate counts of entering and exiting vehicles. And cities often price on-street parking significantly lower than off-street parking, leading drivers rationally to choose to cruise (Millard-Ball, 2014).

Intelligent transportation systems (ITS), such as smart parking solutions, are being used to help relieve such congestion and help drivers navigate this situation by providing predictive and real-time information on current and future traffic and parking availability (Fabusuyi et al., 2014). A well-known smart parking solution to the problem of 'cruising for parking' was SF*park* (SFpark San Francisco Metropolitan Authority, 2014). SF*park* was a Federal Highway Administration

(FHWA) funded parking management project in San Francisco, California, from 2009 to 2014. The City of San Francisco installed close to 12,000 magnetometer sensors in 8000 parking spaces, creating a wireless sensor network to collect and distribute information on parking availability (Chatman & Manville, 2014). Although conceived as a demonstration project, the problems with replicating the program elsewhere include the enormous cost, which is beyond the reach of most cities, the limited life cycle of fixed sensors, and the variable pricing mechanism used by SF*park* (Fabusuyi & Hampshire, 2018).

Apart from the demonstration program, other approaches to managing on-street parking include sensing using smartphones (Nawaz et al., 2013) or through sensors attached to vehicles (Mathur et al., 2010), which require drivers to opt into the program. Crowdsensing, via on-vehicle sensors, is another opt-in system since it depends entirely on the willingness of drivers to attach the sensors to their vehicles and to provide information for others to use (Liao et al., 2016). Other forms of sensing include a combination of on-vehicle sensors and crowdsourcing (Roman et al., 2018). User opt-in is required in this type of project as well.

Detecting parking occupancy is also carried out via vehicle-to-infrastructure (V2I) communication. An example is a novel on-street parking management system that detects parking occupancy using low-cost Bluetooth beacon transmitters installed in vehicles that are connected to receivers located near on-street parking spots (Chen et al., 2019). Some of these solutions are being advocated by the private sector or by the private sector in partnership with universities. These include the use of radar sensors mounted on streetlights by Siemens called the Integrated Smart Parking Solution (Siemens, Integrated Smart Parking Solution, 2015). Here, the radar sensors are mounted on streetlights to scan a bigger area than can be done from a single vehicle. The radar sensors monitor both traffic flows and parking spots to help provide information for drivers searching for parking.

Ford Motor Company and Georgia Institute of Technology (Georgia Tech) have collaborated to build a system that gives information about on-street parking availability using sonars and radars on existing cars (Ford Motor Company, Parking Spotter, 2015). This technology is available to Ford drivers as a paid service. Other systems use vehicles' pre-installed parking sensors to better classify on-street areas into either legal or illegal parking spots to give more accurate information to drivers searching for parking (Coric & Gruteser, 2013). Using data from parking meters or kiosks is yet another method of collecting valuable parking information for drivers, given that parking meters or kiosks are responsible for 95% of on-street paid parking spots (Yang & Qian, 2017). This transaction data has been used with other data to provide not just the real-time information on parking availability but also to forecast short-term on-street parking occupancy. Yang et al. (2019), for example, use a deep neural network-based model that uses parking meter transactions along with traffic and weather conditions to predict parking occupancy. An earlier approach (Yang & Qian, 2017) simulated individual payment and parking behavior using a probabilistic payment model, which, when accumulated over all transactions, creates time-varying occupancy estimations with a low error rate.

The use of parking spaces or meters to estimate parking occupancy is, however, complicated by the fact that drivers can neglect to pay for parking, pay more than required for their time spent, or pay less than required. We can use the parking meter payments in a queueing process model of on-street parking to create a queueing inference engine (QIE) (Bertsimas & Servi, 1992; Larson, 1990; Mandelbaum & Zeltyn, 1998). However, our approach builds on literature in three significant ways. First, our underlying queueing model has no restriction on the inter-arrival time distribution. Second, our parking meter data makes up a subset of service commencement times, in contrast to QIE, whose observations are the complete set of departure times. Additionally, we only observe a noisy estimate of the service times, namely the payment amounts. Finally, we introduce a completely different framework based on simulating the queuing process and Markov chain Monte Carlo methods.

Our approach is innovative given that the authors know of no existing work that applies a QIE approach for estimating on-street parking spaces using parking meter transaction data. Our approach improves on the San Francisco Municipal Transportation Authority (SFpark San Francisco Metropolitan Authority, 2014; Demisch, 2016) regression-based model for estimating hourly parking occupancy using payment data. First, our model provides predictions at the time scale of the observed payments, while SFMTA's regression-based approach is at the hourly level. Secondly, our framework learns the model parameters on the fly, while previous methods require the user to tune the regression approach for each geographic region. Thus, the ease of replicating our approach is afforded to users as they can deploy our method more broadly compared to the regression-based approach.

3 Models

3.1 Parking Model

The unit of analysis is a single on-street parking block with s parking spaces. We model parking occupancy at the block level as a *GI*/*GI*/*s* stochastic queueing model. We define $\{\alpha_j\}$ to be a sequence of random independent inter-arrival times of the drivers with a general distribution and $\{\nu_j\}$ as a sequence of random independent parking (service) times with a general distribution. These inter-arrival and service times are the random input primitives to the queue. The queue transforms the random input primitives into arrival times, service start times, and departure times. The physics of the queue determines the values of these outputs. The queue transforms the random primitives according to the set of equations (see (Sutton et al., 2011; Kiefer & Wolfowitz, 1955; Krivulin, 1994)).

$$a_j = a_{j-1} + \alpha_j$$

$$b_{ji} = \max\{d_j' | a_j' < a_j, p_j' = i\}$$

$$c_j = b_{ji}$$

$$p_j = \text{argmin} b_{ji}$$

$$u_j = \{ a_j, c_j \}$$

$$d_j = v_j + u_j \tag{1}$$

where the outputs are the sequences of arrival times $\{a_j\}$ and departure times $\{d_j\}$. Intermediate variables assist the transformation; u_j denotes the service time for the j-th arrival, c_j represents the first moment when a parking space becomes available for the j-th arrival who is assumed to occupy one of the s available parking spaces. This transformation is one-to-one and invertible (Sutton et al., 2011). Additionally, the likelihood of any set of N arrival and departure times is:

$$p(a,d) = \prod_{j=1}^{N} h_\alpha \left(a_j - a_{j-1}\right) \cdot h_\nu \left(\text{d}_j - u_j\right) \tag{2}$$

where h_α is the density of the inter-arrival times with parameters θ_1 and h_ν is the density of the service times with parameters θ_2. This product form of the likelihood is due to the independence of the inter-arrival times and the service times. These equations correspond to drivers parking at available spaces in the order they arrive. We can extend our results to a random order of service by slightly modifying the queue input and output Eqs. (6). For ease of presentation, we focus on the first-come-first-served (FCFS) service discipline. Using this transformation, we construct the number of parked cars, the number of cars searching for parking, and the time needed to find an available parking space.

3.2 *Payment Observation Model*

We now introduce payments into the parking model. We assume that a driver pays immediately after parking. The driver pays at an automated payment station that aggregates payments for all s spaces on the block. After the k-th driver parks, the time remaining on the meter is:

$$m_k = \left(m_{k-1} + \beta_k - \left(\tau_k - \tau_{k-1}\right), 0\right)$$

where β_k is the amount paid time by the k-th parking driver and τ_k is the time of the k-th payment. The payment amount relates to the true parking time ν through a general joint distribution. For the sake of simplicity, we assume that the payment amount is a mixture of two distributions. The first component of the mixture is an exponential distribution with a mean value equal to the true parking time. The second component of the mixture is a point mass at zero, which represents the probability, p, of a driver not paying for parking.

The observations consist of the time of each payment $\tau_{1:K}$ and the time remaining on the meter $m_{1:K}$. We focus on analyzing a batch of observations $y_{1:K} = (\tau, m)_{1:K.}$

3.3 State Space

An important feature of the model is that not all parking drivers pay for parking. Therefore, the number of payments does not correspond to the number of arriving vehicles. We now introduce a random variable to track the number of vehicles arriving. Let the random variable $J(k)$ be the number of drivers arriving before the k-th observed payment. Since the k-th payment occurs at time τ_k, the number of arrivals up to the k-th payment satisfies:

$$J(k) = \left(\sum_{i=J(k-1)+1}^{n} 1_{\{\beta_i > 0\}} \leq 1\right) \tag{3}$$

where $J(k) = J(k - 1) + x_k$ and x_k is a geometric random variable with rate $P(\beta_k = 0)$.

The number of arrivals before the kth payment, $J(k)$, evolves according to a Markov chain at payment times. It is a counting process with independent increments. The number of arrivals until k payments has a negative binomial distribution with parameter $1 - p$ and k.

The state space is the total number of arrivals and the corresponding collection of arrival times, service start times, and departures times of all customers in the queue at the time of the k-th payment.

$$x_k = \left(j(k), a_{1:j(k)}, u_{1:j(k)}, d_{1:j(k)}\right)$$

The state has two components: the number of arrivals before the kth payment and the queueing quantities for these arrivals. The dimension of x_k grows randomly with k due to the unknown number of arrivals between payments.

4 Inference

Let θ be the parameters of the inter-arrival time, service time, and payment distributions in the parking and payment models. Our goal is to estimate

$$p(y_{1:K}) = \frac{p_\theta(x_{1:K}, y_{1:K}) \bullet p(\theta)}{p(y_{1:K})} \tag{4}$$

where $p(\theta)$ is the prior distribution on the model parameters

$$p_\theta(x_{1:K}, y_{1:K}) = p_\theta(x_{1:K}) \bullet p_\theta(x_{1:K})$$

$$= \mu_\theta(x_1)\prod_{k=2}^{K} f_\theta(x_{k-1})\cdot\prod_{k=1}^{K} g_\theta(x_k)$$

where the process $\{x_n; n \geq 1\}$ has an initial density $\mu_\theta(\cdot)$ and the payments are independent of each other, given the state and they have a common density $g_\theta(x)$ where x is the state of the system. The transition probability is

$$f_\theta(x_{k-1}) = p^{j(k)-j(k-1)-1}(1-p)\cdot\prod_{j=j(k-1)}^{j(k)} h_\alpha(a_j - a_{j-1})\cdot h_v(d_j - u_j)$$

This joint density $p_\theta(x_{1:K}, y_{1:K})$ has the form of a Feynman-Kac model. The rich literature on Feynman-Kac models (see Del Moral, 2000) provides the theoretical foundation for the proposed computational algorithms.

4.1 Particle Filter

The *GI*/*GI*/*s* queue transformation constrains the likely values of the random primitives given the payment observations. First, we consider the case where the parameters of the inter-arrival, service time, and payment distributions are known. Our goal is to compute the posterior distribution of the state given payment observations, $p_\theta(y_{1:K})$. The traditional Kalman filter (Rudolph Emil Kalman, 1960) does not apply to this goal because the dynamics of the state are neither linear nor do they have a Gaussian distribution. The queueing inference engine techniques (Bertsimas & Servi, 1992; Larson, 1990; Mandelbaum & Zeltyn, 1998) do not apply to the goal either, due to the general nature of the distribution of the random primitives.

The solution is to simulate sample trajectories from $p_\theta(y_{1:K})$ using Sequential Monte Carlo (SMC), also known as a particle filter (Doucet & Johansen, 2009; Gordon et al., 1993). The particle filter approximates the posterior distribution over state trajectories using N particles. The particle filter proceeds in three steps. First, for each particle, we simulate a one-step state transition. Next, we compute an importance weight for each particle based on the likelihood of observing the current payment given the state of the particle. Finally, we propagate, kill, or replicate each particle in proportion to its importance weight. Particles with larger weights have a higher likelihood of replicating, and those with smaller weights tend to die out. In this way, the surviving particles approximate samples from the posterior distribution $p_\theta(y_{1:K})$. For any number of particles, the particle filter produces an unbiased estimate of the posterior distribution (Doucet & Johansen, 2009). It also produces an unbiased estimate of the likelihood of the observations as shown in Eq. 5.

$$\hat{p}(y_{1:K}) = \prod_{k=1}^{K}\left(\frac{1}{N}\sum_{n=1}^{N} w_k^{(n)}\right) \tag{5}$$

where $w_k^{(n)}$ is the importance weight of the n-th particle at step k.

The two ingredients of the particle filter are the state transition probabilities and the observation likelihood model. The particle filter requires only the ability to simulate the one-step transitions of the state. The observation likelihood model is more complex to compute than the state transitions.

4.2 Approximate Bayesian Computation (ABC)

In this section, we construct the observation likelihood, $g(x_k)$, of the k-th observation given the state. The observations consist of the payment time and time remaining on the meter just after the kth payment. The time of the kth payment must correspond to either an arrival time or a departure time of a previous arrival. However, the structure of the departure times is complex and does not preserve arrival order. Further, the departure times depend on residual service times that are difficult to compute for general service distributions. The amount of the k -th payment depends on the service time of the corresponding driver, who need not be the kth arrival. These factors make the likelihood challenging to compute in closed form; therefore, we simulate an estimate of the likelihood density.

For each particle, we simulate H realizations of the kth payment given the state x_k. This simulation is fast due to the sample path construction of the queue and payment models. A particle receives a high weight if the simulated kth payments are "close" to the observed kth payment and receives a low weight otherwise. This is called approximate Bayesian computation (ABC) (Beaumont et al., 2002; Marjoram et al., 2003). Jasra et al. (2012) and Calvet et al. (2015) successfully use ABC in the particle filtering context. We approximate the likelihood density function by using a kernel-based distance metric in the observation space

$$\widetilde{g_\epsilon}(y_k) = \frac{1}{H \cdot \epsilon} \sum_{h=1}^{H} K\left(\frac{y_k - \widetilde{y_k^{(h)}}}{\varepsilon} \right) \tag{6}$$

where ϵ is the bandwidth of the kernel, a H is the number of simulations.

4.3 Particle Marginal Metropolis-Hastings (PMMH)

We turn to jointly estimating both model parameters. We are interested in the parameters of the inter-arrival time distribution (λ), the service time distribution (μ), and payment probability (p), and we implement the Particle Markov Chain Monte Carlo method (PMCMC) (Andrieu et al., 2010). This combines two powerful methods: Markov Chain Monte Carlo and the particle filter. Specifically, we use the Particle Marginal-Metropolis Hastings (PMMH) method. The outputs of PMMH are an estimate of the data likelihood, $\widehat{p_\theta}(y_{1:K})$, and samples from the posterior distribution $p(y_{1:K})$.

The key idea of the approach is that the particle filter generates state proposals within the Metropolis-Hastings algorithm. The particle filter also generates an estimate of the likelihood $p_\theta(y_{1:k})$ which is part of the proposal acceptance probability. We accept joint state and parameters proposals that explore parts of the posterior distribution that are more likely. In essence, we are searching over time, the evolution of the state space and parameters that are the most likely given the observed sequence of payments.

Here is how it works. The input to the method is a batch of payments observations, $y_{1:K}$. First, we generate proposed values for the parameters using a Gaussian random walk proposal mechanism. Next, we run the particle filter to generate a proposed state trajectory $x_{1:K}$ under the posterior distribution, and we generate a likelihood estimate $\widehat{p_\theta}(y_{1:K})$ under the proposed parameters. We accept the proposed parameters and state trajectory according to an acceptance probability that depends on the likelihood and the parameter proposal distribution. We iterate this process, which results in unbiased estimates of the joint distribution over the parameters and the state trajectory. The estimates are consistent for any number of particles. The output of PMMH is a distribution of accepted state proposal-parameter pairs. Each accepted proposal of the MCMC step consists of both a state trajectory and corresponding model parameters.

We modify the PMMH algorithm by plugging in the ABC particle filter to generate the state proposals. In their discussion of Andrieu et al. (2010), Cornebise and Peters (2009) advocate this approach when the likelihood is intractable. Our main result shows that the ABC-PMMH method generates unbiased estimates of the target joint distribution as the number of particles and simulated ABC observations go to infinity.

Theorem 1 *The ABC-PMMH converges to the correct posterior distribution when the number of particles grows appropriately with the size of the kernel density estimator, that is*

$$\left| loglog\, p_\theta^{ABC} - loglog\, p_\theta \right| \to 0$$

where the number of particles, N scales with the number of simulated H payments.

$$N = O\left(\left(\frac{H}{loglog\, H}\right)^{-2/5}\right)$$

and the bandwidth of the kernel density estimator decreases with H according to $\varepsilon = c \cdot \left(\frac{loglog\, H}{H}\right)^{1/5}$

Proof The Particle Marginal Metropolis-Hasting acceptance probability depends on the likelihood estimate generated by the particle filter

$$p_\theta(y_{1:T}) = \prod_{t=1}^{T}\left(\frac{1}{N}\sum_{n=1}^{N} g_\theta(x_t^n)\right)$$

where T is the number of steps (payments) and N is the number of particles. This is an unbiased estimator of the likelihood (Del Moral & Miclo, 2000) for any number of particles.

However, the expression for the density function g is intractable. Therefore, we estimate the density g using a Kernel density estimator with bandwidth ϵ thus:

$$\widehat{g_\varepsilon}(y) = \frac{1}{H\cdot\varepsilon}\sum_{h=1}^{H} K\left(\frac{y_h - y}{\varepsilon}\right)$$

where $(y_h)_{h=1}^{H}$ are simulated from true distribution g. The corresponding observation likelihood estimate generated by the particle filter is

$$loglog\ p_\theta^{ABC} = \sum_{t=1}^{T}\sum_{n=1}^{N}\frac{1}{\varepsilon NH}\sum_{h=1}^{H} K\left(\widetilde{y_t^{n,h}};, y_t;, \varepsilon^2\right)$$

If $\left|loglog\ p_\theta^{ABC} - loglog\ p_\theta\right| \to 0$, then the posterior distribution converges to the correct posterior distribution by applying Theorem 4 of Andrieu et al. (2010). The difference between the true and approximate log likelihood is bounded above by this expression:

$$\left|loglog\ p_\theta^{ABC} - loglog\ p_\theta\right| \le \sum_{t=1}^{T}\frac{1}{N}\sum_{n=1}^{N}\left|\frac{1}{H\cdot\varepsilon}\sum_{h=1}^{H} K\left(\widetilde{y_k^{n,h}};, y_k;, \varepsilon^2\right) - g_\theta\left(y_k | x_k^n\right)\right|$$

$$\le \sum_{t=1}^{T}\frac{1}{N}\sum_{n=1}^{N}\sup_{y_k}\left|\frac{1}{H\bullet\varepsilon}\sum_{h=1}^{H} K\left(\widetilde{y_k^{n,h}};, y_k;, \varepsilon^2\right) - g_\theta\left(y_k | x_k^n\right)\right|$$

$$\le T\cdot\delta\cdot\left(\frac{loglog\ H}{H}\right)^{2/5}$$

The last inequality follows from the uniform convergence results for kernel density estimators.

Theorem 2 Fan and Yao (2003): If the If the density g has a bound second derivative and the bandwidth is chosen so that $\varepsilon = c\cdot\left(\frac{loglog\ H}{H}\right)^{1/5}$, *then*

$$\sup_y\left|\hat{g}_\varepsilon(y) - g(y)\right| \le \delta\bullet\left(\frac{loglog\ H}{H}\right)^{2/5}$$

We adopted the PMMH method for two reasons. First, the approach allows for flexibility in model specification and only requires the ability to simulate sample paths of the model. Secondly, PMMH handles state spaces with random dimensions. Our state space is of an unknown dimension due to the unknown number of arrivals between payments. In the literature, this problem is called trans-dimensional (Green, 2003). One way to perform estimation on trans-dimensional problems is the Reversible Jump Markov Chain Monte Carlo (RJMCMC) method (Green, 1995). This popular method requires the development of problem-specific and often complex proposal moves (Karagiannis & Andrieu, 2013). There are several papers that use the PMCMC methods in trans-dimensional problems. For example, PMCMC was used to construct phylogenetic trees (Persing et al., 2015), and to estimate volatility in asset pricing models (Bauwens et al., 2014).

5 Numerical Experiments

In this section, we test the performance of the PMMH approach to jointly estimating parking occupancy and the model parameters. First, we validate the approach by estimating parking occupancy and the model parameters using simulated data from a *GI*/*GI*/*s* queue assuming all customers pay for parking ($p = 1$). Next, we jointly estimate the parking occupancy and the model parameters under partial payment compliance. Finally, we apply the method to parking meter payment data from SF*park*, a large-scale parking intervention policy in San Francisco.

5.1 Simulated Data

First, we assume that the drivers pay for parking 100% of the time. To validate the estimation technique, we simulate 40 payments using the parking and payment models (Sect. 3.1) with parameters $(\lambda, \mu, p) = (0.752, 5.0, 1.0)$ and with 7 spaces. We perform the inference by simulating a candidate set of state trajectories and parking model parameters (λ, μ). Next, we run a particle filter to generate $N = 600{,}000$ samples of the state from the posterior distribution, and an estimate of the data likelihood. The proposed state is a sample from the N particles and the model parameters. We accept or reject the proposal with a probability proportional to the estimated likelihood of the observed payments. We iterate this procedure until we accept a predefined number of proposals in excess of a burn-in period.

The output of the PMMH is a joint posterior distribution of the model parameters and state trajectories. Figure 2 is the contour plot of the accepted model parameters (λ, μ). From this joint distribution, we select the most likely parameter-trajectory pair, $\left(\underline{\lambda}, \underline{\mu}\right) = \left(0.935, 5.195\right)$, with the largest value of the data likelihood. Figure 3 plots the mean and confidence intervals of accepted trajectories generated by the PMMH algorithm. The true path is computed from the parking queueing model corresponding to the 40 observed payments. We find that the true trajectory falls well within the 95% confidence interval of the estimated trajectory distribution. The corresponding root mean square error is 1.12 cars.

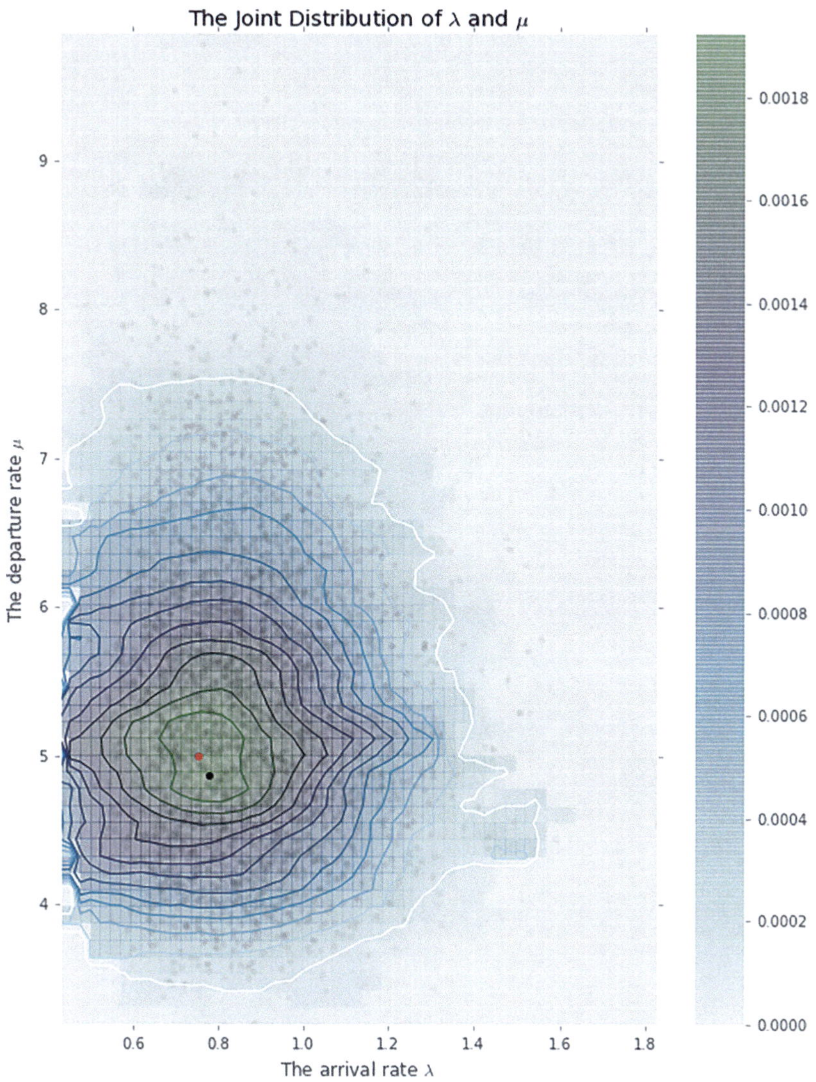

Fig. 2 Joint distribution of the model parameters accepted by the PMMH under the model parameters $(\lambda, \mu, p) = (0.752, 5.0, 1.0)$ and with 7 spaces

Next, we assume that 80% of drivers pay for their parking while 20% do not. We simulate a new sequence of 40 payments from a model with the same parameters $(\lambda, \mu, p) = (0.752, 5.0, 0.8)$ and 7 parking spaces. From the generated contour plot of the accepted model parameters, we select the most probable parameter-trajectory pair, $\left(\underline{\lambda},\underline{\mu},p\right) = \left(0.891, 4.975, .808\right)$, which has the largest value of the data likelihood. Figure 4, constructed using a similar approach in Fig. 3, plots the mean and confidence intervals of accepted trajectories generated by the PMMH algorithm.

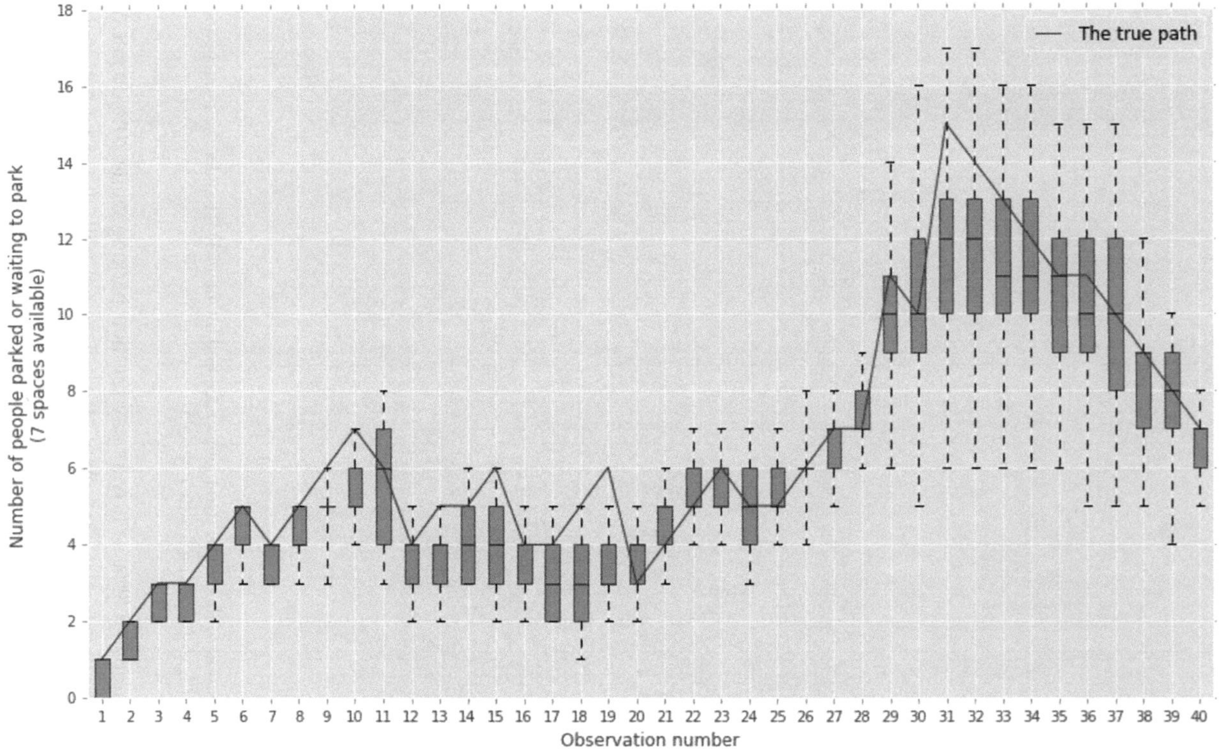

Fig. 3 Plot of true number of people parked versus our belief of the number of people parked on a simulated block. The solid line is the parking occupancy trajectory at 40 simulated payment times under the model parameters $(\lambda, \mu, p) = (0.752, 5.0, 1.0)$ and with 7 spaces. The estimated parking occupancy trajectory using PMMH, with 600,000 particles, is shown as a box-stem plot at the payment times. The most likely value of the parameters from the estimation is $(\lambda, \mu, p) = (0.935, 5.195, 1.0)$ The root mean square difference of the median of our belief against the true path is 1.12 cars

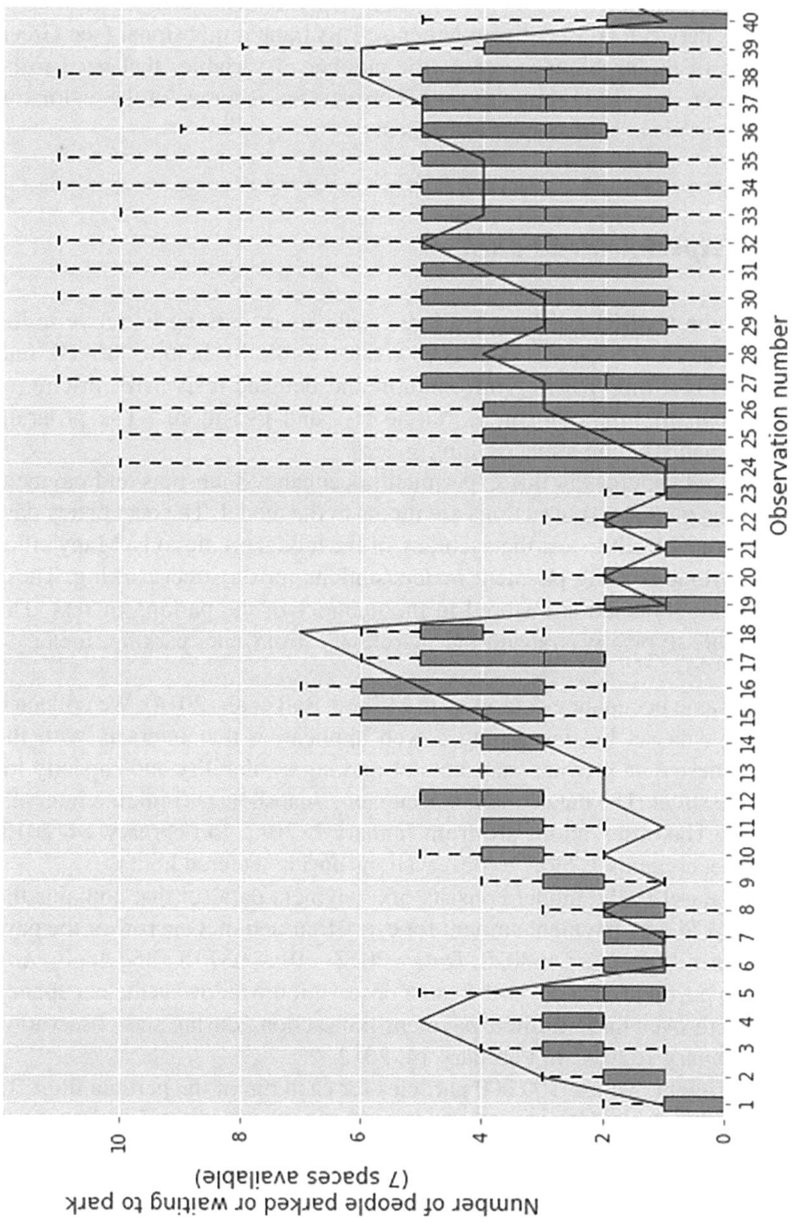

Fig. 4 The solid line is the parking occupancy trajectory at 40 simulated payment times under the model parameters $(\lambda, \mu, p) = (0.752, 5.0, 0.8)$ and with 7 spaces. The estimated parking occupancy trajectory using PMMH, with 600,000 particles, is shown as a box-stem plot at the payment times. The most likely value of the parameters from the estimation is $(\lambda, \mu, p) = (0.891, 4.975, 0.808)$ The root mean square difference of the median of our belief against the true path is 1.65 cars

The true path is computed from the parking queueing model corresponding with the 40 observed payments.

There is a noticeable decrease in estimation accuracy caused by the increased number of model parameters. When $p < 1$, the particle filter estimates the number of arriving drivers between consecutive payments. This trans-dimensional (see Green (1995)) feature of the problem increases the number of variables that need to be inferred. This increases the complexity and decreases the accuracy of the estimator. The resulting root mean square error is 1.65 cars.

5.2 *Field Experiment: SFpark*

We now apply the PMMH to SF*park*, a large-scale smart parking initiative in the City of San Francisco. SF*park*'s goal is to reduce traffic by helping drivers find parking. It uses real-time parking information and demand-responsive pricing to achieve this goal. SF*park*'s slogan is "circle less and live more." The program includes more than 7000 on-street parking spaces.

We use two data sources in this experiment: occupancy snapshots and payment transactions. The payment transactions are inputs to the model. The occupancy data set serves as ground truth to test the accuracy of the model predictions. Many cities now have electronic parking payment meters/stations for on-street parking. These payment data are collected and stored in the memory of the parking meters. The parking data are typically transmitted wirelessly from the parking meter to the vendor.

We use the same occupancy dataset as in Millard-Ball et al. (2014). We collected the occupancy data set by developing a web application that interacts with the SF*park* API. These data provide snapshots of parking availability and capacity for each side of the street. The data contains occupancy snapshots at 5-minute intervals for 340 blocks. The observations are from January 1, 2012, to February 14, 2012. The API data set contains 1,730,770 observations during metered hours.

The input dataset to the model consists of a payment data set that contains the date, time, block id, and payment amount for every transaction. One row of the payment has the form $\{'blockId' : 468022, 'Date' : 2012-01-03T15:35:46.46, 'Amt' : 1.25\}$. These payment data cover the same observation window as the occupancy data. There were over three million payment transactions during the observation period from January 1, 2012, to February 14, 2012.

In this experiment, we use 160,000 particles for each run of the particle filter. To speed up the Metropolis-Hastings algorithm, we use an adaptive random walk mechanism to generate parameter proposals (Peters et al., 2010). The burn-in period of the adaptive Metropolis-Hastings step was 200 accepts, with 3800 accepts after burn-in. We selected the 2200 and 2400 blocks of Mission Street in San Francisco for the inference. Figure 5 shows a plot of the estimated and true occupancy trajectory over a 2 h time window on the 2200 block of Mission Street in San Francisco on January 12, 2013. Figure 6 shows a plot of the estimated and true occupancy

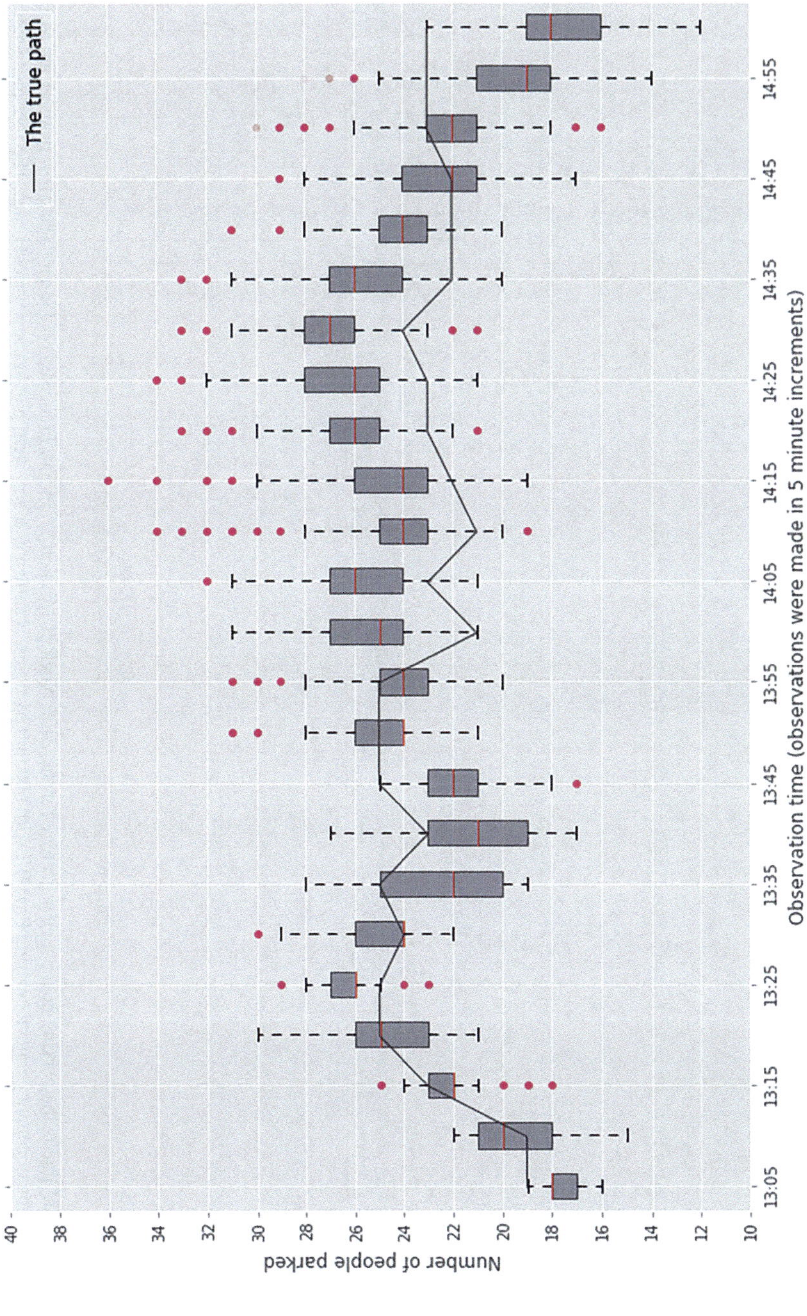

Fig. 5 The solid line is the parking occupancy trajectory for the 2200 block of Mission St. in San Francisco on January 12, 2013 from 1 pm–3 pm. The estimated parking occupancy trajectory using PMMH, with 600,000 particles, is shown as a box-stem plot at the payment times. The root mean square difference of the median of our belief against the true path is 2.54 cars

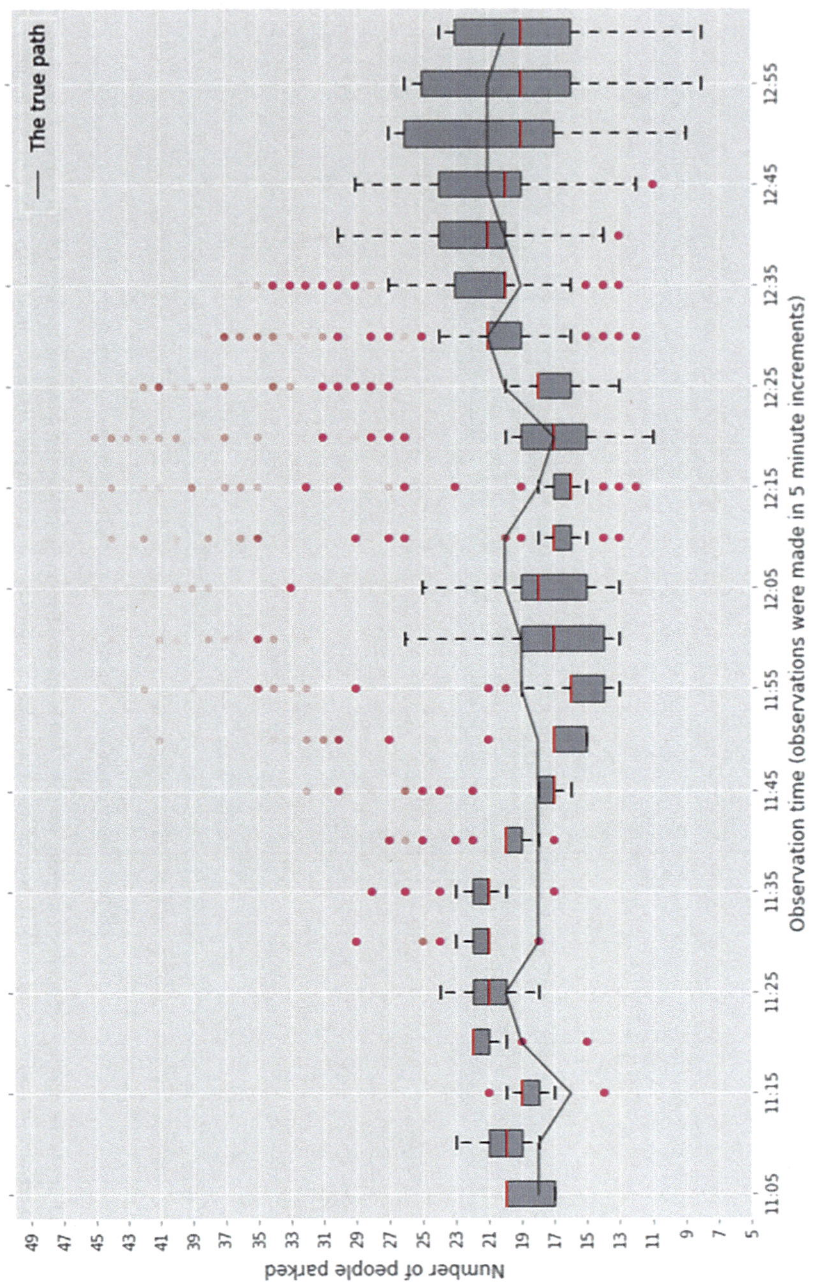

Fig. 6 The solid line is the parking occupancy trajectory for the 2400 block of Mission St. in San Francisco on January 18, 2013, from 11 am to 1 pm. The estimated parking occupancy trajectory using PMMH, with 600,000 particles, is shown as a box-stem plot at the payment times. The root mean square difference of the median of our belief against the true path is 2.01 cars

trajectory over the 2 h time window on the 2400 block of Mission Street in San Francisco on January 18, 2013. As expected, the mean squared error is larger for the field test than the simulated data. However, the performance of our approach is superior to SF*park*'s Sensor Independent Rate Adjustment (SIRA) regression approach.

5.3 *Sensor Independence Rate Adjustment Approach*

SF*park* developed a method to estimate the hourly parking occupancy rate from hourly payment rate data (SFpark San Francisco Metropolitan Authority, 2014; Demisch, 2016). Occupancy rate is the total occupied time divided by the sum of occupied and unoccupied time. The payment rate is the total paid time divided by the sum of the total paid time and unpaid time. It is based on multiple linear regression, where dummy variables correspond to the various neighborhoods in San Francisco. SF*park* currently uses SIRA to generate monthly occupancy estimates. These occupancy estimates influence the monthly changes in parking prices.

Our method produces occupancy estimates at a finer time resolution than the SIRA method. In addition to parking occupancy, we estimate the arrival rates and parking times, which SIRA does not offer. We also ran tests of our method against SF*park* SIRA method. During the same time and location that we used above, the true hourly average occupancy was 84%. Our method estimated the hourly average occupancy at 88%, while SIRA estimated an hourly average occupancy of 105%.

6 Discussion and Conclusion

As cities and urban areas become more data-rich, examples of data being repurposed or given parallel lives will become more ubiquitous; we just demonstrated one. In addition, our numerical methods show that revisions to the estimates could be made in real time, thus allowing the dynamic pricing of parking spaces in a manner that reflects both the temporal and spatial dimensions of the demand. We have demonstrated the effectiveness of our approach in achieving SF*park*'s occupancy objective, given the increased reliability in our estimates compared to SIRA's. We have also shown how our approach could be replicated nationwide—a feature that delivers on a core objective of SF*park* as a demonstration project. This could be achieved at minimal cost compared to the invasive and expensive price tags typically associated with in-ground sensors and the reduction in reliability when their batteries start to fail.

The ability to modify parking rates, particularly in real time, has implications for alleviating parking search-related congestion. Although urban planners know of the areas where there is scarce parking, they do not have a good way of quantifying the magnitude of the problem. Once these neighborhoods are identified, we can look to a variety of remedies, such as: changing parking rates to incentivize drivers to choose off-street instead of searching for on-street parking; rezoning areas so that more off-street parking is available; increasing off-street parking with direct investments; or implementing congestion pricing in neighborhoods with high rates of cruising.

All four of the remedies mentioned above can be viewed as policies that either explicitly or implicitly change the price of parking. Changing the price of on-street parking is the most direct way to reduce drivers' incentive to search for on-street parking and is the path that policy makers in SF*park* took. Investing directly in off-street parking and rezoning for more off-street parking increases the off-street parking supply and should reduce the price of off-street parking. These approaches are often difficult to implement in practice, and their effects are difficult to estimate beforehand. Using congestion pricing in high traffic areas functions as an implicit increase in the price of both on-street and off-street parking. Congestion pricing has other benefits since it reduces traffic overall. However, it is difficult to implement in a targeted manner.

To conclude, we introduce an inference framework for estimating parking occupancy from parking meter payment data. Given a historical record of parking payment transactions, our method provides estimates of second-by-second parking occupancy. This framework allows cities to use existing infrastructure (parking meters) to estimate parking occupancy without installing specialized parking sensors. The inference methodology is an extension of a simulation-based approach called Particle Markov Chain Monte Carlo method that includes an approximate Bayesian computation step to compute the likelihood weights. We show that this method yields unbiased asymptotic estimates of the posterior distribution.

The framework has the virtue of being unsupervised; that is, no training set of ground truth parking occupancy data is required to perform the inference. The inference relies solely on the underlying model for the parking search process. Although we employed the *GI*/*GI*/*s* queue with first-come first-served, the PMMH framework supports any simulated parking search model. Beyond introducing the inference framework and establishing its basic theoretical properties, subsequent efforts may focus on developing the most appropriate parking search queuing model. This methodology can serve as the basis for a tool that can potentially enable a parking information system for citizens and provide policy makers and city planners the ability to evaluate the impact of parking policy interventions such as pricing modifications, information provision, and time limit changes.

Acknowledgement This research was supported by the National Science Foundation under Grant No. 1435164.

References

Andrieu, C., Doucet, A., & Holenstein, R. (2010). Particle markov chain Monte Carlo methods. *Journal of the Royal Statistical Society: Series B (Statistical Methodology), 72*(3), 269–342.

Bauwens, L., Dufays, A., & Rombouts, J. V. K. (2014). Marginal likelihood for markov switching and change-point garch models. *Journal of Econometrics, 178*, 508–522.

Beaumont, M. A., Zhang, W., & Balding, D. J. (2002). Approximate Bayesian computation in population genetics. *Genetics, 162*(4), 2025–2035.

Bertsimas, D. J., & Servi, L. D. (1992). Deducing queueing from transactional data: The queue inference engine, revisited. *Operations Research, 40*(3), S217–S228.

Calvet, L. E., & Czellar, V. (2015). Accurate methods for approximate Bayesian computation filtering. *Journal of Financial Econometrics, 13*(4), 798–838. https://doi.org/10.1093/jjfinec/nbu019

Chatman, D. G., & Manville, M. (2014). Theory versus implementation in congestion priced parking: An evaluation of SF*park*, 2011–2012. *Research in Transportation Economics, 44*, 52–60.

Chen, H. T., Lin, P. Y., & Lin, C. Y. (2019). A smart roadside parking system using bluetooth low energy Beacons. In L. Barolli, M. Takizawa, F. Xhafa, & T. Enokido (Eds.), *Web, artificial intelligence and network applications. WAINA 2019. Advances in intelligent systems and computing* (p. 927). Springer.

Coric, V., & Gruteser, M. (2013). Crowdsensing maps of on-street park-ing spaces. In *Proceedings of the 2013 IEEE international conference on distributed computing in sensor systems (DCOSS '13)* (pp. 115–122). IEEE Computer Society. https://doi.org/10.1109/DCOSS.2013.15

Cornebise, J. & Peters, G. W. (2009), *Comments on" particle markov chain monte carlo" by c. andrieu, a. doucet and r. hollenstein*, arXiv preprint arXiv:0911.3866.

Del Moral, P., & Miclo, L. (2000). Branching and interacting particle systems. Approximations of Feynman-Kac formulae with applications to non-linear filtering. *Seminaire de probabilities de Strasbourg, 34*, 1–145.

Demisch, A. (2016). Demand-responsive pricing on the cheap: Estimating parking occupancy with meter payment data. *Transportation Research Record: Journal of the Transportation Research Board., 2543*(2016), 125–133. https://doi.org/10.3141/2543-14

Doucet, A., & Johansen, A. M. (2009). A tutorial on particle filtering and smoothing: Fifteen years later. *Handbook of Nonlinear Filtering, 12*, 656–704.

Fabusuyi, T., & Hampshire, R. (2018). Rethinking performance-based parking pricing: A case study of SF*park*. *Transportation Research Part A: Policy and Practice, 115*, 90–101.

Fabusuyi, T., Hampshire, R., Hill, V., & Sasanuma, K. (2014). Decision analytics for parking availability in downtown Pittsburgh. *Interfaces, 44*(3), 286–299.

Fan, J., & Yao, Q. (2003). *Nonlinear time series: Nonparametric and parametric methods*. Springer Science and Business Media.

Ford Motor Company, Parking Spotter. (2015). Retrieved January 12, 2021, from https://drivemyfamily.com/2015/03/24/fords-parking-spotter-technology-quickly-evolving/

Gordon, N. J., Salmond, D. J., & Smith, A. F. M. (1993). Novel approach to nonlinear/non-gaussian bayesian state estimation. *IEEE Proceedings, 140*, 107–113.

Green, P. J. (1995). Reversible jump markov chain Monte Carlo computation and bayesian model determination. *Biometrika, 82*(4), 711–732.

Green, P. (2003). Trans-dimensional markov chain Monte Carlo. In *Oxford statistical science series* (pp. 179–198). Oxford Academic.

Jasra, A., Singh, S. S., Martin, J. S., & McCoy, E. (2012). Filtering via approximate bayesian computation. *Statistics and Computing, 22*(6), 1223–1237.

Karagiannis, G., & Andrieu, C. (2013). Annealed importance sampling reversible jump MCMC algorithms. *Journal of Computational and Graphical Statistics, 22*(3), 623–648.

Kiefer, J., & Wolfowitz, J. (1955). On the theory of queues with many servers. *Transactions of the American Mathematical Society, 78*(1), 1–18. https://doi.org/10.2307/1992945

Krivulin, N. K. (1994). A recursive equations-based representation for the g/g/m queue. *Applied Mathematics Letters, 7*(3), 73–77.

Larson, R. C. (1990). The queue inference engine: Deducing queue statistics from transactional data. *Management Science, 36*(5), 586–601.

Liao, R., Roman, C., Ball, P., Ou, S., & Chen, L. (2016). *Crowdsourcing on-street parking space detection*. Cornell University. Retrieved October 14, 2020, from arXiv.org/abs/1603.00441

Mandelbaum, A., & Zeltyn, S. (1998). Estimating characteristics of queueing networks using transactional data. *Queueing Systems, 29*(1), 75–127.

Marjoram, P., Molitor, J., Plagnol, V., & Tavare, S. (2003). Markov chain Monte Carlo without likelihoods. *Proceedings of the National Academy of Sciences, 100*(26), 15324–15328.

Mathur, S., Jin, T., Kasturirangan, N., Chandrasekaran, J., Xue, W., Gruteser, M., & Trappe, W. (2010). ParkNet: Drive-by sensing of road-side parking statistics. In *MobiSys'10—Proceedings of the 8th international conference on mobile systems, applications, and services* (pp. 123–136). https://doi.org/10.1145/1814433.1814448

Millard-Ball, A. (2014). Rachel R Weinberger, and Robert C Hampshire, is the curb 80% full or 20% empty? Assessing the impacts of San Francisco's parking pricing experiment. *Transportation Research Part A: Policy and Practice, 63*, 76–92.

Nawaz, S., Efstratiou, C., & Mascolo, C. (2013). Park sense: A smartphone-based sensing system for on-street parking. In *Proceedings of the 19th annual international conference on mobile computing and networking* (pp. 75–86).

Persing, A., Jasra, A., Beskos, A., Balding, D., & De Iorio, M. (2015). A simulation approach for change-points on phylogenetic trees. *Journal of Computational Biology, 22*(1), 10–24.

Peters, G. W., Hosack, G. R., & Hayes, K. R. (2010). *Ecological non-linear state space model selection via adaptive particle markov chain monte carlo (adpmcmc)*, arXiv preprint arXiv:1005.2238.

Pierce, G., & Shoup, D. (2013). Getting the prices right: An evaluation of pricing parking by demand in San Francisco. *Journal of the American Planning Association, 79*(1), 67–81.

Roman, C., Liao, R., Ball, P., Ou, S., & de Heaver, M. (2018). Detecting on-street parking spaces in smart cities: Performance evaluation of fixed and mobile sensing systems. *IEEE Transactions on Intelligent Transportation Systems, 19*(7), 2234–2245.

Rudolph Emil Kalman. (1960). A new approach to linear filtering and prediction problems. *Journal of Fluids Engineering, 82*(1), 35–45.

Schaller, B. (2021). *In a reversal 'car-rich' households are growing, despite uber and lyft urban car ownership is growing*. Retrieved January 12, 2021, from https://www.bloomberg.com/news/articles/2019-01-07/

SFpark San Francisco Metropolitan Authority. (2014). *Sensor Independent Rate Adjustments (SIRA) methodology and implementation plan.*

Shoup, D. C. (2006). Cruising for parking. *Transport Policy, 13*(6), 479–486.

Siemens, Integrated Smart Parking Solution. (2015). https://www.mobility.siemens

Sutton, C., Jordan, M. I., et al. (2011). Bayesian inference for queueing networks and modeling of internet services. *The Annals of Applied Statistics, 5*(1), 254–282.

Yang, S., & Qian, Z. S. (2017). Turning meter transactions data into occupancy and payment behavioral information for on-street parking. *Transportation Research Part C: Emerging Technologies, 78*, 165–182.

Yang, S., Ma, W., Pi, X., & Qian, Z. S. (2019). A deep learning approach to real-time parking occupancy prediction in transportation networks incorporating multiple spatiotemporal data sources. *Transportation Research Part C: Emerging Technologies, 107*, 248–265.

Urban Design Analytics in the Age of Data Science and AI: Discovery, Measurement, Design, and Prediction

Stephen Law, Demin Hu, Xuhui Lin, Xiaowei Gao, Zicheng Fan, Yao Shen, and Tao Yang

Abstract How can data science and artificial intelligence (AI) methods be used in the urban design process? In this chapter, we argue that data science can be integrated at every stage of the urban design process, assisting rather than replacing human experts. We illustrate this process through four sequential stages: discovering knowledge from contextual data, digital reconstruction for augmented design, applying generative methods in design development, and using causal methods and simulations for design prediction and evaluation. To demonstrate, we present a case study on street frontage analysis, focusing on data science methods such as image classification and image generation to transform the urban design workflow. We conclude by discussing opportunities and challenges, such as the need to collect multi-modal urban design data and the importance of responsible urban design education and practice.

Keywords Urban design · Urban analytics · Machine learning · Active frontage · Artificial intelligence

S. Law (✉) · D. Hu · X. Lin · X. Gao
University College London, London, UK
e-mail: stephen.law@ucl.ac.uk; Demin.hu.22@ucl.ac.uk; xuhui.lin.16@ucl.ac.uk; xiaowei.gao.20@ucl.ac.uk

Z. Fan
National University Singapore, Singapore, Singapore
e-mail: zicheng.f@u.nus.edu

Y. Shen
Tongji University, Shanghai, China
e-mail: 18093@tongji.edu.cn

T. Yang
Tsinghua University, Beijing, China
e-mail: yangtao128@mail.tsinghua.edu.cn

R. Goodspeed et al. (eds.), *Digital-Era Urban Transformations*, The Urban Book Series, https://doi.org/10.1007/978-3-031-98300-9_9

1 Background

Urban design is an interdisciplinary and iterative process that shapes towns and cities, streets and places (Carmona, 2009). This process typically begins with analysing wider contexts (social, economic, and environment), followed by understanding a site, then developing designs, and lastly evaluating the design before finalizing it. At its core, the urban design process is contextual, recursive, and collaborative in nature, requiring a high degree of continuous feedback and design development with different stakeholders and practitioners. A key question that arises is: how can data science and AI methods[1] be used in the urban design process? We argue that data science can be integrated into every stage of the design process. We also argue that human experts cannot be removed from the loop. The majority of these advanced systems, in particular agentic large language models, consist of opaque machinery that is prone to implicit biases. Thus, without expert guidance within the design process, these systems run the risk of generating designs that do not necessarily align with successful urban design principles and contextual sensitivities. We refer to this concept or notion as 'machine-in-the-urban-design-loop' or 'machine-assisted-urban-design'.[2] We illustrate this process in Fig. 1 through four stages: (1) discovering knowledge from a vast array of contextual data, (2) digital reconstruction for site measurements and augmented urban design, (3) generative methods in design ideation, and (4) causal methods and simulations for design predictions and evaluation. To demonstrate, we will present a case study on street frontage analysis. We will conclude by discussing some opportunities and challenges for integrating these technologies into the urban design process in practice.

[1] artificial intelligence method refers in this context to the narrow usage of Machine Learning and Deep Learning techniques.

[2] This notion has been presented by the main author in the Design Computation Input Output Conference (DCIO2021) as an analogy to 'human-in-the-loop'(Zanzotto, 2019).

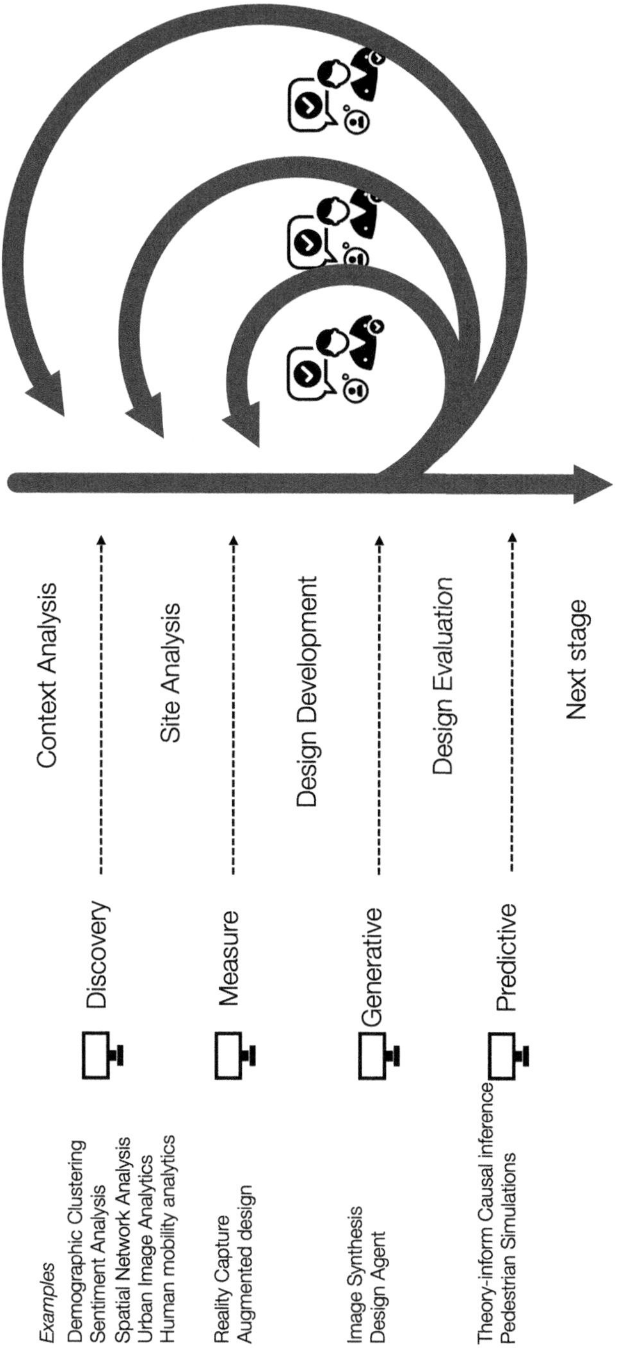

Fig. 1 Urban design analytics in the age of data science and AI

2 Machine-Assisted Urban Design Analytics

2.1 Knowledge Discovery for a more Dynamic Understanding of Cities

Knowledge discovery refers to the process of generating new insights from databases (Fayyad et al., 1996). In urban design, modern data science methods enable the translation of both structured data in tabular format (e.g. census) and unstructured data such as imagery (eg. remote sensing imagery, street views), point cloud (eg. LiDAR), geometry (eg. graphs and polygons) and sequential data (eg. text data and video data) into new insights. Early examples of these techniques include applying unsupervised methods such as K-means clustering for geodemographic classification (Harris et al., 2005), the use of graph theory to link spatial configuration with human behaviour in space syntax (Hillier & Hanson, 1984), applying urban modelling to estimate commuting flows (Batty, 1976), and the use of geospatial analysis methods to calculate accessibility potential (Hansen, 1959) and retail catchment area (Huff, 1963). In remote sensing, we have also seen the use of multi-spectral imagery to calculate indices describing buildings, vegetation, or water (Tucker, 1979; Pettorelli et al., 2005). Overlaying different layers of contextual information is a crucial tool for evidence-based urban design (Karimi, 2012).

Recent advancements in machine learning allow for more enriched knowledge discovery. For instance, applying machine vision methods on street-level imagery (Biljecki & Ito, 2021) to quantify and map objects in urban spaces such as street enclosures (Law & Neira, 2019), building age (Hoffmann et al., 2019), urban greenery (Li et al., 2015), and skyview/solar irradiance (Fujiwara et al., 2024). Additionally, these methods help assess the perceptions of our environment (Liang et al., 2024) such as visual walkability (Wang et al., 2024a), scenic quality (Seresinhe et al., 2015) and safety (Naik et al., 2014). Moreover, similar approaches have also been applied to retrieve measurements from high-resolution satellite imagery or 3D geometric point cloud data, such as vitality (Scepanovic et al., 2021), pollution (Du et al., 2024), and tree height (Francis et al., 2023).

Another research direction focuses on leveraging human mobility data (Pappalardo et al., 2023) to analyse the impact of public space design on pedestrian flow (Hillier & Hanson, 1984) and social interactions (Whyte, 1980). The increasing availability of location-based human mobility data from mobile devices now affords more opportunities for smarter digital urban design (Batty, 2013), offering deeper insights into social-spatial segregation (Liao et al., 2025), pedestrian navigation patterns (Miranda et al., 2021), pedestrian road safety (Gao et al., 2024), and active travel behaviour (Breum et al., 2022).

Finally, we have also seen extensive use of Natural Language Processing (NLP) methods to mine geographic textual data. One example is the research of (Hu et al., 2019), who analysed neighbourhood blog sentiment data to mine opinions about different topics (e.g. transport) in New York City. Another is the use of Latent Dirichlet Allocation (LDA) to identify urban area types in London (Lansley & Longley, 2016; Lin et al., 2025) or the dominant characteristics of Airbnb customers

(Wang et al., 2023a). Building on these approaches, we have recently seen the use of Large Language Models (LLMs) and Vision Language Models (VLMs) to forecast mobility trajectories (Wang et al., 2023b), extract building information (Zeng et al., 2024) and in developing foundation models for urban representation learning (Wang et al., 2024b) and urban region understanding (Xiao et al., 2024).

The information extracted from novel data sources can complement existing data layers to provide richer contextual insights in the urban design process and building/city information systems BIM/CIM. Furthermore, with this wealth of multi-modal high-frequency data, we have the potential to move beyond surface-level observations to a more dynamic understanding of urban development. Such an approach could transform traditional urban design into more adaptive-aware urban design practices. For example, to: (1) Identify long-term patterns in urban growth and functional changes, (2) Uncover the mechanisms behind the emergence of new urban functions, and (3) Analyse the complex interactions between various urban systems over time.

2.2 *More Affordable and Accessible Digital Reconstruction*

The second topic concerns the use of advanced close range remote sensing and neural rendering methods for site measurements and augmented urban design. In the past, digital reconstruction required substantial effort and costs in surveying the site in great detail. With technological advancements, two opportunities have emerged. The first is the increasing availability of high-resolution 2D images and 3D point cloud data from sensors on close-range consumer handheld mobile devices (Guan et al., 2015). This can be seen in Fig. 2, which shows handheld LiDAR scans of a street in Pompeii and a room in a historic Kyoto Machiya. A research

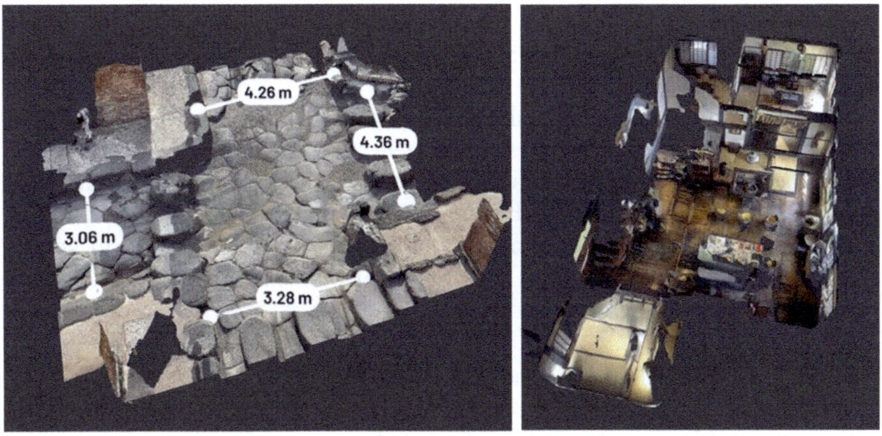

Fig. 2 Examples of Digital Reconstruction. Left. Measuring a street junction in Pompeii. Right. The layout of a historic home in Kyoto. Taken by the author

application of these low-cost sensors is the terrestrial scanning of forest inventory (Mokros et al., 2021). The second is advances in neural rendering, where highly accurate 3D environments can be reconstructed from a sparse set of 2D images (Mildenhall et al., 2020; Kerbl et al., 2023). Two technologies that have arisen in the past 3 years include Neural Radiance Fields (NeRFs) and 3D Gaussian Splatting (3DGS).

NeRF proposed by (Mildenhall et al., 2020) is a novel view synthesis methods that use volumetric rendering to learn a neural scene representation from posed images. Camera poses are typically estimated using traditional Structure-from-Motion (SfM) tools such as COLMAP. On the other hand, 3DGS uses many 3D Gaussians to approximate the scene that has been optimized over multiple images (Kerbl et al., 2023). The ease in close-range, handheld mobile reality capture and advanced neural rendering allows urban designers to quickly capture the reality of the site. This then allows urban designers to visualize their own design. For example, by coupling with augmented reality (e.g. Oculus[3], Apple Vision Pro[4]) and augmented design tools (e.g. UnrealEngine5[5], Dreams on PS5[6]), these technologies can allow designers to situate themselves for immersive site analysis and 3D urban design (Gillespie et al., 2022). The training of future urban designers to use these technologies will be an important facet for urban design pedagogy.

2.3 *Design Development with Generative Methods*

The third topic is the application of generative models in urban design development. Generative computer-aided urban design isn't new and is a well-studied topic in architecture computation, which can be categorized into rule-based methods and optimization-based ones (Koenig et al., 2020; Wortmann, 2017). For example, a procedural model such as CityEngine employs a set of predefined rules to automate city and environment generation. These models are scalable, interpretable, and can easily be tied to explicit urban design rules and guidelines in practice. Other examples include the use of evolutionary genetic algorithms to optimize layout design (Vermeulen et al., 2015).

Aside from the traditional methods, there is now a growing interest in applying vision machine learning techniques such as Generative Adversarial Networks(GANs) and Denoising Diffusion methods such as Latent Diffusion in design ideation. The former, GAN synthesises images by improving the image quality iteratively through a discriminator (Goodfellow et al., 2020). Examples include its use to generate street network patches such as StreetGAN (Hartmann et al., 2017), building

[3] https://www.oculus.com/experiences/quest/?locale=en_GB.

[4] https://www.apple.com/uk/apple-vision-pro/.

[5] https://www.unrealengine.com/en-US/unreal-engine-5.

[6] https://www.playstation.com/en-gb/games/dreams/.

footprints such as GANmapper (Wu & Biljecki, 2022), floor plans such as ArchiGAN (Chaillou, 2020), and street scenes such as UrbanCounterfactual (Law et al., 2023). The latter, Denosing Diffusion (Rombach et al., 2022) methods, such as Latent Diffusion, produce synthetic imagery with similar or higher fidelity than GANs but with more diversity and control. These latent diffusion models map an image to the latent space using an encoder-decoder, where a diffusion model is then learnt by denoising added noise iteratively. Recent development has allowed these methods to gain better control by conditioning these models with text and images (Zhang et al., 2023).[7] A key challenge with image synthesis methods, such as GANs and Denoising Diffusion, is that these pixel-based approaches differ from how a designer designs. As a result, the generated images can often be difficult to control and may produce implausible or inconsistent outcomes. To enhance the generative design process, other research directions have also been explored. One approach is stroke-based rendering, which trains neural networks to draw in a manner that more closely mimics human behavior (Ha & Eck, 2017). Another approach involves the use of reinforcement learning (Han et al., 2021), where design agents are trained to generate options by optimising specific rewards, such as solar irradiation. We envisage novel methods that combine old technology, like procedural modelling, with new ones, such as generative models, to become an integral part of the urban design process. Additionally, the potential use of Agentic Large Language Models (Agent-LLMs) is emerging, enabling the generation of multiple agents capable of retrieving information and simulating various stakeholders/parties (Park et al., 2023), thereby aiding in design decisions (Zhou et al., 2024). The seamless integration of these advanced holistic design tools in human-computer interactive design is increasingly becoming a key research direction. However, it is important to acknowledge the risks associated with end-to-end automation, such as over-reliance (Buçinca et al., 2021) and the implicit biases of these models. To reduce such risks, expert guidance is necessary to ensure that generated designs adhere to established urban design principles and contextual needs.

2.4 *Design Evaluation Using Theory-Driven Causal Models and Simulations*

The fourth and final topic focuses on design prediction and evaluation. While one could develop a supervised predictive ML model that uses various urban parameters to predict the socioeconomic vitality of a new design, some might argue that these parameters may not be inherently causal, leading to potential misinterpretations or misspecifications. This raises concerns about the validity of simple ML-based models when used in complex urban environments for decision making. To address this, a more cautious approach involving theory-driven structural causal models and

[7] https://github.com/comfyanonymous/ComfyUI.

simulations (Pearl, 2009) could be applied in urban design. One example is to link the causal mechanism back to urban design theory. A classic example is space syntax research (Hillier & Hanson, 1984), which theorized the association between spatial configuration and pedestrian movement (Hillier et al., 1993; Penn et al., 1998), to land-use (Hillier, 1997) and land value via a feedback loop as seen in Fig. 3. This urban configuration model has been operationalized through street network analysis and applied in evidence-based urban design projects (Karimi, 2012), including the redesign of Trafalgar Square.[8] Another empirical approach for design evaluation is to build bottom-up Agent-Based Models (ABM) (Crooks et al., 2018; Gilbert, 2019) to both validate traditional theories and propose new ones. In urban ABMs, pedestrian agents can be simulated within an urban environment overtime, following a set of behavioural rules like cognition (Turner & Penn, 2002), that govern their interactions with other agents and with the environment (Crooks et al., 2018). These simulation models make behavioural assumptions within enclosed systems for scenario testing that can potentially better isolate and thus test the effect of a particular design intervention. An urban planning example is the use of agent computing to simulate the effect of the new transport infrastructure, such as the London Elizabeth Line, on housing affordability (Guerrero & Law, 2024) or in simulating the effect of visibility has on pedestrian navigation in complex 3D environments (Gath-Morad et al., 2022). Limitations of these approaches include complex model assumptions that could lead to difficulty in model validation and interpretation.

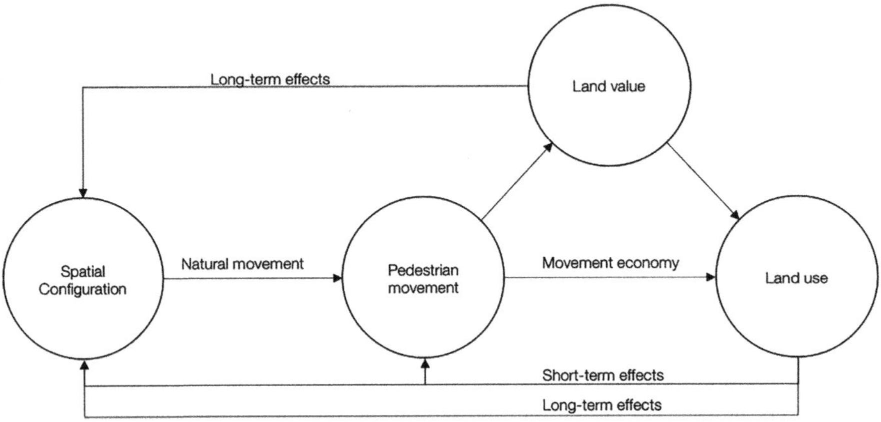

Fig. 3 An example of a theoretically-inspired causal graph linking spatial configuration to pedestrian movement, land use, and land value based on space syntax theory

[8] www.spacesyntax.com and www.fosterandpartners.com.

3 The Case Study of Street Frontage Image Classification and Generation

This section presents two related case studies that illustrate how data science can be applied to urban design analytics, from discovery to design ideation. In the first case study, Convolutional Neural Networks (CNNs) were employed to detect street frontage quality at the scale of Greater London. A CNN is a deep learning model that automatically learns hierarchical features from images using convolutional operations, enabling tasks like object detection, classification, and segmentation. These methods enable urban designers to identify areas that need design interventions. The second case study explores the efficacy of generative models in the urban design ideation phase. By simulating potential improvements, this approach aids in reimagining urban environments that can be used for participatory co-design with stakeholders.

3.1 Street Frontage Net

The quality of street frontages is an important factor in urban design, as it represents the interface between public and private space. The more doors and windows facing a street, the more 'eyes-on-the-streets,' which contributes to the perception of safety (Jacobs, 1961). Despite being advocated in planning guidance, empirical research on this topic has been limited as the collection of street frontage data is costly and time-consuming. One approach to collecting this data more efficiently is to cast this as an image classification problem in computer vision (Krizhevsky et al., 2012). In previous research, we applied deep learning methods in urban frontage classification (Law et al., 2017, 2020) through the creation of the Street-Frontage-Net model (SFN) to classify a street scene into four categories: blank, single-side active, both sides active, and non-urban. The SFN consists of a CNN encoder that extracts image features and a frontage classifier. To optimize the SFN, we fine-tuned four CNN encoder architectures—ResNet50, VGG16, Inception, and Xception, which were state-of-the-art machine vision models in 2018. Our results showed that VGG16 image embeddings achieved the highest accuracy of 92.7% when fine-tuned on a combination of Google Street View images and synthetic images generated using ESRI CityEngine. Accuracy in this context is an evaluation metric that measures the proportion of correctly classified instances out of all instances in a dataset.

After training, the trained SFN model can be used to infer the probability that a street image has active frontages on both sides of the street. This street-level active frontage probability can then be plotted (Fig. 4) to help identify areas in red that have an abundance of active frontages and areas in blue that potentially need intervention. Furthermore, the model has demonstrated its ability to generalize beyond

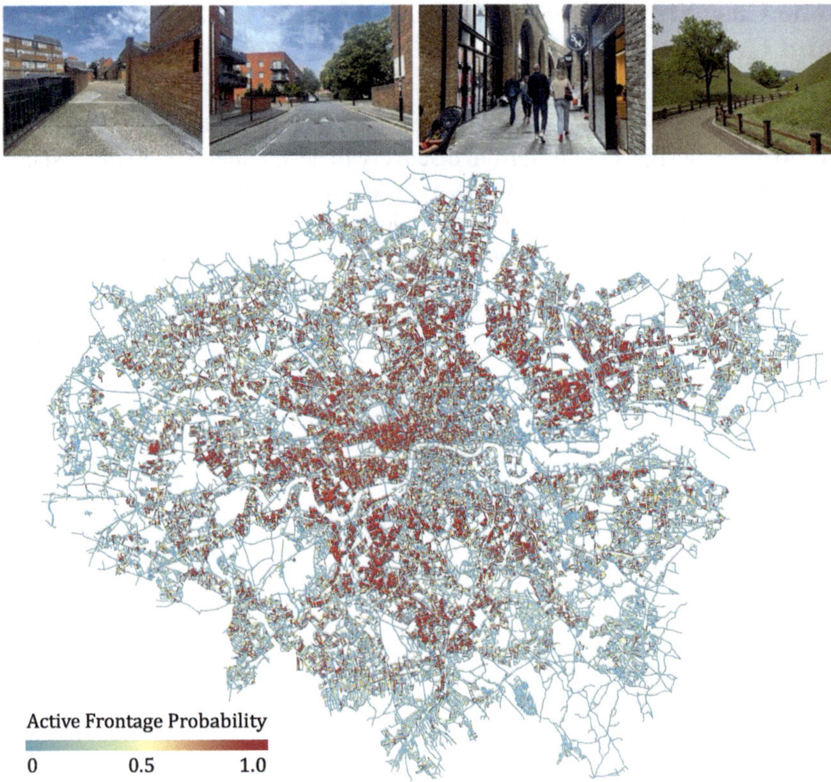

Fig. 4 Results from the StreetFrontageNet classifier. Top: From left to right, blank frontage, single-side active frontage, both-side active frontage, non-urban frontage. Bottom. Active frontage probability map of London, where red denotes a higher probability of active frontages and blue denotes a lower probability of active frontage as inferred from the StreetFrontageNet model. For more on StreetFrontageNet see Law et al. (2017, 2020). Images produced by the author

its training environment. When tested on images from Paris, NYC, Barcelona, San Francisco, Hong Kong, Kyoto and Tokyo, the SFN model retained a robust accuracy for cities that are both geographically closer such as Paris (77.26%) but also culturally closer such as NYC (73.30%) but not robust for cities that are both geographically and culturally far away such as Hong Kong (67.78%) and Tokyo (52.20%). The implications of SFN's success are many-fold. In the next subsection, we will demonstrate how generative methods can be used to visualize/ideate such interventions.

3.2 Street Frontage Generative Net

Beyond the detection of street frontage quality, data science can also be used in generating street frontage scenarios—counterfactuals of how urban streetscapes might look under different design interventions. Image generative models such as Diffusion, GANs, and FLUX offer an opportunity for exploring different possibilities in urban design ideation. For example, in a neighborhood where blank walls dominate the streetscape, a generated image might replace those walls with storefronts, cafes, or windows, creating a more lively and pedestrian-friendly environment. Urban planners can then use these synthetic images to present to stakeholders for advocating changes.

To evaluate its effectiveness in generating plausible counterfactuals, this chapter includes a small demonstration using a variant of the Denoising Diffusion model, a generative AI model that synthesises high-quality images from text prompts by iteratively denoising added noise of an image using a neural network. Specifically, we used a popular and more efficient variant of Stable Diffusion called SDXL (Rombach et al., 2022; Podell et al., 2023), to study whether a pre-trained generative model can synthesize street frontage images with active or blank using text-prompts for a curated image set.

For this test, we curated a small dataset of architectural façades from ten cities namely ['Hong Kong','Seoul','Tokyo','Kyoto','Shanghai','San Francisco', 'Paris', 'Amsterdam', 'Vienna','London'] that totalled 120 images.[9] We then employed a standard SDXL image inpainting pipeline, given an input architectural facade image, a segmentation mask focusing on the street façade, and a text prompt, to generate a counterfactual image based on the inputs.[10] We generated a pair of images from each of the original images in the curated dataset using the following text prompts = ['a closed and boarded up building','a building with lively storefronts']. After that, we manually evaluated the extent to which the two synthetic images aligned with the contrastive text prompts. The results showed that 85/120 of the generated images were closely aligned with the text prompts, while 35/120 were not.

To investigate a little deeper, we then visualized images that successfully aligned with the text prompts alongside those that did not. In each set, the leftmost image is the original, the middle image corresponds to the text prompt 'a closed and boarded-up building' (representing a blank frontage), and the rightmost image corresponds to the prompt 'a building with lively storefronts' (representing active frontages).

---

[9] Images taken by the authors.

[10] For image consistencies, we used the same random-seed.

Fig. 5 Successful examples of generated active-blank street frontage pairs. From top-to-bottom: (**a**) London, (**b**) Kyoto, (**c**) Hong Kong. Images produced by the author

A case is considered successful when the generated image pairs more accurately reflect the contrastive text prompts. Ideally, images generated from the active frontage prompt should depict a street scene with a vibrant storefront at the ground level, while those generated from the inactive frontage prompt should appear blank or boarded up. These visual comparisons help assess the model's performance qualitatively.

First, we look at a couple of successful cases. Figure 5a shows a pub in London, where the blank frontage image depicts a less active facade compared to its counterpart, which shows a shop. Figure 5b shown in the second row, illustrates a traditional Machiya in Kyoto.[11] Similarly, the blank frontage image appears less active, while the active frontage image shows clearly a shop. Figure 5c, which sits in the bottom row, features a red-brick historic building in Hong Kong. The blank frontage version shows a boarded-up entrance, while the active frontage counterpart transforms the original entrance into a shop. Overall, this set of images appears generally plausible, with no significant contextual changes. However, some realism issues remain, particularly relating to scale inconsistencies in some of the objects.

Next, we look at a couple of failure cases. Figure 6a, in the first row, shows a traditional machiya in Kyoto, where both the blank and active frontages images contain no windows or shopfront. The only minor difference is that the image on the

[11] A Machiya is a traditional wooden townhouse commonly found in Japan. Their typology is similar to shophouses found in Southeast Asia and Guangdong in China.

Fig. 6 Failure examples of generated active-blank street frontage pairs. From top-to-bottom; (**a**) Kyoto, (**b**) London, (**c**) London. Images produced by the author

right has a sign at the front, which is seemingly more like a shop. Figure 6b, in the second row, presents a modern building in London, where the blank frontage image clearly lacks windows. While the active frontage image introduces activity in the upper floor, while leaves the ground floor unchanged. Figure 6c shows a commercial building in London, where the blank frontage image correctly depicts boarded-up windows, and the active frontage image correctly adds a row of shops. However, since the shops appear closed, we classified this case as incorrect. These results show that the generative model is sometimes imperfect in making these transformations, highlighting the need for careful fine-tuning and potential integration with other generative models to produce more accurate and realistic urban images.

These early results highlight some promises and limitations of using generative vision models to synthesize street frontage design. With further fine-tuning, this technology can be employed to explore the implications of various urban design interventions. For instance, elements such as green spaces, street furniture, or pedestrian zones can be integrated into the counterfactual scenarios, allowing planners to visualise the cumulative impact of different urban design interventions. Such visualizations allow stakeholders from the community to imagine and thus more easily engage in the urban design process. While generative models show promise in synthesizing street frontage designs, they face limitations such as contextual unawareness, data biases, and challenges in stakeholder interpretations, which must be addressed for effective integration into urban planning.

4 Key Challenges

We have described how data science methods can be used in the urban design process, and a case study demonstrating both in detecting and designing street frontage. We conclude by describing a couple of potential challenges and opportunities. We summarise these challenges into three areas namely: lack of multi-modal data, the potential misuse and over-reliance of these methods, and finally a limited understanding of how urban design affects us holistically. The first challenge is the lack of multi-modal data. Despite the growing availability of large scale urban design image dataset such as the original facades dataset (Isola et al., 2017), the global streetscape dataset (Hou et al., 2024), or the 3D geometric dataset such as ABC (Koch et al., 2019), there is still a gap in the availability of multi-modal urban design process data. This is important to improve our understanding of how and why urban design ideas emerge collaboratively and collectively. A second challenge is the need to equip future generations of urban designers with the technical and analytical abilities to responsibly use modern data science techniques. It is obvious that there are clear benefits in how GenerativeAI can improve productivity, but there are also visible risks of over-reliance on AI, which can lead to a decline in system-2 critical thinking (Buçinca et al., 2021; Essel et al., 2024). Therefore, it is essential not only to teach how to use these tools effectively but also to emphasize their responsible and transparent use, fostering critical and creative thinking in both education and practice in the long term. The final challenge—and opportunity—lies in the limited understanding of how urban design influences our health and well-being holistically. While empirical research has provided evidence of the impact of urban design on well-being, the underlying neurophysiological mechanisms remain largely underexplored. Neuroarchitecture is an emerging field that investigates how architectural environments affect neurological functions and cognition (Chatterjee et al., 2021; Valentine, 2023), with early efforts extending into urban planning and design. Research in this area has the potential to reveal the neurological processes and pathways through which our urban environments shape health and well-being. A deeper understanding of these mechanisms, supported by multi-modal physiological sensing (e.g., EEG, fMRI), could potentially inform and impact urban design practices, leading to healthier and liveable cities.

4.1 Concluding Remarks

To conclude, this chapter offered some insights into how data science and AI methods can be applied to urban design analytics, not as replacements for human expertise but as tools to augment and enhance decision-making. These applications include using data mining techniques for insight discovery, leveraging highly accurate close-range remote sensing and neural rendering methods for measurements and augmented urban design, to the use of generative models for automated design ideations, and lastly applying theoretical and causal models to predict and evaluate design. We have described two of these processes through the case study of street frontage classification and generation, and highlighted a couple of challenges and opportunities ahead. We emphasized the need for expert involvement, as AI models have implicit biases and lack contextual awareness. The concept of 'machine-assisted urban design' underscores the necessity of keeping human designers in the loop to ensure AI-generated solutions align with sound urban design principles and needs. As urban design is both an art and a science, it is critical for designers to learn how to use these tools responsibly, ensuring the scientific evidence is applied to create more beautiful and liveable places.

Acknowledgements We appreciate the editor's thoughtful feedback and attention to detail, which greatly improved the work.

References

Batty, M. (1976). *Urban modelling*. Cambridge University Press.

Batty, M. (2013). *The new science of Cities*. MIT Press.

Biljecki, F., & Ito, K. (2021). Street view imagery in urban analytics and gis: A review. *Landscape Urban Planning, 215*, 104217.

Breum, S. M., Kostic, B., & Szell, M. (2022). Computational desire line analysis of cyclists on the dybbølsbro intersection in copenhagen. *arXiv Preprint*

Buçinca, Z., Malaya, M. B., & Krzysztof, G. Z. (2021). To trust or to think: cognitive forcing functions can reduce overreliance on AI in AI-assisted decision-making. In *Proceedings of the ACM on Human-computer Interaction 5, no. CSCW1* (pp. 1–21).

Carmona, M. (2009). *Public places urban spaces: The dimensions of Urban Design*. Routledge.

Chaillou, S. (2020). Archigan: Artificial intelligence x architecture. In *Architectural intelligence* (pp. 117–127).

Chatterjee, A., Coburn, A., & Weinberger, A. (2021). The neuroaesthetics of architectural spaces. *Cognitive Processing, 22*, 1–6.

Crooks, A., Malleson, N., Manley, E., & Heppenstall, A. (2018). *Agent based modelling and geographical information systems: A practical primer*. Sage.

Du, S., Zhang, Y., Sun, W., & Liu, B. (2024). Quantifying heterogeneous impacts of 2d/3d built environment on carbon emissions across urban functional zones: A case study in Beijing, China. *Energy Buildings, 319*, 114513.

Essel, H. B., Vlachopoulos, D., Essuman, A. B., & Amankwa, J. O. (2024). Chatgpt effects on cognitive skills of undergraduate students: Receiving instant responses from ai-based conversational large language models (llms). *Computers and Education Artificial Intelligence, 6*, 100198.

Fayyad, U., Piatetsky-Shapiro, G., & Smyth, P. (1996). From data mining to knowledge discovery in databases. *AI Magazine, 17*(3), 37–37.

Francis, J., Disney, M., & Law, S. (2023). Monitoring canopy quality and improving equitable outcomes of urban tree planting using lidar and machine learning. *Urban Forestry Urban Greening, 89*, 128115.

Fujiwara, K., Ito, K., Ignatius, M., & Biljecki, F. (2024). A panorama based technique to estimate sky view factor and solar irradiance considering transmittance of tree canopies. *Building and Environment, 266*, 112071.

Gao, X., Jiang, X., Haworth, J., Zhuang, D., Wang, S., Chen, H., & StephenLaw. (2024). Uncertainty-aware probabilistic graph neural networks for road-level traffic crash prediction. *Accident Analysis & Prevention, 208*, 107801.

Gath-Morad, M., Melgar, L. E. A., Conroy-Dalton, R., & Holscher, C. (2022). Beyond the shortest-path: Towards cognitive occupancy modeling in bim. *Automation in Construction, 135*, 104131.

Gilbert, N. (2019). *Agent-based models*. Sage Publications.

Gillespie, D., Qin, Z., & Tsigkari, M. (2022). Augmenting design: Extending experience of the design process with glaucon, an experiential collaborative xr toolset. In *Design modelling symposium Berlin* (pp. 332–342).

Goodfellow, I., Pouget-Abadie, J., Mirza, M., Bing, X., Warde-Farley, D., Ozair, S., Courville, A., & Bengio, Y. (2020). Generative adversarial networks. *Communications of the ACM, 63*(11), 39–144.

Guan, H., Yongtao, Y., Ji, Z., Li, J., & Zhang, Q. (2015). Deep learningbased tree classification using mobile lidar data. *Remote Sensing Letters, 6*(11), 864–873.

Guerrero, O. A., & Law, S. (2024). The spatial structure of housing affordability and the impact of public infrastructure. *Journal of Simulation, 18*, 1–15.

Ha, D., & Eck, D. (2017). A neural representation of sketch drawings. *arXiv preprint arXiv:1704.03477*

Han, Z., Yan, W., & Liu, G. (2021). A performance-based urban block generative design using deep reinforcement learning and computer vision. In *Proceedings of the 2020 DigitalFUTURES: The 2nd International Conference on Computational Design and Robotic Fabrication (CDRF 2020)* (pp. 134–143). Springer.

Hansen, W. (1959). How accessibility shapes land use. *Journal of the American Institute of Planners, 25*(2), 73–76.

Harris, R., Sleight, P., & Webber, R. (2005). *Geodemographics, GIS and neighbourhood targeting* (Vol. 7). Wiley.

Hartmann, S., Weinmann, M., Wessel, R., & Klein, R. (2017). Streetgan: Towards road network synthesis with generative adversarial networks. In *Proceedings of international conference on computer graphics, visualization and computer vision 2017*.

Hillier, B. (1997). Cities as movement economies. In *Intelligent Environments* (pp. 295–344). North-Holland.

Hillier, B., & Hanson, J. (1984). *The social logic of space*. Cambridge University Press.

Hillier, B., Penn, A., Hanson, J., Grajewski, T., & Jianming, X. (1993). Natural movement: Or, configuration and attraction in urban pedestrian movement. *Environment and Planning B: Planning and Design, 20*(1), 29–66.

Hoffmann, E. J., Wang, Y., Werner, M., Kang, J., & Zhu, X. X. (2019). Model fusion for building type classification from aerial and street view images. *Remote Sensing, 11*(11), 1259.

Hou, Y., Quintana, M., Khomiakov, M., Yap, W., Ouyang, J., Ito, K., Wang, Z., Zhao, T., & Biljecki, F. (2024). Global streetscapes—A comprehensive dataset of 10 million street-level images across 688 cities for urban science and analytics. *ISPRS Journal of Photogrammetry and Remote Sensing, 215*, 216–238.

Hu, Y., Deng, C., & Zhou, Z. (2019). A semantic and sentiment analysis on online neighborhood reviews for understanding the perceptions of people toward their living environments. *Annals of the American Association of Geographers, 109*(4), 1052–1073.

Huff, D. L. (1963). *A probabilistic analysis of shopping center trade areas* (Vol. 39). Land.

Isola, P., Zhu, J.-Y., Zhou, T., & Efros, A. A. (2017). Image-to-image translation with conditional adversarial networks. In *Proceedings of the IEEE conference on computer vision and pattern recognition* (pp. 1125–1134).

Jacobs, J. (1961). *The death and life of great American cities*. Random House.
Karimi, K. (2012). A configurational approach to analytical urban design: 'Space syntax' methodology. *Urban Design International, 17*(4), 297.
Kerbl, B., Kopanas, G., Leimkuhler, T., & Drettakis, G. (2023). 3d gaussian splatting for real-time radiance field rendering. *ACM Transactions on Graphics, 42*(4), 139–131.
Koch, S., Matveev, A., Jiang, Z., Williams, F., Artemov, A., Burnaev, E., Alexa, M., Zorin, D., & Panozzo, D. (2019). Abc: A big cad model dataset for geometric deep learning. In *Proceedings of the IEEE/CVF conference on computer vision and pattern recognition* (pp. 9601–9611).
Koenig, R., Miao, Y., Aichinger, A., Knecht, K., & Konieva, K. (2020). Integrating urban analysis, generative design, and evolutionary optimization for solving urban design problems. *Environment and Planning B: Urban Analytics and City Science, 47*(6), 997–1013.
Krizhevsky, A., Sutskever, I., & Hinton, G. E. (2012). Imagenet classification with deep convolutional neural networks. *Advances in Neural Information Processing Systems, volume, 25*, 84.
Lansley, G., & Longley, P. A. (2016). The geography of twitter topics in London. *Computers, Environment and Urban Systems, 58*, 85–96.
Law, S., & Neira, M. (2019). An unsupervised approach to geographical knowledge discovery using street level and street network images. In *Proceedings of the 3rd ACM SIGSPATIAL international workshop on AI for geographic knowledge discovery* (pp. 56–65).
Law, S., Shen, Y., & Seresinhe, C. (2017). An application of convolutional neural network in street image classification: The case study of London. In *Proceedings of the 1st workshop on artificial intelligence and deep learning for geographic knowledge discovery* (pp. 5–9).
Law, S., Seresinhe, C. I., Shen, Y., & Gutierrez-Roig, M. (2020). Streetfrontage-net: Urban image classification using deep convolutional neural networks. *International Journal of Geographical Information Science, 34*, 1–27.
Law, S., Hasegawa, R., Paige, B., Russell, C., & Elliott, A. (2023). Explaining holistic image regressors and classifiers in urban analytics with plausible counterfactuals. *International Journal of Geographical Information Science, 37*(12), 2575–2596.
Li, X., Zhang, C., Li, W., Ricard, R., Meng, Q., & Zhang, W. (2015). Assessing street-level urban greenery using google street view and a modified green view index. *Urban Forestry & Urban Greening, 14*(3), 675–685.
Liang, X., Chang, J. H., Gao, S., Zhao, T., & Biljecki, F. (2024). Evaluating human perception of building exteriors using street view imagery. *Building and Environment, 263*, 111875.
Liao, Y., Gil, J., Yeh, S., Pereira, R. H. M., & Alessandretti, L. (2025). Socio-spatial segregation and human mobility: A review of empirical evidence. *Computers, Environment and Urban Systems, 117*, 102250.
Lin, X., Yang, T., & Law, S. (2025). From points to patterns: An explorative poi network study on urban functional distribution. *Computers, Environment and Urban Systems, 117*, 102246.
Mildenhall, B., Srinivasan, P. P., Tancik, M., Barron, J. T., Ramamoorthi, R., & Ng, R. (2020). Nerf: Representing scenes as neural radiance fields for view synthesis. In *European conference on computer vision* (pp. 405–421). Springer.
Miranda, A. S., Fan, Z., Duarte, F., & Ratti, C. (2021). Desirable streets: Using deviations in pedestrian trajectories to measure the value of the built environment. *Computers, Environment and Urban Systems, 86*, 101563.
Mokros, M., Miita, T., Singh, A., Tomastik, J., Chuda, J., Wezyk, P., Kuzelka, K., Surovy, P., Klimanek, M., Zieba-Kulawik, K., & Bobrowski, R. l. (2021). Novel low-cost mobile mapping systems for forest inventories as terrestrial laser scanning alternatives. *International Journal of Applied Earth Observation and Geoinformation, 104*, 102512.
Naik, N., Philipoom, J., Raskar, R., & Hidalgo, C. (2014). Streetscore-predicting the perceived safety of one million streetscapes. In *Proceedings of the IEEE conference on computer vision and pattern recognition workshops* (pp. 779–785).
Pappalardo, L., Manley, E., Sekara, V., & Alessandretti, L. (2023). Future directions in human mobility science. *Nature Computational Science, 3*(7), 588–600.
Park, J. S., O'Brien, J., Cai, C. J., Morris, M. R., Liang, P., & Bernstein, M. S. (2023). Generative agents: Interactive simulacra of human behavior. In *Proceedings of the 36th annual acm symposium on user interface software and technology* (pp. 1–22).
Pearl, J. (2009). *Causality*. Cambridge University Press.

Penn, A., Hillier, B., Banister, D., & Jianming, X. (1998). Configurational modelling of urban movement networks. *Environment and Planning B: Planning and Design, 25*(1), 59–84.

Pettorelli, N., Vik, J. O., Mysterud, A., Gaillard, J.-M., Tucker, C. J., & Stenseth, N. C. (2005). Using the satellite-derived ndvi to assess ecological responses to environmental change. *Trends in Ecology Evolution, 20*(9), 503–510.

Podell, D., English, Z., Lacey, K., Blattmann, A., Dockhorn, T., Muller, J., Penna, J., & Rombach, R. (2023). Sdxl: Improving latent diffusion models for high-resolution image synthesis. *arXiv preprint*

Rombach, R., Blattmann, A., Lorenz, D., Esser, P., & Om-mer, B. (2022). High-resolution image synthesis with latent diffusion models. In *Proceedings of the IEEE/CVF conference on computer vision and pattern recognition* (pp. 10684–10695).

Scepanovic, S., Joglekar, S., Law, S., & Quercia, D. (2021). Jane jacobs in the sky: Predicting urban vitality with open satellite data. *Proceedings of the ACM on Human-Computer Interaction, 5*(CSCW1), 1–25.

Seresinhe, C., Preis, T., & Moat, S. (2015). Quantifying the impact of scenic environments on health. *Scientific Reports, 5*(1), 1–9.

Tucker, C. J. (1979). Red and photographic infrared linear combinations for monitoring vegetation. *Remote Sensing of Environment, 8*(2), 127–150.

Turner, A., & Penn, A. (2002). Encoding natural movement as an agent-based system: An investigation into human pedestrian behaviour in the built environment. *Environment and Planning B: Planning Design, 29*(4), 473–490.

Valentine, C. (2023). Architectural allostatic overloading: Exploring a connection between architectural form and allostatic overloading. *International Journal of Environmental Research and Public Health, 20*(9), 5637.

Vermeulen, T., Knopf-Lenoir, C., Villon, P., & Beckers, B. (2015). Urban layout optimization framework to maximize direct solar irradiation. *Computers, Environment and Urban Systems, 51*, 1–12.

Wang, J., Chow, Y. S., & Biljecki, F. (2023a). Insights in a city through the eyes of airbnb reviews: Sensing urban characteristics from homestay guest experiences. *Cities, 140*, 104399.

Wang, X., Fang, M., Zeng, Z., & Cheng, T. (2023b). Where would i go next? large language models as human mobility predictors. *arXiv preprint arXiv:2308.15197.*

Wang, M., Haworth, J., Chen, H., Liu, Y., & Shi, Z. (2024a). Investigating the potential of crowd-sourced street-level imagery in understanding the spatiotemporal dynamics of cities: A case study of walkability in inner London. *Cities, 153*, 105243.

Wang, X., Cheng, T., Law, S., Zeng, Z., Yin, L., & Liu, J. (2024b). Multimodal contrastive learning of urban space representations from poi data. *arXiv preprint arXiv:2411.06229.*

Whyte, W. (1980). *Social life of small urban spaces. Conservation Foundation.*

Wortmann, T. (2017). Model-based optimization for architectural design: Optimizing daylight and glare in grasshopper. *Technology—Architecture+Design, 1*, 176.

Wu, A. N., & Biljecki, F. (2022). Ganmapper: Geographical data translation. *International Journal of Geographical Information Science, 36*(7), 1394–1422.

Xiao, C., Zhou, J., Xiao, Y., Huang, J., & Xiong, H. (2024). Refound: Crafting a foundation model for urban region understanding upon language and visual foundations. In *Proceedings of the 30th ACM SIGKDD conference on knowledge discovery & data mining* (pp. 3527–3538).

Zanzotto, F. M. (2019). Human-in-the-loop artificial intelligence. *Journal of Artificial Intelligence Research, 64*, 243–252.

Zeng, Z., Goo, J. M., Wang, X., Chi, B., Wang, M., & Boehm, J. (2024). Zero-shot building age classification from facade image using gpt-4. *arXiv preprint*, arXiv:2404.09921.

Zhang, L., Rao, A., & Agrawala, M. (2023). Adding conditional control to text-to-image diffusion models. In *Proceedings of the IEEE/CVF international conference on computer vision* (pp. 3836–3847).

Zhou, Z., Lin, Y., Jin, D, & Li, Y. (2024). *Large language model for participatory urban planning. arXiv preprint arXiv:2402.17161.*

Fingerprinting NYC's Scaffolding Problem with Longitudinal Dashcam Data

Dorin Ruinsky Shapira, Matt Franchi, and Wendy Ju

Abstract Scaffolds, also called sidewalk sheds, are intended to be temporary structures to protect pedestrians from construction and repair hazards. However, some sidewalk sheds are left up for years; this trend recently motivated New York City's (NYC) mayor to issue a 'Get Sheds Down' plan that notably took down a scaffold in Harlem that had been up for *21 years*. Today, there are over 9000 active permits for scaffolds in NYC. This research uses computer vision to develop a longitudinal map of scaffolding. Using a dataset of 29,156,833 images recorded from vehicle-attached dashcams between August 2023 and January 2024 in NYC, we develop an algorithm to track scaffolding presence over time and match detected scaffolds to reported permit locations. We identify 850,766 images of scaffolding, tagging 5156 active sidewalk sheds and estimating 529 unpermitted sheds. We discuss implications for urban tech, governmental inspection processes, and out-of-distribution evaluations outside NYC.

1 Introduction

Scaffolds, or colloquially 'sidewalk sheds,' are temporary structures used in the construction, maintenance, and repair of buildings, bridges, and other human-made structures. The use of scaffolds facilitates the safety of workers and citizens walking near the area (Rubio-Romero et al., 2013).

New York City (NYC) Mayor Eric Adams has declared a war on longstanding sidewalk sheds (Heyward, 2023). Why? Because many sidewalk sheds are left up for long periods of time, even decades. NYC Local Law 11 mandates that the

D. R. Shapira
Technion - Israel Institute of Technology, Haifa, Israel
e-mail: dorinruinsky@campus.technion.ac.il

M. Franchi (✉) · W. Ju
Cornell University, Ithaca, NY, USA
e-mail: mattfranchi@cs.cornell.edu; wendyju@cornell.edu

R. Goodspeed et al. (eds.), *Digital-Era Urban Transformations*, The Urban Book Series, https://doi.org/10.1007/978-3-031-98300-9_10

facades for all buildings over 6 stories high must be inspected every five years, any deficiencies must be repaired, and the public must be protected during the process. Many building owners find it easier to leave the scaffolds up than to make necessary repairs (Calder, 2023).

The visual prevalence of scaffolding in the urban landscape raises concerns about its impact. One issue associated with scaffolding is its detrimental impact on the urban landscape; it obstructs side views, obscures the facades of storefronts, and disrupts the overall aesthetic appeal of the surroundings (Deacon, 2013). It creates accessibility challenges for the mobility-impaired (Achtemeier, 2019), and helps give cover to illicit activity (Calder, 2023; McGeehan, 2024).

According to NYCOpenData, there are currently more than 9000 active permits for scaffolds in New York City. However, some of the scaffolds around the city are likely unpermitted, including cases where the permit has expired. For the city to 'Get Sheds Down', as the Mayor has pledged to do, it must find ways to detect where the scaffolds are rather than just relying on official records.

In this paper, we describe a project that uses computer vision to 'fingerprint', or sense at fine spatiotemporal granularity, New York City's scaffolding problem. We use longitudinal analysis of data from the networked dashcams that are installed in many of the taxis and rideshare vehicles that traverse the city's different neighborhoods. This makes it possible to profile not only where sidewalk sheds are now, but also how long different sheds have been in place. This data medium is relatively novel (Madhavan & Darrell, 2017; Franchi et al., 2023), and we introduce custom methods for longitudinal analysis that harness the temporal density afforded by crowdsourced dashcam.

We observe well-known hotspots of scaffolding in the Upper West Side (West Side Rag, 2022) and Upper East Side (Upper East Side, 2023) neighborhoods of Manhattan, as well as in outer-borough neighborhoods like Brooklyn's Brownsville and Carroll Gardens. We also see new scaffolds tagged over time in our five-month dataset, as ridesharing vehicles newly make trips to less-visited parts of the city. By training a set of YOLOv7 (Wang et al., 2022) and YOLOv11 (an architecture built on top of YOLOv10 (Wang et al., 2024)) object detection models to recognize scaffolds and developing a custom algorithm to confirm scaffolds with time, we establish the capability to track how long scaffolds are up in different neighborhoods, without actively sending out inspectors. We validate our findings via active sidewalk shed permit data published on the NYCOpenData platform by the NYC Department of Buildings (NYC OpenData, 2024).

The primary contribution of this work is to affirm networked dashcam as a medium that elevates the state-of-the-art for passive urban sensing. Vehicles that record images in our dataset aren't sent out to actively monitor sidewalk sheds; they instead gather snaps of scaffolding passively throughout each day's scores of trips.

Beyond helping the municipal government and urban citizens with the challenge of infrastructure that overstays its welcome, this project highlights the way that dashcam data can make it tractable to canvas sidewalk level activity throughout the city, making it easier to address the different issues encountered in different boroughs throughout this city of over 8.5 million people.

2 Related Work

2.1 Street-Level Image Datasets

Streetview imagery datasets, such as Google Street View (GSV), Microsoft Bing Maps, Baidu Total View, and Tencent Street View, have gained prominence in urban analytics. These datasets have been employed to conduct diverse research tasks, ranging from tree counting (Yi, 2019; Aikoh et al., 2023; Liu et al., 2020), utility pole identification (Zhang et al., 2018), and traffic sign detection (Yifan et al., 2018; Balali et al., 2015) to bike rack (Maddalena et al., 2020) and maintenance hole mapping (Boller et al., 2019). More recently, crowdsourced dashcam data has become available (Madhavan & Darrell, 2017). Here, cameras are affixed to the dashboards of ridesharing vehicles, allowing for coverage proportional to ridesharing popularity in an area. Hence, ridesharing-heavy areas like New York City have extensive coverage (Lam et al., 2021).

The primary advantage of crowdsourced dashcam is its much higher temporal density; GSV and similar static scene datasets offer a rich picture, but dashcam makes rich timelapses possible (Franchi et al., 2023). Dashcam data has distinctive attributes that suit it for automated object detection in complex, detail-rich environments; it is frame-by-frame, cost-effective to obtain, variable in camera perspective, and capable of precisely capturing infrastructure, anonymized pedestrians, other vehicles, and small fixtures (Franchi et al., 2024, 2025).

2.2 Social Science from Street-Level Images

One of the earliest studies utilizing GSV for novel social science purposes was Odgers et al.'s survey of children's neighborhoods. This work used image samples from over 1000 neighborhoods, asking coders to rate them for signs of physical disorder, physical decay, and safety. They found GSV to be a reliable and cost-effective tool for measuring both negative and positive characteristics of local neighborhoods (Odgers et al., 2012).

Lu et al. investigated whether the streets of Hong Kong included enough greenery to encourage people to exercise, using GSV images as an analysis medium (Yi, 2019). Helbich et al. similarly looked for correlations between greenness visible in GSV and mental health in the Netherlands (Helbich et al., 2021). Here, a low, static sampling density restricts longitudinal analysis and restrains specific applications.

Gebru et al. investigated if information in GSV correlated with demographic information (Gebru et al., 2017). Their analysis involved the detection and classification of cars appearing in specific neighborhoods, finding that it was possible to correlate a neighborhood's vehicle makes and models with the voting preferences of that neighborhood's residents.

2.3 Fingerprinting Governmental Agencies and Services

A growing number of social scientists are combining pedestrian perspectives, publicly available datasets, and other novel data sources to address questions regarding the efficiency, productivity, and equity of governmental agencies and services. During the COVID-19 pandemic, (Chowdhury et al., 2021) used dashcam data to sense social distancing compliance in New York City. In (Franchi et al., 2023), the authors use Nexar dashcam data to detect the presence of New York Police Department (NYPD) vehicles during 2020. Through data fusion with a publicly accessible American Community Survey (ACS) demographics dataset, the authors uncovered significant spatial and demographic disparities over how likely different residents were to encounter policing vehicles. This paper formatively shows how high-frequency dashcam data, like that offered by Nexar, can be used to study governmental services.

Prior to the above work, numerous researchers had used Google Street View, a resource offering similar visual perspectives but with a much lower sampling frequency, in pursuit of auditing city services (Li et al., 2015; Rundle et al., 2011). Agostini et al. (2023), Liu et al. (2023), and Minkoff (2016) show that 311, New York City's governmental tool for sourcing infrastructure complaints from citizens, suffers from underreporting. Similar findings were generated outside of New York in other similar reporting systems (Pak et al., 2017; Kontokosta & Hong, 2021; Clark et al., 2013). Agostini et al. (2023) audits this phenomenon and attempts to mitigate it by developing a Bayesian model that leverages spatial correlation between street flooding reports. This work, while not utilizing image data, still shows that new computational approaches can produce more equitable government services.

Semantically further, yet still relevant, is work that audits environmental (Apte et al. 2017; Hacker et al., 2022); physical (Hipp et al., 2022; Hara et al., 2013); or demographic (Gebru et al., 2017; Yin et al., 2015; Klemmer et al., 2021) characteristics of urban environments.

2.4 Sidewalk Scaffolding

As components of the built environment are ultimately meant to facilitate safety, scaffolding is subject to strict inspection standards, both at a federal level from the Occupational Health and Safety Administration (OSHA) and at a local level from the New York City Administrative Code (NYCAC). Sect. 3314.4.3 of the NYCAC allocates specific provisions for scaffolding inspection. Our analysis deals primarily with *supported scaffolds*, which are constructed over sidewalks to protect pedestrians from debris that may fall from work occurring above the sidewalk (NYC Department of Buildings, 2015). *Suspended scaffolds* are placed much higher and hang from overhead support structures on roofs or building setbacks (NYC Department of Buildings, 2023); they generally are not visible from the fixed vantage point of a vehicular dashcam camera. For the remainder of the paper, we use 'scaffolding' for simplicity, but only consider supported scaffolds in our analysis.

Presently, scaffolding inspection is a manual process, with audits done on downstream reports rather than the observations themselves. There are three types of inspections for supported scaffolds:

1. Per 3314.4.3.3 (NYC Department of Buildings, 2022), installation inspections occur after installation of a supported scaffold. This inspection verifies that all components are in safe condition and, if designed, are installed in accordance with the design drawings.
2. Per 3314.4.3.5 (NYC Department of Buildings, 2022), before every construction shift or other event that might affect structural integrity, a scaffold must be inspected by a competent supervisor who can verify that all components of the scaffold remain safe to use.
3. Per 3314.4.3.6 (NYC Department of Buildings, 2022), inspections must be made following a repair or adjustment to a scaffold at a site. The completion of these inspections blocks further work on the site.

Researchers and journalists have described the implications of widespread scaffolding for accessibility and quality of life. In the Minneapolis-St. Paul metro area, Jacob Achtemier describes a survey of nineteen normally-sighted, low vision, and blind pedestrians, stating that 'construction emerged as a primary, universal theme for explaining safety challenges and mobility efficiency with low vision and blindness', with sidewalk scaffolding being a prominent obstacle experienced (Achtemeier, 2019).

Leslie Deacon studied the landscape of sidewalk planning in the East Village neighborhood of Manhattan and offers a blunt description of the sidewalk sheds permanent throughout the neighborhood: 'Pedestrians must navigate these tunneled passages along entire blocks near Astor Place and other areas scattered throughout the Village, removing themselves from the typical experience of the neighborhood's sidewalks and shifting into bland segments suffering placelessness' (Deacon, 2013). In a 2020 New York Times article called 'Our Lives, Under Construction', Dina Seiden, a Brooklyn-based author and comedian, says, 'People hate their fellow pedestrians in a scaffolding confinement more profoundly than they do once liberated' (Green, 2020).

Overall, existing research finds scaffolding to be a detriment to both accessibility and quality of life, and journalistic coverage is largely negative (Green, 2020; Yglesias, 2022), dubbing scaffolding a 'scourge' of the urban landscape (New York Daily News Editorial Board, 2023).

3 Data

3.1 Dashcam Image Dataset

We source a dashcam imagery dataset of NYC street scenes from Nexar, a company specializing in manufacturing cloud-connected dashboard cameras for consumer vehicles. In New York City, Nexar dashcams are often equipped on ridesharing vehicles and sometimes freight. All images are obfuscated between image capturing and later, on-premises downloading. Faces and license plates are blurred, and vehicle dashboards are cropped out via a letterboxing process.

We assemble the dataset in-house, developing a high-speed, parallelized utility that downloads image data and metadata for each frame, and organizes data by the day of capture and h3–06[1] index.

We collect data between August 11th, 2023, and January 10th, 2024, with a period of sustained density during the entire month of November 2023. Our aim is to craft a dataset appropriate for examining both short-term changes in sidewalk sheds and longitudinal changes over longer periods. We hope that with a five-month range, we craft a dataset where at least some scaffolds are completely put up or taken down within the period of coverage.

All images are sized at 720p, 1280 pixels by width and 720 pixels by height. The aforementioned metadata includes a unique per-image frame ID, the UTC-localized timestamp when the image was taken, GPS coordinates for the capture location (accurate even in skyscraper-dense areas of Manhattan), and camera heading. Camera heading refers to the camera's orientation angle along the Z-axis in relation to global coordinates, allowing us to determine which direction a vehicle was headed at the time of image capture.

3.2 *NYC OpenData*

The Department of Buildings (DOB) maintains a comprehensive database listing all scaffolds that either currently possess or have previously obtained permit approval for their construction. This data includes essential information, such as the scaffold's coordinates, permit expiration date, permit issue date, the presence of permit renewals, and the permit type. Additional details, such as the construction firm and building obligations, although present in the database, are not within the scope of this paper.

As of January 22, 2024, shortly after the end of our data collection period, there were 8364 active sidewalk sheds in New York City, with an average age of 513 days. Scaffolds have been up for as long as 6478 days—equivalent to over 17 years.

We visualize a basic 'scaffolding impact factor', computed by taking the summed age of all scaffolds in a census area *G*. For Fig. 1, we group by Neighborhood Tabulation Area (NTA),[2] and normalize by the spatial area of each NTA. We see

[1] H3, or Hexagonal Hierarchical Spatial Index, is a grid system developed by Uber (uber/h3, 2024) that segments areas of the Earth into identifiable grid cells. Nexar utilizes H3 for frame indexing, and so we work with that until merging in less arbitrary geographies like NYC boroughs, etc. The h3–06 level of the index covers a fairly large area, with an average area of 13.95 mile^2.

[2] A Neighborhood Tabulation Area (NTA) is a moderately-sized statistical geography used to report the Decennial Census and American Community Survey (ACS). They are created by the NYC Department of City Planning to report populations at a mid-size granularity, as well as for easier visual recognition.

Fig. 1 Citywide Distribution of Scaffolding Impact Factor, computed as the summed age (in days) of all reported scaffolds in a neighborhood N

scaffolding impact highest in some of Manhattan's most famous, longstanding residential neighborhoods (which aligns with journalistic coverage), including the Upper West Side (West Side Rag, 2022) and Upper East Side (Upper East Side, 2023). The commercially dominated Midtown district of Manhattan is also one of the most scaffolding-dense areas of the city. Interestingly, the Brownsville NTA in Brooklyn also emerges as an outlier in the scaffolding impact distribution. Lastly, many of the NTAs in Upper Manhattan and the South Bronx have higher-than-average levels of scaffolding impact.

4 Deep Learning Model

We develop a deep learning approach for in-the-wild scaffold detection in urban environments. Our methodology follows the standard supervised-learning paradigm of four steps:

1. Compile and annotate a training set.
2. Compile and annotate a test set.
3. Train a classifier on the training set.
4. Final model evaluation on the test set.

4.1 Training Dataset

To put together the training dataset, we aim to identify images containing scaffolds. Unlike Franchi et al., 2023, where the geospatial distribution of objects to detect was unknown, we can utilize public data published by the NYC Department of Buildings (NYC Department of Buildings, 2024; NYC OpenData, 2024) to filter for images near reported scaffolding sites. We apply a filter to include only current active permits, denoted as having expiration dates later than 2023. We curate a dataset comprising only those images captured within a 100-m radius of an active permit location. This fairly coarse 100-m threshold is necessary to retrieve all qualifying images from larger buildings, as the NYC DOB stores permit coordinates at the centroid of the affected building.

While we acknowledge the potential for distribution shift using this approach (compared to a truly random sample of the image distribution), we rely on the standardized appearance of scaffolding to mitigate any possible performance losses. Using Nexar metadata described in Sect. 3.1 and the public datasets detailed in 3.2, we calculate the proximity to the nearest permit location for each coordinate. To provide further assurance of ground truth signal in the images, we determine whether the camera's heading was oriented toward the permit area.

Fig. 2 illustrates the steps for collecting the training dataset. The first step is indicated by a black pin, denoting the permit location; the building with scaffolding is also highlighted in red. The white pin marks the coordinates of the ego vehicle when the image is captured. In the first step (left subfigure), the algorithm checks if the permit is within a 100-m radius. In the second step (right subfigure), α denotes the camera heading angle, or the camera's orientation angle relative to true north.

To enable the trained model to learn from various scaffold representations in NYC's different areas, we matched the distribution of annotated data with the permit distribution across NYC boroughs. We accomplished this by grouping images by borough and randomly selecting them from each borough, aiming to maintain a distribution similar to that of permits. The image distribution across boroughs is presented in Table 1. Here, we compare the reported distribution of permits from NYC OpenData to the distribution of scaffolding images in our training set, stratified by NYC borough. The dissimilarity between the distributions was evaluated using KL divergence, yielding a value of 0.0038 (0.3%).

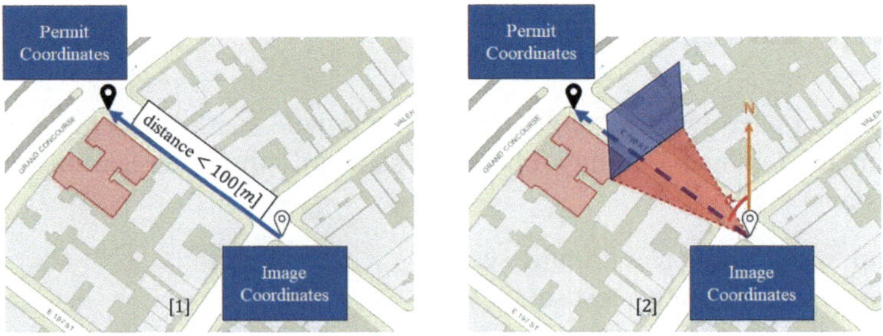

Fig. 2 (1) Spatial proximity check (2) Ego Vehicle-Scaffold angle check

Table 1 Distribution of permits across NYC boroughs

Borough	Permit distribution	Dashcam distribution
Manhattan	0.529	0.538
Brooklyn	0.220	0.221
Bronx	0.150	0.135
Queens	0.088	0.098
Staten Island	0.005	0.011

We chose 2214 images for annotation. We annotate images using Label Studio (HumanSignal, 2025), an open-source data labeling tool. We frame annotation as a rectangular object detection task. 1015 scaffolds were annotated from those images.

4.2 *Model Training*

We utilized well-established models from the You Only Look Once (YOLO) suite, achieving the best validation performance (mAP 0.85) from the YOLOv7-E6E object detector model. Within the YOLOv11 suite, training on the YOLO11-S(mall) weights yielded the highest mAP, 0.84. Since YOLOv7 slightly outperforms YOLOv11 and we are not operationally sensitive to inference speed, we opt to use the trained YOLOv7-E6E model for inference on our entire dataset. We initialized this model using pre-trained weights from the MS-COCO dataset, which encompasses 80 object categories, including people, cars, buses, traffic lights, and more. We fine-tune these weights on our image dataset, using the full 720p image resolution and a batch size of 2 during the training process.

We start with two classes. Initially, the annotations distinguished between white and green scaffolds, shown in Fig. 3, but we observed that only 7.7% of the images contained white scaffolds. Due to their similar structural characteristics and concerns over an imbalanced dataset, this low prevalence prompted us to combine white and green scaffolds into a single scaffold class. Interesting future work might result from a stratified study, as white scaffolds are installed by the company Urban Umbrella that specifically caters to high-wealth clientele (Bloomberg.com, 2022).

4.3 *Model Validation*

We generated our test set from random samples of Nexar images, selecting data from 15 days in August 2023. The resulting dataset, which underwent manual annotation, consisted of 2006 images. Within this dataset, we annotated 241 scaffolds.

The application of our trained model to this dataset produced adequate results as depicted in Fig. 4. Following the standard supervised learning paradigm, at the confidence threshold that maximizes the F1-Score, 0.79, we observe a recall of 0.78 and a precision of 0.79.

Fig. 3 White and Green Scaffolds

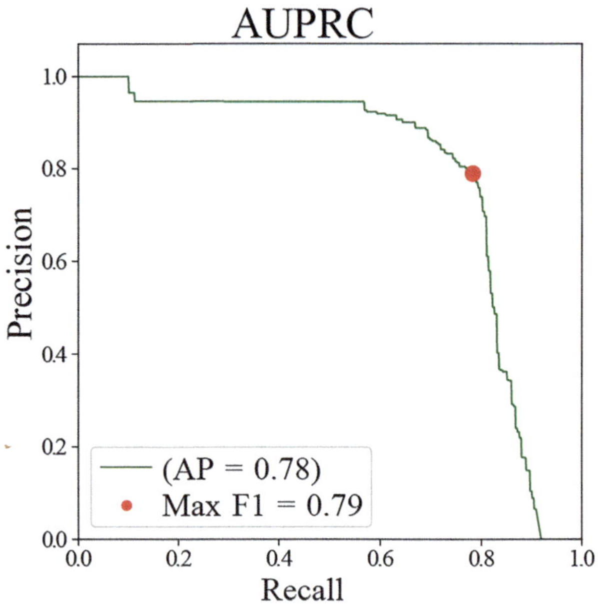

Fig. 4 Precision-Recall curve for the final classifier, evaluated on the test set. We report an average precision of 0.78 and an F1-score of 0.79

4.4 Model Detections

In Fig. 5, we display examples of true positive detections, false positive detections of bridges and balconies as scaffolds, and false negatives, where the model misses at least one scaffold present in an image. Aware of false positives, where the model incorrectly identifies scaffolds when none exist, we aim to improve the

Fig. 5 Examples of scaffolding detections. (a) True positives. (b) False positives. (c) False negatives

model's performance. We move to analyze repeated detections in these areas to make an informed judgment about whether they are true positives or false negatives.

Across the period of coverage, we detect 850,766 scaffolds in our images. We split these detections by month in Table 2.

Table 2 Scaffold detections by month/year

Period	# of Scaffolding detections	# Days of coverage
Aug 2023	187,682	18
Sep 2023	84,336	5
Oct 2023	285,127	13
Nov 2023	776,496	30
Dec 2023	164,661	8
Jan 2024	8919	1

5 Improving Model Accuracy

With 75 unique days of NYC image data spanning from August 2023 to January 2024, we are enabled to develop a tagging algorithm specifically designed to scrutinize and verify detected objects. This verification process is critical to ensure the accuracy of each detection. Given the objective of confirming scaffolds through multiple detections, the algorithm is tailored to prioritize maximum precision in the computer vision model. The rationale behind this approach is grounded in the high likelihood of accurately identifying scaffolds as true positives from repeating images of the same area.

Our goal is to develop a highly accurate model, and while this may result in a lower recall rate, it is a calculated trade-off. Despite this emphasis on precision, it is necessary to maintain a balance and not excessively compromise recall. We have opted for a confidence threshold of 0.85 to strike this balance. This threshold is carefully chosen to enhance precision while mitigating the impact on recall as much as possible. This approach ensures that while we prioritize accuracy, we also maintain a reasonable level of recall, thereby achieving a more refined and effective algorithm. For the 85% threshold, $r = 56.76\%$ and $p = 93.29\%$.

5.1 *Arranging the Detections*

Given the variation in detection angles and distances, a standardized method is needed to compile images showing the same scaffold. To achieve this, we implement a grid-based layout, mapping each coordinate to a grid system covering New York City. This grid is composed of rectangular 80 ft by 80 ft cells.

Our system connects with the YOLOv7 detection algorithm's output. We create a data frame for every processed image that includes the image's coordinates, capture time, and a flag indicating whether a scaffold was detected.

We adjust each coordinate by extending it 60 ft in the camera's direction. This adjustment is based on an analysis of various images, observing the distance between the scaffolds and the camera, which was noted to range from 0 to 120 ft.

The algorithm sorts all detections in chronological order. It then associates every row to its respective grid area, compiling a list that indicates whether a scaffold was detected. After processing the entire data frame, we accumulate a chronologically ordered list of scaffold detections or absences for each grid point.

The algorithm reviews each grid point by examining the last 20 detections. The number 20 is chosen so there will be enough images for the algorithm to see the area's history while balancing the algorithm's run time. From those images, we need to choose a threshold that will decide the number of detected scaffold images in the list to determine if this area contains a scaffold.

5.2 Confidence Threshold Selection

Let *A* denote the set of all scaffolds in NYC. Let *B* denote the set of scaffolds detected by our classifier. Let *C* denote the set of scaffolds tagged by the algorithm over time.

Given recall *r* and precision *p*, we define the following probabilities: – $P(B|A) = r$ is the probability of computer vision detecting a scaffold given that the scaffold exists (relevant), which equals 56.76% – $P(A|B) = p$ is the probability of a scaffold being relevant given that it was detected by computer vision, which equals 93.29%. The precision and recall curves are shown in Fig. 6, and maps of scaffolds detected are shown in Fig. 7.

The threshold for the algorithm detection over time is denoted by TH, and the sum of detections C from time 1 to 20 is represented by:

$$\sum_{i=1}^{20} C_i = B_i \geq TH$$

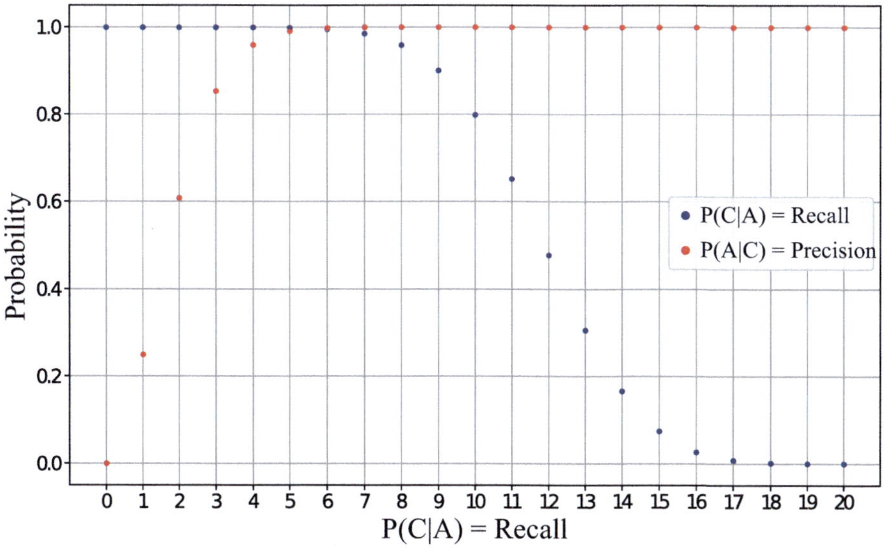

Fig. 6 Precision curve and recall curve of the tagging algorithm. Precision and recall curves. *The graph shows precision (blue line) and recall (red line) curves as functions of the threshold value*

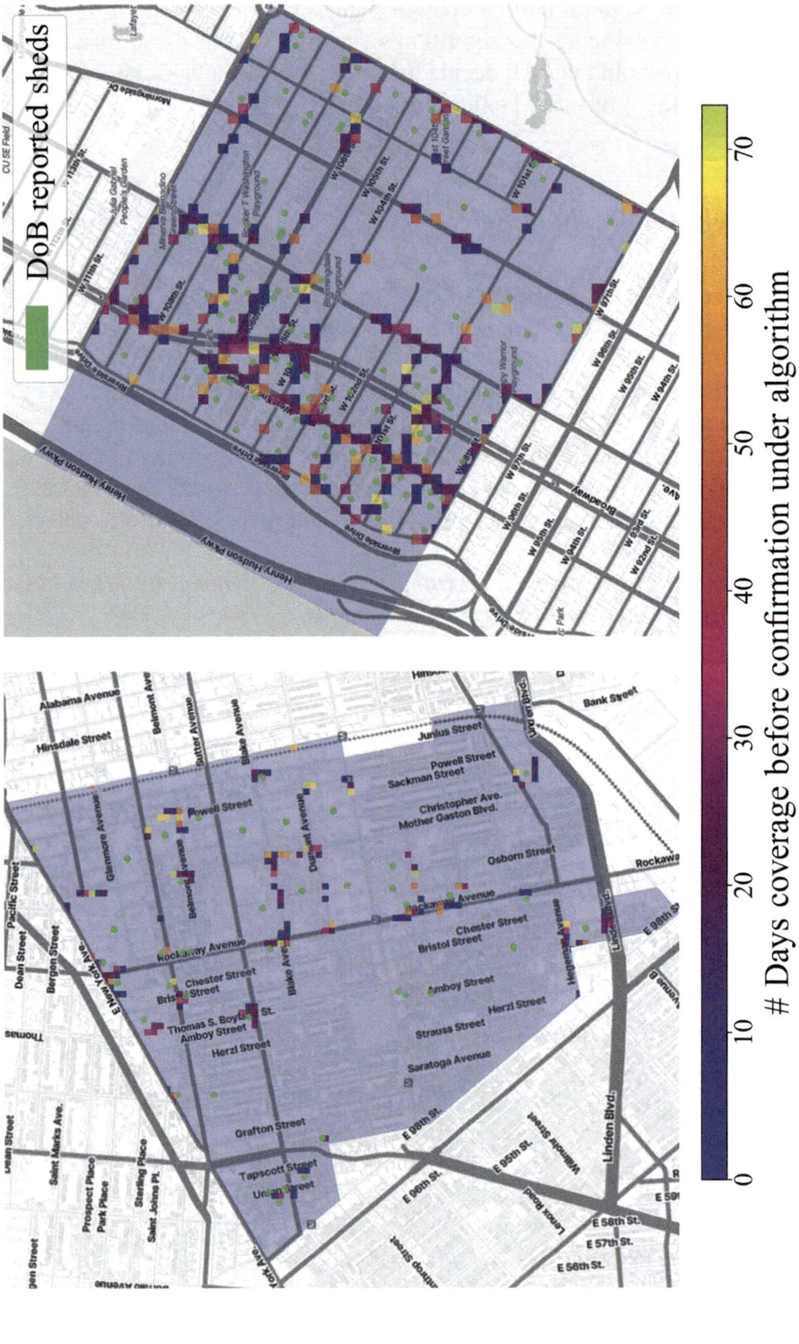

Fig. 7 Zoom-in of two neighborhoods with high scaffolding impact factor. Neighborhood comparison shows Brownsville and the Upper West Side, with DOB-reported scaffolds (in green) and our detections (in gradient). *Note that DoB-reported locations are often aligned to the building a scaffold belongs to, hence reported areas sometimes being in the interior of a block*

Table 3 Repeated detections threshold in tagging algorithm vs. recall and precision

Threshold (TH)	Recall (%)	Precision (%)
5	99.91	99.10
6	99.59	99.84
7	98.56	99.98

See the resulting recall and precision performance at different threshold values in Table 3.

5.3 Evaluation on DoB Active Shed Data

We access a dataset of active sidewalk sheds in the city as ground truth for our model detections. This evaluation is the primary way we evaluate our model's accuracy. The estimand (Lundberg et al., 2021) is two-fold: (1) What proportion of the known active scaffolds in the city can we confirm? (2) What proportion of known active scaffolds are out of range in our data distribution?

By combining these two evaluations, we get a ceiling of performance, and then a representation of how close we are to that ceiling. Furthermore, we also investigate how many of our detected scaffolds align with a known permit, giving us an idea of how many unpermitted scaffolds there are in the city.

5.3.1 Tagged Scaffolds

Tagged scaffolds are those sidewalk sheds confirmed by our tagging algorithm through longitudinal detections. To identify the number of tagged scaffolds, we align the raw coordinates from each positive image to an 80 ft. × 80 ft. grid system. Upon finding that this small cell size results in multiple cells per scaffold, we group cells around known active sheds together, resulting in a 320 ft. × 320 ft. area centered at each known coordinate. At this point, we can answer question (1) and say there are 5685 scaffolds detected in our dashcam dataset of New York City.

5.3.2 Performance Ceiling

To determine a performance ceiling, we compute the number of known scaffolds that lie outside the coverage of our dataset. We proxy the same algorithm used to tag/confirm a scaffold; so, if there are less than 20 images within 120 ft[3] of a known sidewalk shed, we count that shed as out-of-scope. After this analysis, we find that

[3] We use 120 ft as this is the maximum distance between ego-vehicle and scaffold that we observed during annotation.

2512 scaffolds lie outside of scope. And so, as of January 10, 2024, 2512 out of 8336 scaffolds are invisible in our dataset, or 30.1%.

As expected, this varies across each borough's scaffolding distribution; by borough: 19.4% in Manhattan, 34.5% in Brooklyn, 40.0% in The Bronx, 43.5% in Queens, 74.2% of sidewalk sheds in Staten Island are undetectable due to lack of dashcam data

5.3.3 Checking for Permits

To estimate the distribution of scaffolds without permits in our dataset, we evaluate tagged scaffolds that lie outside of any 320x320 ft. area cast by an active scaffold. We find that out of our 5685 sidewalk shed detections, 529, or 9.3%, are unpermitted.

5.3.4 Summary

We are satisfied with this model‘s performance. Out of 8336 known sidewalk sheds:—2512 are undetectable due to lack of coverage—529 are predicted to be unpermitted—5156 are confirmed by our model.

This means that our model missed 668 scaffolds, or 8.0%, an unknown proportion of which are due to our coarse 320 × 320 ft. detection region around each known shed.

6 Discussion

This analysis takes several important steps towards using large-scale, crowdsourced dashcam data for the sensing of the urban built environment. We prove it is possible to sense scaffolding both statically and longitudinally, establishing a framework to aggregate detections over time to maximize accuracy. This approach is especially well-suited for analyses that sense static or relatively immobile objects.

6.1 Tradeoffs

We made several design decisions with tradeoffs during this project:

1. Some parameters of the tagging algorithm were chosen arbitrarily, like the 20-day rolling window. You would adjust this time window based on the object to be detected and the frequency of change.
2. We select a confidence threshold of 0.85 for our supervised classifier, electing for higher precision and lower recall; this threshold is dependent on use case and individual model performance, and should not be blindly utilized in other projects.

3. We opt for a coarse, 320 ft. × 320 ft. detection region around each permit, which groups the too-small 80 ft. × 80 ft. cells together at the expense of noise in areas with highly dense scaffolding distributions like the Upper West Side neighborhood of Manhattan.

6.2 Future Work

We detail several areas of future work that would strengthen the results of this project and explore our model's generalization ability.

First, we would like to evaluate our scaffolding classifier on scaffolds from other cities around the United States, and other cities around the world. A limitation of the police vehicle classifier in (Franchi et al., 2023) is that it would not generalize well, due to the unique branding placed on police vehicles in New York City. We anticipate, from visual inspection of scaffolds in other cities, that the distribution of scaffolding appearances is more homogeneous than that of police vehicles, and so classifier generalization is likely better. That said, we would still like to prove this empirically, as models that generalize well are quite valuable.

Second, we would like to collaborate more directly with a city agency or journalistic organization in conducting further scaffolding permit compliance analysis. The problem of scaffolding has earned much contemporary news coverage (Mayor Adams Takes Down City's Longest-Standing Sidewalk Shed, 2023; McGeehan, 2024; Siff, 2023), encouraging public discourse. Downstream usage of this project's dataset by an entity like NYC's Department of Buildings could yield the ability to:—Pinpoint unpermitted scaffolds that should have been permitted—Confirm unpermitted scaffolds that do not require permitting, for reasons like:—Emergency work (NYCAC Sect. 28-105.4.1) (Project Requirements for Owner—Scaffold—Buildings, n.d.) - Design and permit requirements (NYCBC 3314.2) (NYC. New York City Building Code, n.d.)

Third, we would like to investigate the process of model distillation, in pursuit of real-world applications of our classifier. By distilling our model's weights into a smaller model, we could enable real-time inference on edge devices like networked dashcams (Cob-Parro et al., 2021; Sarwar Murshed et al., 2022; Koubaa et al., 2022), installed on either private or public, city-owned fleets. On-the-edge inference is not only less storage intensive but also preserves privacy by never transferring captured images outside of the imaging device (Khanna, 2021; Liu et al., 2020). In this type of work, utilizing images of public street scenes and computer vision, it is imperative to consider privacy as highly as possible (Upmanyu et al., 2009; Xiang, 2022; Das et al., 2017).

6.3 Limitations

We acknowledge several limitations of our work:

1. The grid system we used to map detections to unique scaffolding instances is a coarse 320 ft by 320 ft, and leaves room for further refinement.
2. Our coverage of the city is limited by the distribution of ridesharing drivers in the city; as such, our ability to detect scaffolds in outer boroughs like Staten Island is diminished.
3. While we consider this a deployment of our research, it is a retrospective deployment. Real-time, continuous monitoring of scaffolding deployments in New York City would require much closer collaboration with both our data provider and city agencies.
4. There is bias present in the data; a randomly sampled image from our dataset would *not* equate to a randomly sampled image from a hypothetical vehicle in NYC traffic. While this bias is not as pertinent when considering questions of object presence, instead of object probability, we have developed statistical upweighting methods for dense dashcam data in our other work (Franchi et al., 2023).
5. This approach does not automate the scaffolding inspection process in New York City, due to complex, varied policies and requirements for manual inspections.

6.4 *Impacts*

To the best of our knowledge, this is the first project studying longitudinal scaffolding distributions at a city-wide scale. Not only can we assert 'yes, there is a scaffold here', but we can also track scaffolds as they are constructed and taken apart over time. As mentioned earlier, this work does not automate the scaffolding inspection process required by NYC policy. However, this work would easily enable targeted inspections for agencies and the city government, filtering the over 8000 scaffolds present in New York City into a manageable subset (for example, the 529 scaffolds we identify without an associated permit).

Three primary groups are impacted by this work:

1. Urban planners and data scientists are presented with evidence of a promising new data medium, and are also given tools to bootstrap their own analyses.
2. NYC agencies are given a new way to monitor the built environment, yielding productivity and efficiency gains.
3. The general public is given further awareness of the extent of the city's scaffolding problem, and hopefully, this work adds to the momentum of 'getting sheds down!'

Dataset Release

As this dashcam dataset was acquired under a research evaluation license, we are restricted from publishing the raw data. However, we have published a visually-equivalent 2.4-million image dataset of NYC dashcam images, taken during 2020; available at Harvard Dataverse. This analysis would be replicable with that dataset.

Code Release

We release code to replicate the analysis in this paper at this repository: https://github.com/IRL-CT/urban-fingerprinting

Acknowledgements The authors would like to thank Ilan Mandel, Maria Teresa Perreira, Emma Pierson and Eilam Shapira for helpful discussions surrounding the ideation and framing of this work. We thank Nexar, Inc., for providing us with dashcam image data for this project.

Appendix

A1. Algorithm 1

Algorithm 1 Finding The Nearest Permit

Input: $[X, Y]$: Permit coordinates, $[x_i, y_i]$: Image coordinates, CAM_DEG: Camera heading, TH_DIST: Threshold distance, TH_DEG: Threshold angle.
Output: List of images associated with permits for annotation.

$A = [X, Y] - [x_i, y_i]$
if $\|A\| \leq TH\_DIST$ **then**
  if $|\text{atan2}(A)| \leq CAM\_DEG + TH\_DEG$ **then**
    $Dataset \leftarrow [x_i, y_i]$
  end if
end if

See Fig. 8.

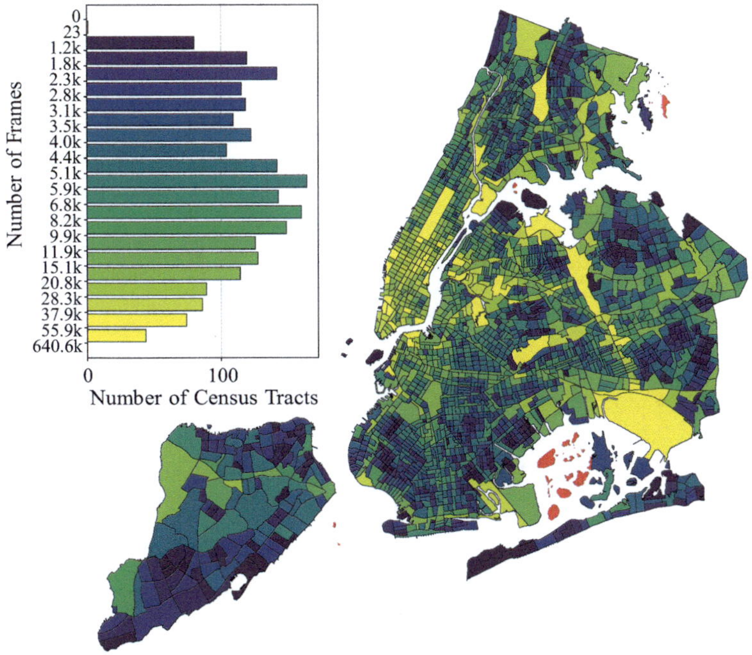

Fig. 8 Spatial density of our dashcam image dataset, shown by Census tract

A2. Algorithm 2

Algorithm 2 Over Time Detection

Input: YOLOv7 Detection, $[x_i, y_i]$: Image coordinates, α: Camera heading, t: capture time.
Output: $[x^{20X20}, y^{20X20}]$ grid.

for each image in YOLOv7 output **do**
  Extract: $[x_i, y_i]$, t, and detection flag
  $[x_i', y_i'] \leftarrow [x_i, y_i] + \alpha \times 20\text{m}$
  Adjusted coordinates to grid $([x_i^{20X20}, y_i^{20X20}])$
  grid$([x_i^{20X20}, y_i^{20X20}]) \leftarrow t, l.append$(detection flag)
end for
for $[x_i^{20X20}, y_i^{20X20}]$ in grid **do**
  last_20 = grid$([x_i^{20X20}, y_i^{20X20}]).l[: -20]$
  if sum(last_20$(True)$)$/20 \geq 30\%$ **then**
    grid$([x_i^{20X20}, y_i^{20X20}]) \leftarrow$ 'containing a scaffold'
  else
    grid$([x_i^{20X20}, y_i^{20X20}]) \leftarrow$ 'not containing a scaffold'
  end if
end for
return Updated grid with scaffold detection results

See Fig. 9.

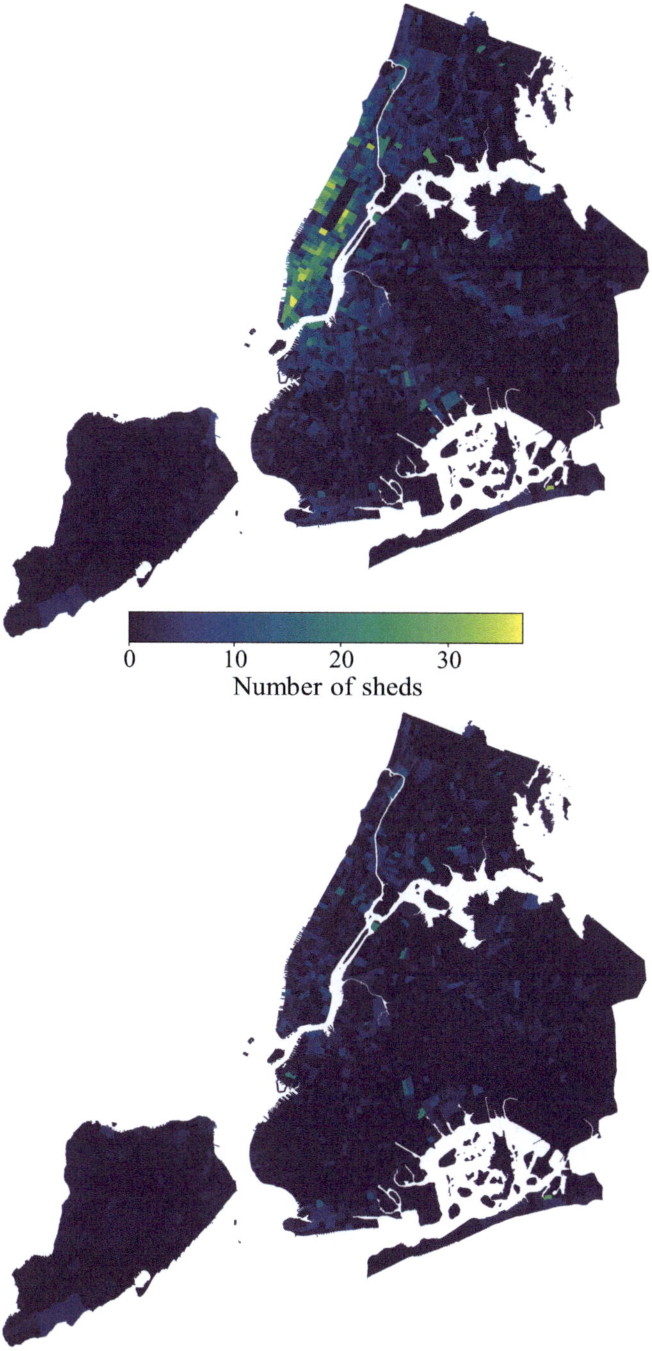

Fig. 9 Active, DOB-reported sheds (top) vs. sheds that are out-of-scope from detection via our dashcam dataset (bottom), summed by Census tract. As visible, coverage is lacking on the fringes of the city's outer boroughs, which makes sense given the ego-vehicles of our dashcams: ride-sharing vehicles

References

Achtemeier, J. D. (2019). *The impact of weather conditions and infrastructure design on the mobility of people with impaired vision*. Master's thesis, University of Minnesota, United States. ISBN: 9781687924490.

Agostini, G., Pierson, E., & Garg, N. (2023). A Bayesian spatial model to correct under-reporting in urban crowdsourcing. arXiv:2312.11754 [cs, stat].

Aikoh, T., Homma, R., & Abe, Y. (2023). Comparing conventional manual measurement of the green view index with modern automatic methods using google street view and semantic segmentation. *Urban Forestry & Urban Greening, 80*, 127845.

Apte, J. S., Messier, K. P., Gani, S., Brauer, M., Kirchstetter, T. W., Lunden, M. M., Marshall, J. D., Portier, C. J., Vermeulen, R. C. H., & Hamburg, S. P. (2017). High-resolution air pollution mapping with Google Street View cars: Exploiting big data. *Environmental Science & Technology, 51*(12), 6999–7008.

Balali, V., Depwe, E., & Golparvar-Fard, M. (2015). *Multi-class traffic sign detection and classification using Google Street view images.*

Boller, D., De Vitry, M. M., Wegner, J. D., & Leitão, J. P. (2019). Automated localization of urban drainage infrastructure from public-access street-level images. *Urban Water Journal, 16*(7), 480–493.

Calder, R. City Council seeking to overhaul scaffolding abuse in NYC, March 2023.

Chowdhury, T., Ding, Q., Mandel, I., Wendy, J., & Ortiz, J. (2021). Tracking urban heartbeat and policy compliance through vision and language-based sensing. In *Proceedings of the 8th ACM International Conference on Systems for Energy-Efficient Buildings, Cities, and Transportation, BuildSys '21* (pp. 302–306). Association for Computing Machinery.

Clark, B. Y., Brudney, J. L., & Jang, S.-G. (2013). Coproduction of Government Services and the New Information Technology: Investigating the Distributional Biases. *Public Administration Review, 73*(5), 687–701. eprint: https://onlinelibrary.wiley.com/doi/pdf/10.1111/puar.12092

Cob-Parro, A. C., Losada-Gutierrez, C., Marron-Romera, M., Gardel-Vicente, A., & Bravo-Munoz, I. (2021). Smart video surveillance system based on edge computing. *Sensors, 21*(9), 2958. Number: 9 Publisher: Multidisciplinary Digital Publishing Institute.

Das, A., Degeling, M., Wang, X., Wang, J., Sadeh, N., & Satyanarayanan, M. (2017). Assisting users in a world full of cameras: A privacy-aware infrastructure for computer vision applications. In *2017 IEEE Conference on Computer Vision and Pattern Recognition Workshops (CVPRW)* (pp. 1387–1396) ISSN: 2160-7516.

Deacon, L. A. (2013). *Planning sidewalks: Implications of regulating sidewalk space in the East Village*. PhD thesis, Columbia University.

Franchi, M., Dey, D., & Wendy, J. (2024). Towards Instrumented Fingerprinting of Urban Traffic: A Novel Methodology using Distributed Mobile Point-of-View Cameras. In *Proceedings of the 16th International Conference on Automotive User Interfaces and Interactive Vehicular Applications, AutomotiveUI '24* (pp. 53–62). New York, NY, USA, Association for Computing Machinery.

Franchi, M., Parreira, M. T., Bu, F., & Ju, W. (2025). The robotability score: Enabling harmonius robot navigation on urban streets. In *Proceedings of the 2025 SIGCHI Conference on Human Factors in Computing Systems*. ACM.

Gebru, T., Krause, J., Wang, Y., Chen, D., Deng, J., Aiden, E. L., & Fei-Fei, L. (2017). Using deep learning and Google Street View to estimate the demographic makeup of neighborhoods across the United States. *Proceedings of the National Academy of Sciences, 114*(50), 13108–13113.

Green, P. (2020). Our Lives, Under Construction. *The New York Times*.

Hacker, K. P., Greenlee, A. J., Hill, A. L., Schneider, D., & Levy, M. Z. (2022). Spatiotemporal trends in bed bug metrics: New York City. *PLOS ONE, 17*(5), e0268798. Public Library of Science.

Hara, K., Le, V., & Froehlich, J. (2013). Combining crowdsourcing and google street view to identify street-level accessibility problems. In *Proceedings of the SIGCHI conference on human factors in computing systems* (pp. 631–640). ACM.

Helbich, M., Poppe, R., Oberski, D., Zeylmans, M., & van Emmichoven, and Raoul Schram. (2021). Can't see the wood for the trees? An assessment of street view and satellite derived greenness measures in relation to mental health. *Landscape and Urban Planning, 214*, 104181.

Heyward, G. (2023). Mayor Adams mounts effort to take down NYC scaffolding. *Gothamist*. Section: News.

Hipp, J. R., Lee, S., Ki, D., & Kim, J. H. (2022). Measuring the Built Environment with Google Street View and Machine Learning: Consequences for Crime on Street Segments. *Journal of Quantitative Criminology, 38*(3), 537–565.

HumanSignal. Label Studio Documentation, 2025.

Khanna, S. (2021). Identifying privacy vulnerabilities in key stages of computer vision, natural language processing, and voice processing systems. *International Journal of Business Intelligence and Big Data Analytics, 4*(1), 1–11. Number: 1.

Klemmer, K., Neill, D. B., & Jarvis, S. A. (2021). Understanding spatial patterns in rape reporting delays. *Royal Society Open Science, 8*(2), rsos.201795.

Kontokosta, C. E., & Hong, B. (2021). Bias in smart city governance: How sociospatial disparities in 311 complaint behavior impact the fairness of data-driven decisions. *Sustainable Cities and Society, 64*, 102503.

Koubaa, A., Ammar, A., Kanhouch, A., & Al-Habashi, Y. (2022). Cloud Versus Edge Deployment Strategies of Real-Time Face Recognition Inference. *IEEE Transactions on Network Science and Engineering, 9*(1), 143–160.

Lam, C. T., Liu, M., & Hui, X. (2021). The geography of ridesharing: A case study on New York City. *Information Economics and Policy, 57*, 100941.

Li, X., Zhang, C., Li, W., Ricard, R., Meng, Q., & Zhang, W. (2015). Assessing street-level urban greenery using Google Street View and a modified green view index. *Urban Forestry & Urban Greening, 14*(3), 675–685.

Liu, Z., Wu, Z., Gan, C., Zhu, L., & Han, S. (2020). DataMix: Efficient privacy-preserving edge-cloud inference. In A. Vedaldi, H. Bischof, T. Brox, & J.-M. Frahm (Eds.), *Computer Vision – ECCV 2020* (Lecture Notes in Computer Science) (pp. 578–595). Springer.

Liu, Z., Bhandaram, U., & Garg, N. (2023). Quantifying spatial under-reporting disparities in resident crowdsourcing. *Nature Computational Science*, 1–9. Nature Publishing Group.

Lundberg, I., Johnson, R., & Stewart, B. M. (2021). What is your estimand? Defining the target quantity connects statistical evidence to theory. *American Sociological Review, 86*(3), 532–565. SAGE Publications Inc.

Maddalena, E., Ibanez, L.-D., & Simperl, E. (2020). Mapping points of interest through˜ street view imagery and paid crowdsourcing. *ACM Transactions on Intelligent Systems and Technology, 11*(5), 1–28.

Madhavan, V., & Darrell, T. (2017). *The bdd-nexar collective: A large-scale, crowsdsourced, dataset of driving scenes*. Master's thesis, EECS Department, University of California.

Matt, F., Zamfirescu-Pereira, J. D., Wendy, J., & Pierson, E. (June 2023). Detecting disparities in police deployments using dashcam data. In *Proceedings of the 2023 ACM Conference on Fairness, Accountability, and Transparency, FAccT '23* (pp. 534–544). Association for Computing Machinery.

Mayor Adams Takes Down City's Longest-Standing Sidewalk Shed, Scaffolding, December 2023.

McGeehan, P. (2024). New York City's Everlasting Scaffolding. *The New York Times*.

Minkoff, S. L. (2016). NYC311: A tract-level analysis of citizen–government contacting in New York City. *Urban Affairs Review, 52*(2), 211–246. SAGE Publications Inc.

New York Daily News Editorial Board. Shedding the sheds: NYC embarks on crusade to combat scourge of scaffolding, 2023.

NYC Department of Buildings. Code Notes - Supported Scaffolds. Technical report, July 2015.

NYC Department of Buildings. Scaffolding Inspections. NYC Building Code, November 2022.

NYC Department of Buildings. Suspended Scaffold Tips—Buildings, 2023.
NYC Department of Buildings. Active Sidewalk Shed Permits, February 2024.
NYC OpenData. DOB Permit Issuance I NYC Open Data, January 2024.
NYC. New York City Building Code.
Odgers, C. L., Caspi, A., Bates, C. J., Sampson, R. J., & Moffitt, T. E. (2012). Systematic social observation of children's neighborhoods using Google Street View: a reliable and cost-effective method. *Journal of Child Psychology and Psychiatry, and Allied Disciplines, 53*(10), 1009–1017.
Pak, B., Chua, A., & Moere, A. V. (2017). FixMyStreet Brussels: Socio-Demographic Inequality in Crowdsourced Civic Participation. *Journal of Urban Technology, 24*(2), 65–87. arXiv:1708.02274 [cs].
Project Requirements for Owner - Scaffold - Buildings.
Rubio-Romero, J. C., Gamez, M. C. R., & Carrillo-Castrillo, J. A. (2013). Analysis of the safety conditions of scaffolding on construction sites. *Safety Science, 55*, 160–164.
Rundle, A. G., Bader, M. D. M., Richards, C. A., Neckerman, K. M., & Teitler, J. O. (2011). Using Google Street view to audit neighborhood environments. *American Journal of Preventive Medicine, 40*(1), 94–100.
Sarwar Murshed, M. G., Murphy, C., Hou, D., Khan, N., Ananthanarayanan, G., & Hussain, F. (2022). Machine learning at the network edge: A survey. *ACM Computing Surveys, 54*(8), 1–37.
Scaffolding as Status Symbol? One Company Has High Hopes. *Bloomberg.com*, February 2022.
Siff, A. Tired of Scaffolding? NYC Has New Plan to Get Rid of Long-Standing Sidewalk Sheds, 2023.
uber/h3, January 2024. original-date: 2017-12-21T01:38:35Z.
Upmanyu, M., Namboodiri, A. M., Srinathan, K., & Jawahar, C. V. (2009). Efficient privacy preserving video surveillance. In *2009 IEEE 12th International Conference on Computer Vision* (pp. 1639–1646). IEEE.
Upper East Side, Midtown Have The Most Sidewalk Sheds Citywide, July 2023. Section: Real Estate.
Wang, C.-Y., Bochkovskiy, A., & Liao, H.-Y. M. (2022). *YOLOv7: Trainable bag of freebies sets new state-of-the-art for real-time object detectors*. arXiv:2207.02696 [cs]
Wang, A., Chen, H., Liu, L., Chen, K., Lin, Z., Han, J., & Ding, G. (2024). *YOLOv10: real-time end-to-end object detection* (p. arXiv:2405.14458).
West Side Rag. The omnipresence of scaffolding and its impact on City Lives; why and what is being done?, 2022. Section: NEWS.
Xiang, A. (2022). *Being 'seen' vs. 'mis-seen': Tensions between privacy and fairness in computer vision.*
Yglesias, M. (2022). New York City shouldn't be covered in sidewalk sheds.
Yi, L. (2019). Using Google Street View to investigate the association between street greenery and physical activity. *Landscape and Urban Planning, 191*, 103435.
Yifan, L., Jiaming, L., Zhang, S., & Hall, P. (2018). Traffic signal detection and classification in street views using an attention model. *Computational Visual Media, 4*(3), 253–266.
Yin, L., Cheng, Q., Wang, Z., & Shao, Z. (2015). 'Big data' for pedestrian volume: Exploring the use of Google Street View images for pedestrian counts. *Applied Geography, 63*, 337–345.
Zhang, W., Witharana, C., Li, W., Zhang, C., Li, X., & Parent, J. (2018). Using deep learning to identify utility poles with crossarms and estimate their locations from Google Street view images. *Sensors, 18*(8), 2484. Number: 8 Publisher: Multidisciplinary Digital Publishing Institute.

Development of a Method for Identifying the Presence of Front Roads Based on Building Coverage Ratio

Takahito Shimizu, Kotaro Mizutani, Yuki Akiyama, Hiroki Baba, and Chiaki M. Akiyama

Abstract This study aims to develop a new method for identifying the presence of front roads (roads located in close proximity to buildings that ensure daily traffic and safety functions, such as evacuation and emergency response during disasters) based on the Building Coverage Ratio. This enables the evaluation of building–road proximity relationships, which have been difficult to assess in previous studies. Our newly developed method identifies the presence of front roads by verifying the intersections between polygons representing site areas generated from buildings and road polygons. In addition, we improved the method using field survey data, allowing it to identify the presence of front roads with high reliability. These results contribute to the development of a method to easily identify the proximity between buildings and roads over a wide area.

Keywords Front roads · Buildings · Proximity · Digital road map · Japan

---

T. Shimizu (✉) · Y. Akiyama
Department of Urban and Civil Engineering, Faculty of Engineering, Tokyo City University, Tokyo, Japan
e-mail: g2581624@tcu.ac.jp; akiyamay@tcu.ac.jp

K. Mizutani
Graduate School of Integrative Science and Engineering, Tokyo City University, Tokyo, Japan
e-mail: kmizu@tcu.ac.jp

H. Baba
Faculty of Science and Engineering, Chuo University, Tokyo, Japan
e-mail: hbaba893@g.chuo-u.ac.jp

C. M. Akiyama
Reitaku University, Chiba, Japan
e-mail: chakiyam@reitaku-u.ac.jp

R. Goodspeed et al. (eds.), *Digital-Era Urban Transformations*, The Urban Book Series, https://doi.org/10.1007/978-3-031-98300-9_11

1 Introduction

In urban planning, the arrangement of roads and buildings is widely recognized to play a critical role in traffic flow, infrastructure development, and disaster prevention planning. Similarly, in the real estate market, location conditions often influence land and housing prices, with physical attributes, such as road width and setback distance, frequently reflected in economic value.

Against this backdrop, studies on the locational relationship between buildings and roads have been conducted in various fields, with findings from the transportation, disaster prevention, and real estate sectors further emphasizing its importance. For instance, in the real estate field, Abdulla et al. (2023) revealed a positive correlation between road width and land value. Similarly, Souza (2009) revealed that building setback distances and frontage widths significantly influence housing prices. In the field of disaster prevention, Osaragi and Oki (2012) revealed that narrow roads are common in densely built-up areas in Japan, making evacuation difficult owing to collapsed buildings during disasters. In the transportation field, Raji and Attah (2017) revealed that many land developers in Nigeria construct buildings that occupy as much as 70% of the plot, leading to poor setbacks, limited car accessibility, and inadequate parking spaces. Shaikh et al. (2019) and Kumar and Pushplata (2015) reported similar findings in Pakistan and India, respectively. Considering these previous studies, the locational relationship between buildings and roads, particularly the proximity of each building to a road that ensures daily transportation and safety functions, such as evacuation and emergency activities during disasters, is of critical importance.

In Japan, to ensure daily transportation and facilitate rescue and evacuation during disasters, the Building Standards Act specifies that a building site must have at least 2 m of frontage on a road that is at least 4 m wide. However, compliance with these requirements does not necessarily ensure adequate traffic functionality or safety. For instance, Minai et al. (2016) pointed out that in flag-shaped lots, the width of the site portion connecting the building to the road is narrow, barely allowing vehicle passage, which can delay emergency activities. In addition, densely built areas with old buildings and narrow roads are common in Japan. Numerous studies have indicated that such areas pose significant risks during earthquakes, including evacuation difficulties and the rapid spread of fires (e.g., Ichikawa et al., 2004; Osaragi & Oki, 2012; Yoshihara & Tanaka, 2020). These concerns were brought into sharp focus during the major earthquake, with a magnitude of 7.5, that struck the Noto Peninsula on January 1, 2024. Densely built-up areas were particularly affected during this earthquake, giving rise to serious challenges. The roads were blocked by collapsed structures and debris, rendering them impassable. This obstruction not only hindered evacuation efforts but also made it extremely difficult for emergency vehicles and rescue teams to reach the affected areas, resulting in significant delays in rescue operations (Suppasri et al., 2024).

Thus, understanding the proximity between buildings and roads while considering traffic and safety functions is important in fields such as transportation, disaster prevention, and real estate. In particular, countries such as Japan frequently face natural disasters, including earthquakes, which emphasizes the need for buildings to fulfill traffic and safety functions, further emphasizing the importance of this analysis. However, data on the proximity of roads to individual buildings have not been developed, and no simple method has been established. Therefore, an efficient method using existing data is required to understand the proximity of roads to individual buildings over a wide area.

2 Related Work

Previous studies aimed at analyzing the proximity between buildings and roads can be broadly categorized into two approaches: those utilizing Voronoi divisions and those focusing on the geographic shape of roads.

First, Voronoi diagrams have been widely studied as a method for representing the proximity of spatial objects such as buildings (e.g., Gold, 1989; Jiatian et al., 2009; Usui & Asami, 2010a; Hamaina et al., 2013). Gold (1989) pointed out the effectiveness of Voronoi diagrams as a method for determining the proximity between buildings and roads. Usui and Asami (2010a) used Voronoi diagrams to determine the proximity of buildings and roads. However, they indicated that methods using Voronoi diagrams require considerable calculation time.

Methods that utilize the geographical shape of roads include those of Usui and Asami (2010a, b). Usui and Asami (2010a) proposed a method that uses buffer areas derived from road line data and Usui and Asami (2010b) proposed a method to estimate the optimal buffer distance for determining road proximity by analyzing the number of buildings per unit area.

However, these studies did not compare or verify the accuracy of their results with the actual proximity conditions of buildings obtained through on-site surveys; therefore, the extent to which they reflect reality is unclear. Thus, no method has been established to understand the proximity between buildings and roads in a straightforward manner.

3 Study Objective

In light of the abovementioned problems, this study aimed to develop a simple and highly reliable method for broadly identifying the proximity between buildings and roads. Specifically, we developed a new method to identify the presence of front roads (in this study, we define "Front Roads" as roads close to buildings that guarantee daily transportation and safety functions, such as evacuation and emergency activities during disasters.) for each building, using road link data that represent the road network as lines and building polygon data. In addition, we verified the results of this method using actual data on the presence of front roads collected through field surveys.

The proposed method exhibits four distinct features. First, it addresses the challenges of previous studies, particularly the difficulty of comprehensively and rapidly understanding road proximity conditions on a per-building basis over a wide area. Second, the data used in this study were continuously and frequently updated for all of Japan, thereby ensuring high versatility and immediacy. Third, this is the first study to verify the accuracy of the developed method based on actual information about front roads collected through field surveys, providing high reliability. Finally, because data similar to those used in this study are available in other countries, the method can be applied to regions outside Japan facing similar issues.

By applying this method, the presence of front roads for each building could be identified. This enables more detailed urban analyses, such as identifying areas prone to evacuation difficulties during disasters or regions that are inaccessible to demand-responsive transportation systems.

The study area was Takasaki City, Gunma Prefecture (Fig. 1). Located approximately 100 km northwest of Tokyo, Takasaki City has approximately 370,000 inhabitants. It offers excellent access to the Tokyo metropolitan area, facilitated by bullet train lines and expressways. In addition, the city features a diverse range of areas, including densely populated urban zones with concentrated buildings and roads, relatively low-density suburban neighborhoods, and mountainous regions with sparse roads and buildings. This diverse region serves as an ideal location for validating methods that can be applied in a wider area.

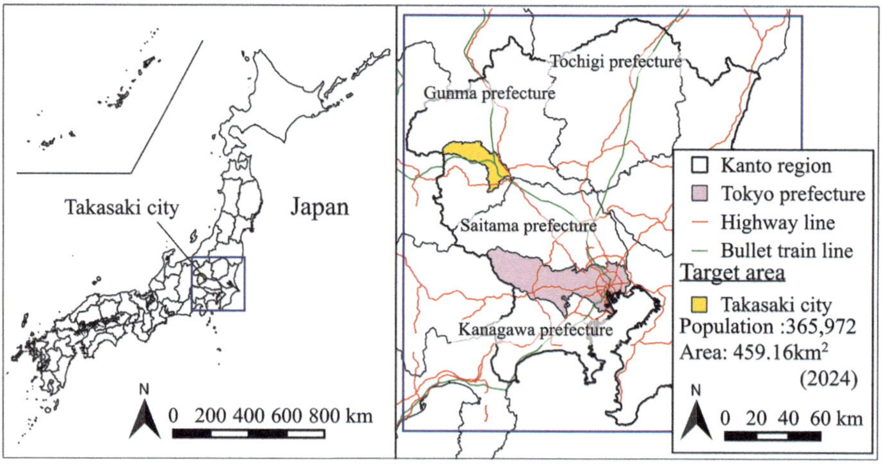

Fig. 1 Overview of the study area

4 Development of a Method for Identifying the Presence of Front Roads

4.1 Data

This study used the data presented in Table 1, which includes three types: road link data, building polygon data, and land use zone polygon data.

4.1.1 Road Link Data

First, for the road link data, we used both the "All Road Link Data" and the "Narrow Road Data" included in the Digital Road Map Database (DRM-DB) provided by the Japan Digital Road Map Association. These datasets were combined, and duplicates were removed to construct a road network database covering the entire country of Japan. In addition, although most road line data in the DRM-DB represent road shapes using centerlines, roads in the real world have two-dimensional shapes that include widths. Therefore, in this study, approximate road surfaces (hereinafter called "Road Polygons") were created by buffering both sides of the road links with half of the lower limit of the road width categories ("less than 3.0 m", "3.0 to 5.5 m", "5.5 to 13.0 m" and "13.0 m or more") based on the attributes of the road network database.

4.1.2 Building Polygon Data

Second, for the building polygon data, we used the "Outline of a Building" from the Basic Map Information provided by the Geospatial Information Authority of Japan. This dataset comprises polygon data representing nearly all building shapes in Japan, including residential, commercial, and industrial facilities. In Japan, the Basic Plans for Housing set the minimum living area standard at 25 m^2 or larger. Consequently, this study targeted building polygons with areas of 25 m^2 or larger.

Table 1 Overview of data

Data name	Data source	Year
All Road Link Data Narrow Road Data	DRM-DB	2023
Outline of a Building	Basic Map Information	2023
Land Use Zone Polygon Data	National Land Numerical Information	2019

4.1.3 Land Use Zone Polygon Data

Finally, for land use zone polygon data, we used the latest version of "Land Use Zones (2019)" from the National Land Numerical Information provided by the Ministry of Land, Infrastructure, Transport, and Tourism (MLIT). Land use zones in Japanese urban planning are designed to promote the systematic development of urban areas by specifying the permitted uses for each region, with restrictions on building types and sizes defined for each use. This dataset includes information such as land use zone classifications, Building Coverage Ratio (BCR) regulations, and Floor Area Ratio (FAR) regulations for each polygon. This study targeted buildings in 10 types of land use zones within Takasaki City, excluding the "Exclusively Industrial Zone" where residential construction is not permitted. Buildings in areas without designated land use zones were also included, with a total of 11 area types. We assigned regional classifications and BCR regulatory values to each building using spatial joins with land use zone polygon data. In areas without specified BCR regulations, we set the BCR regulatory value to 50%.

4.2 Method for Identifying the Presence of Front Roads

In this study, we defined a front road as a road with traffic and safety functions and proposed a method to identify the presence of a front road for each building. Specifically, we first extracted the front road candidates based on the shape of the building polygon. Next, we generated an "Extended Building Polygon" representing the site to which the building belongs and determined whether this polygon intersected with a front-road candidate. The determination process is described in detail below.

4.2.1 Extraction of Candidate Front Roads

This section details the method for extracting the road nearest to each building as a candidate front road. This process is shown in Fig. 2a. First, the nearest road polygon was identified for all vertices of the building polygon, and the distance between each vertex and the nearest road polygon was calculated. The road polygon with the shortest distance was determined to be the road nearest to the building. In addition, a vector (hereinafter called "Nearest Neighbor Vector") was assigned to the building, with the vertex having the shortest distance as the starting point and the closest point on the edge of the nearest road polygon as the endpoint.

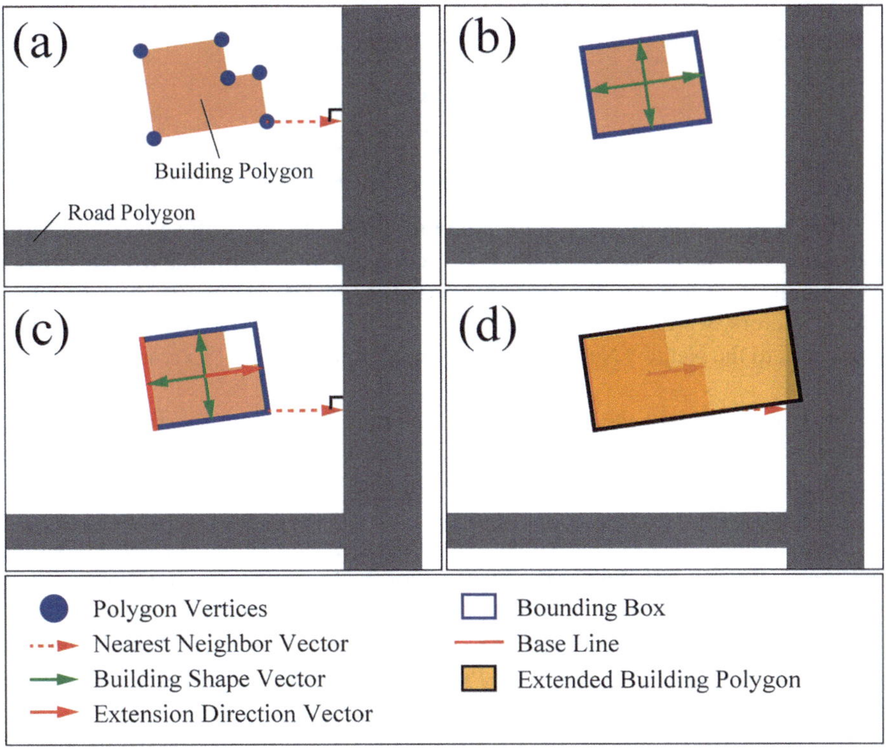

Fig. 2 Details of the method

4.2.2 Extraction of Expansion Direction Candidates for Building Polygons

This section describes the method for extracting candidates in the direction in which the building site extends. This process is shown in Fig. 2b. First, to approximate the diverse shapes of buildings, we generated a minimum rectangle (hereinafter called "Bounding Box") that encompasses each building and provides an approximate representation of its shape. Next, we created four vectors (hereinafter called "Building Shape Vectors") originating from the center of gravity of the Bounding Box and extending towards the midpoints of each side of the rectangle. These vectors represent candidates for the expansion direction of the building polygon.

4.2.3 Generation of Extended Building Polygons

This section details the method for generating the extended building polygons. This process is shown in Fig. 2c, d. First, we calculated the cosine similarity between each of the four Building Shape Vectors and the Nearest Neighbor Vector. Cosine similarity measures the similarity in the directions of two vectors, and for any n-dimensional vectors x and y, it is defined by Eq. (1).

$$\cos(x,y) = \frac{\sum_{k=1}^{n} x_k y_k}{\sqrt{\sum_{k=1}^{n} x_k^2}\sqrt{\sum_{k=1}^{n} y_k^2}} \tag{1}$$

The cosine similarity takes values ranging from 1 to −1, with values closer to 1 indicating that the directions of the two vectors are more similar and values closer to −1 indicating that the directions are nearly opposite. Among the four Building Shape Vectors, we extracted the one with the cosine similarity closest to 1 when compared to the Nearest Neighbor Vector and defined it as the "Extension Direction Vector."

Next, we extracted the edge of the Bounding Box where the Building Shape Vector, with the cosine similarity closest to −1 compared to the Extension Direction Vector, intersected. We defined this edge as the "Base Line" used to extend the building polygon.

Then, to generate an Extended Building Polygon, we calculated its area based on the area of the building polygon. As the area of the Extended Building Polygon corresponds to the site area of the building, we set a parameter α to convert the building area into the equivalent site area. The area of the Extended Building Polygon was calculated using the parameter α and Eq. (2). In Sect. 4.2.3, we defined this parameter α as the upper limit of the BCR for each area.

$$\textit{Extended building polygon area} = \frac{100}{\alpha} \times \textit{Building polygon area} \tag{2}$$

Finally, we extended the building polygon from the Base Line in the direction of the Extension Direction Vector such that the area of the Extended Building Polygon matched the calculated area.

4.2.4 Determining the Presence of a Front Road

According to the Japanese Building Standard Act, a certain distance (usually 50 cm or more) must be maintained from the site boundary when constructing a building. As a result, the actual site extends not only in the direction of the Extension Direction Vector, but also in other directions. However, with this method, the Extended Building Polygon only accounts for the area in the direction of the Extension Direction Vector. Therefore, in reality, if the building has a front road, the Extended Building Polygon and the candidate front road are expected to intersect.

From the above, the presence of a front road for each building was determined by checking the intersection between the Road Polygon and the Extended Building Polygon created in Sect. 4.2.3. If they intersected, the building was classified as having a front road. Otherwise, it was classified as having no front road.

4.3 *Field Survey and Visual Survey*

Validation is necessary to verify the accuracy of this method for identifying the presence of front roads in buildings. For this purpose, we selected sample areas from the study area to collect data on the presence of front roads. Specifically, information on the actual presence of front roads for each building was collected through field and visual surveys of the buildings and their surroundings using Google Street View.

4.3.1 Overview of Field Survey and Visual Survey

This section details the selection of the sample areas for the field survey to collect verification data. First, we classified the small areas within Takasaki City into 13 categories based on the types of land use zones introduced in Sect. 4.1.3 and the upper limits of the BCR. Next, we selected two areas from each category to ensure an unbiased survey and sampled approximately 100–200 buildings in each area (Table 2). Field surveys were conducted in the areas denoted as D, K, T, and X. For the other areas, we conducted visual surveys using Google Street View to examine the presence of front roads for each building. The surveys were conducted based on the criteria listed in Table 3.

4.3.2 Results of Field Survey and Visual Survey

Table 2 presents the results of the field and visual surveys. The lowest percentage of buildings with front roads was found in Area F (Category 1, Exclusive High-Rise Residential Zone, BCR 50%, Kaneko-machi) at 24.7%, whereas the highest was observed in Area G (Category 1, Medium-High Rise Residential Zone, BCR 60%, 2-chome Nakai-machi) at 100%.

Examples of buildings without front roads include cases in which private roads or dead-end streets not recorded in the DRM-DB were extended to residential buildings (Fig. 3(a)). In addition, although rare, instances such as those shown in Fig. 3(b), where a small stream existed between the building and road, were also observed. Narrow roads not included in DRM-DB were present around these buildings, allowing for minimal mobility.

4.4 *Method Improvement Based on Field and Visual Surveys*

We compared the results of the front road determination using the method from Sect. 4.2 with the survey results from Sect. 4.3 for buildings in the study area. This comparison revealed two main issues with the proposed method.

Table 2 List of sample areas and results of field and visual surveys

Land Use Zone (Abbreviation, BCR)	Sample area		Number of buildings	Number of buildings with front road	Buildings with front road [%]
Category I exclusively low-rise residential zone (ELR1, 40)	A	3-chome Kataoka-machi	154	100	64.94
	B	Kaizawa-machi	138	71	51.45
Category I exclusively low-rise residential zone	C	1-chome Yoshiimachi Nanyodai	152	152	100.00
(ELR1, 50)	D	1-chome Shiroyama-machi	164	163	99.39
Category I mid/high-rise oriented residential zone	E	Munadaka-machi	119	70	58.82
(MOR1, 50)	F	Kaneko-machi	146	36	24.66
Category I mid/high-rise oriented residential zone	G	2-chome Nakai-machi	153	153	100.00
(MOR1, 60)	H	Yoshiimachi Yoshii	112	42	37.50
Category II mid/ high-rise oriented residential zone	I	Kaminamie-machi	150	88	58.67
(MOR2, 60)	J	3-chome Yachiyo-machi	147	104	70.75
Category I residential zone	K	Misatomachi kamishiba	156	98	62.82
(R1, 60)	L.	Iizuka-machi	148	77	52.03
Category II residential zone	M	Azuma-chou	134	89	66.42
(R2, 60)	N	Iizuka-machi	122	84	68.85
Quasi-residential zone	O	Shimotoyoka-machi etc.	138	84	60.87
(QR, 60)	P	Ino-machi etc.	117	75	64.10
Neighborhood commercial zone	Q	Yoshiimachi Yoshii	119	86	72.27
(NCZ, 80)	R	Egi-machi	98	73	74.49
Commercial zone	S	Tonya-machi etc.	165	157	95.15
(CZ, 80)	T	Narita-chou	120	77	64.17
Quasi-industrial zone	U	Kamisano-machi	130	85	65.38
(QIZ, 60)	V	Sugaya-machi	166	105	63.25
Industrial zone	W	Oyagi-machi	123	39	31.71
(IZ, 60)	X	Munadaka-machi	195	195	100.00
(Not designated)	Y	Kurabuchimachi sannokura	184	74	40.22
(ND, 50)	Z	Shimano-machi	169	111	65.68

Table 3 Criteria for determining the presence of front roads in the field and visual surveys

Determination	Determination criteria
Front road present	A series of lots with buildings are adjacent to a road existing in the DRM-DB People can enter the lots from the adjacent road
Front road absent	A series of lots with buildings are not adjacent to a road existing in the DRM-DB There is an adjacent road, but people cannot enter the lots from that road due to the presence of walls, hedges, etc.

(a) (b)

Fig. 3 Examples of buildings without front roads

First, the building polygons were insufficiently expanded. This likely resulted from using the upper limit of the BCR as a parameter α in Sect. 4.2.3, although the actual BCR was often lower.

Second, an error in determination occurred when another building or object was located between the building and its candidate front road.

The methods for improving these issues will be detailed in the following sections.

4.4.1 Examining the Optimal Parameter α

To address the first issue identified in Sect. 4.4, we examined the optimal parameter α. The method for determining the optimal parameter α is described below.

First, the value of α was set with an upper limit of the BCR for the area to which each building belonged and a lower limit of 10%. Next, within this range, we varied the value of α in 10% increments and determined the presence of front roads using our method for each value. Then, the determination results for each α value were compared with the actual presence of front roads, as collected through field and visual surveys. Finally, the α value that yielded the highest accuracy was selected as the optimal parameter α.

As an indicator of accuracy, we defined the Accuracy Rate (AR) as the proportion of cases in which our method correctly determined the presence of front roads compared with the actual situation. The AR was calculated using Eq. (3).

$$AR_i = \frac{ac_i}{bc_i} \tag{3}$$

where *ac* is the number of correctly determined cases of the presence of front roads, *bc* is the total number of buildings, and *i* represents any given area.

4.4.2 Addition of Criteria for Determining the Presence of Front Roads

To address the second issue identified in Sect. 4.4, we first extended the Extension Direction Vector from each building and created a new vector (hereinafter called "Intersection Determination Vector") that ends at the road identified as a candidate for the front road. If this Intersection Determination Vector intersected another building, we determined that the building had no front road (Fig. 4).

The final process for determining the presence of front roads, including the improvement of the method, is shown in Fig. 5.

5 Results

5.1 Results of Examining the Optimal Parameter α

Table 4 lists the results of identifying front roads and calculating ARs for the areas described in Sect. 4.3 using the method described in Sect. 4.2.

First, the method identified the presence of front roads with a maximum accuracy exceeding 85% in almost all areas. In addition, more than half of these identifications achieved a high accuracy (over 90%).

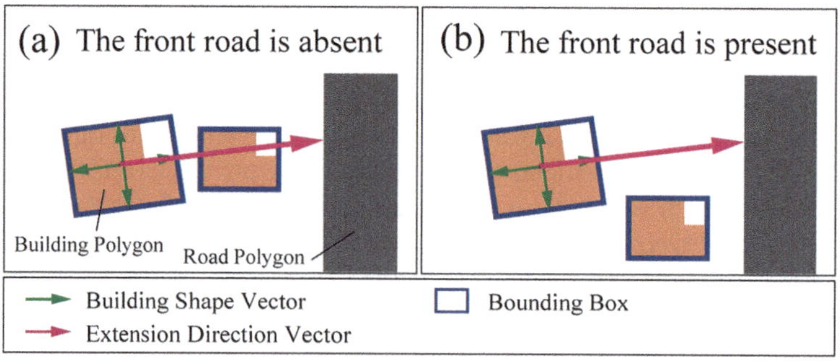

Fig. 4 New determination using the intersection determination vector

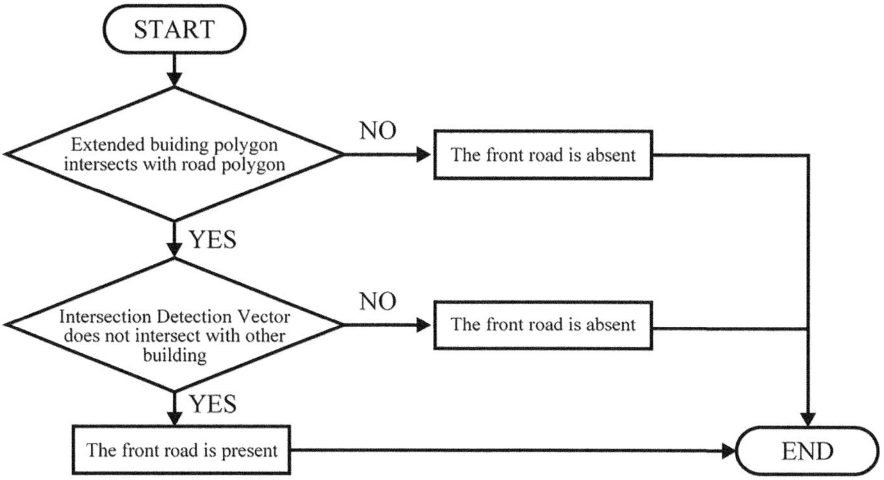

Fig. 5 Final flow for determining the presence of front roads

Observing the average AR for each α, the highest value was 91.18% at $\alpha = 40\%$, followed by 90.49% at $\alpha = 50\%$. The third highest value was 84.74% at $\alpha = 30\%$; however, this represented a relatively large decrease compared to $\alpha = 40\%$ or 50%. The minimum value was 84.24% at $\alpha = 50\%$, followed by 81.88% at $\alpha = 40\%$. These results suggest that α values of 40% or 50% are the most suitable for accurately identifying the presence of front roads using this method.

5.2 *Final Determination Results Using the Intersection Determination Vector*

Table 5 presents the results of identifying the front roads and calculating the ARs using the method described in Sect. 4.2.2.

First, adding the determination using the intersection determination vector enabled the identification of front roads with a maximum accuracy exceeding 90% in almost all areas. In addition, approximately half of the identifications achieved an extremely high accuracy (over 95%).

Observing the average AR for each α, the highest value was 93.41% at $\alpha = 40\%$, followed by 92.83% at $\alpha = 30\%$. For the minimum values, similar to the averages, the highest value was 87.50% at $\alpha = 40\%$, followed by 84.24% at $\alpha = 30\%$.

These results indicate that $\alpha = 40\%$ may be the optimal value for determining the presence of front roads.

Table 4 Accuracy rate with the pre-improvement method

Land use zone (BCR)		Parameter α [%]							
		10	20	30	40	50	60	70	80
ELR1	A.	66.23	70.13	81.17	90.26	–	–	–	–
(40)	B.	52.17	60.14	73.91	81.88	–	–	–	–
ELR2	C.	100.00	100.00	100.00	97.37	95.39	–	–	–
(50)	D.	99.39	99.39	99.39	99.39	87.20	–	–	–
MOR1	E.	59.66	67.23	80.67	94.12	89.92	–	–	–
(50)	F.	34.25	56.85	76.03	86.99	93.84	–	–	–
MOR1	G.	100.00	100.00	100.00	100.00	97.39	69.28	–	–
(60)	H.	40.18	64.29	83.04	94.64	99.11	94.64	–	–
MOR2	I.	59.33	65.33	84.67	91.33	94.67	90.67	–	–
(60)	J.	70.75	71.43	82.31	90.48	86.39	77.55	–	–
R1	K.	64.10	72.44	83.33	92.95	93.59	87.82	–	–
(60)	L.	53.38	72.97	85.81	87.84	84.46	83.11	–	–
R2	M.	67.16	71.64	83.58	88.06	87.31	75.37	–	–
(60)	N.	68.85	77.05	82.79	90.16	89.34	72.95	–	–
QR	O.	61.59	65.94	84.06	87.68	86.23	76.81	–	–
(60)	P.	64.10	66.67	76.92	88.03	88.89	75.21	–	–
NCZ	Q.	71.43	74.79	83.19	86.55	87.39	80.67	72.27	48.74
(80)	R.	75.51	75.51	79.59	86.73	88.78	79.59	62.24	43.88
CZ	S.	95.15	95.76	96.97	98.18	94.55	81.21	61.21	40.00
(80)	T.	65.83	74.17	85.83	95.00	97.50	94.17	92.50	70.83
QIZ	U.	65.38	66.92	76.92	84.62	85.38	76.92	–	–
(60)	V.	63.25	70.48	81.33	90.36	90.36	81.33	–	–
IZ	W.	35.77	52.03	78.86	89.43	86.99	81.30	–	–
(60)	X.	100.00	99.49	98.97	98.97	95.38	77.95	–	–
ND	Y.	48.91	65.76	80.98	88.04	84.24	–	–	–
(50)	Z.	66.86	73.37	82.84	91.72	87.57	–	–	–
Mean		67.28	74.22	84.74	91.18	90.49	80.92	72.06	50.86
Median		65.61	71.54	82.94	90.31	89.12	80.13	67.26	46.31
Maximum		100.00	100.00	100.00	100.00	99.11	94.64	92.50	70.83
Minimum		34.25	52.03	73.91	81.88	84.24	69.28	61.21	40.00

6 Conclusion and Future Works

In this study, we developed a method to identify the presence of front roads (roads in close proximity to buildings that ensure essential daily traffic functions and safety features, such as evacuation and emergency response during disasters) for each building using road link data and building polygon data. Furthermore, by refining the method and comparing its results with those from field and visual surveys, we achieved a highly accurate determination of the actual road adjacency conditions. Our approach offers a straightforward and practical solution for accurately

Table 5 Accuracy rate including determination with intersection determination vector

Land Use Zone (BCR)		Parameter α [%] 10	20	30	40	50	60	70	80
ELR1	A.	85.71	86.36	90.91	91.56	–	–	–	–
(40)	B.	81.16	84.06	86.23	87.68	–	–	–	–
ELR2	C.	99.34	99.34	99.34	96.71	95.39	–	–	–
(50)	D.	99.39	99.39	99.39	99.39	87.20	–	–	–
MOR1	E.	88.24	89.92	94.12	96.64	89.92	–	–	–
(50)	F.	76.03	80.82	86.99	90.41	93.15	–	–	–
MOR1	G.	100.00	100.00	100.00	100.00	97.39	69.28	–	–
(60)	H.	89.29	91.96	93.75	96.43	99.11	94.64	–	–
MOR2	I.	94.00	94.67	95.33	96.00	94.67	90.67	–	–
(60)	J.	93.88	94.56	95.92	93.88	87.76	77.55	–	–
R1	K.	91.03	91.03	92.95	92.95	92.95	87.82	–	–
(60)	L.	83.78	89.19	92.57	91.22	86.49	84.46	–	–
R2	M.	88.81	90.30	93.28	91.79	88.81	75.37	–	–
(60)	N.	88.52	90.98	90.98	89.34	88.52	71.31	–	–
QR	O.	87.68	88.41	92.75	89.13	85.51	76.81	–	–
(60)	P.	78.63	80.34	84.62	89.74	88.03	75.21	–	–
NCZ	Q.	91.60	92.44	92.44	90.76	87.39	80.67	72.27	48.74
(80)	R.	92.86	91.84	89.80	90.82	89.80	80.61	62.24	43.88
CZ	S.	96.36	96.97	97.58	96.36	93.33	80.61	61.21	40.00
(80)	T.	87.50	90.00	93.33	98.33	98.33	94.17	92.50	70.83
QIZ	U.	86.92	87.69	90.00	90.00	85.38	76.92	–	–
(60)	V.	90.36	92.17	95.78	96.39	90.36	81.33	–	–
IZ	W.	82.93	84.55	91.87	94.31	88.62	82.11	–	–
(60)	X.	100.00	99.49	98.97	98.97	95.38	77.95	–	–
ND	Y.	69.02	78.26	84.24	87.50	83.15	–	–	–
(50)	Z.	84.62	86.39	90.53	92.31	87.57	–	–	–
Mean		88.76	90.43	92.83	93.41	90.59	80.97	72.06	50.86
Median		88.67	90.64	92.85	92.63	89.30	80.61	67.26	46.31
Maximum		100.00	100.00	100.00	100.00	99.11	94.64	92.50	70.83
Minimum		69.02	78.26	84.24	87.50	83.15	69.28	61.21	40.00

assessing the proximity of buildings to roads without incurring the high computational costs associated with previous methods.

Based on this study, we suggest the following directions for future research: First, it must be applied in other municipalities to clarify its applicability and challenges. In this study, the method was applied to approximately 3700 buildings in Takasaki City, which significantly increased the sample size compared with that of previous studies. However, it remains unclear whether this method can be applied to all municipalities in Japan. Applying this method to municipalities with diverse geographical conditions will help identify areas where application is difficult and understand the reasons, allowing for further improvements to the method. Next, it is necessary to resolve the misidentifications in the irregular cases described in Sect. 4.3.2. These issues can be addressed by adding further intersection checks using river data, which are available as open data, and may help prevent such misidentifications. By addressing these challenges, we aim to further improve accuracy and promote wider deployment.

The identification of the presence of a front road for each building using this method is expected to contribute to various fields of urban planning. For instance, conventional studies on evacuation during disasters, as previously introduced, often assume that all buildings are accessible by some means, and they fail to consider the accessibility between buildings and roads. However, as indicated in this study, not all buildings necessarily have a front road. Including factors such as the presence of front roads and their widths in evacuation analyses enables a more realistic assessment of the likelihood of evacuation and rescue difficulties during disasters. This perspective also applies to the transportation sector. For instance, in areas where existing public transportation systems are difficult to use, on-demand transportation helps ensure mobility for individuals, such as the elderly, who may have trouble driving. However, some buildings are expected to be inaccessible to vehicles because they lack a front road, which may prevent on-demand transportation. If the presence of front roads for each building can be confirmed, it will become possible to identify buildings that are difficult to access even with the introduction of on-demand transportation, thereby enabling the identification of individuals or areas that truly experience significant challenges in their daily transportation.

In addition, as mentioned earlier, issues related to narrow roads and road adjacency conditions are common challenges worldwide. Thus, it is important to expand the method developed in this study to other countries. To achieve this, the first step would be to examine its applicability in countries with similar datasets, building on the knowledge accumulated in Japan. It is essential to clarify how differences in geographical conditions and legal regulations influence the effectiveness of this method. At the same time, international expansion faces challenges owing to variations in the accuracy and availability of data across countries. Therefore, evaluating the state of data preparation for each region is critical. In addition, this method will likely require improvements and adjustments to address local conditions and challenges. In particular, Asian and African countries, where urbanization has been characterized by imbalances in road development and building placement, face growing challenges related to disaster safety and traffic access (Turok & McGranahan, 2013). The method proposed in this study can significantly contribute to urban planning and disaster response in these regions.

Acknowledgments This study was supported by a grant from the Japan Digital Road Map Association (DRM Association) in 2024. The DRM-DB used in this study was provided by the DRM Association. We express our sincere appreciation for their contribution to this study.

References

Abdulla, H. M., Ibrahim, M. A., & Al-Hinkawi, W. S. (2023). The impact of urban street network on land value: Correlate syntactical premises to the land price. *Buildings, 13*(7). https://doi.org/10.3390/buildings13071610

Gold CM (1989) Spatial adjacency—A general approach. Proceedings of the Auto-Carto, pp. 298–312.

Hamaina, R., Leduc, T., & Moreau, G. (2013). A new method to characterize density adapted to a coarse city model. In *Information Fusion and Geographic Information Systems (IF&GIS 2013)* (pp. 249–263). Springer. https://doi.org/10.1007/978-3-642-31833-7

Ichikawa, F., Sakata, T., & Yoshikawa, T. (2004). An analysis of accessibility to spaces for disaster refuge considering danger of street-blockades caused by collapse of buildings on evacuation routes. *Theory and Applications of GIS, 12*(1), 47–56. https://doi.org/10.5638/thagis.12.47

Jiatian, L., Xiaoqing, Z., Min, S., & Chunxiao, W. (2009). Describing and calculating of geometry adjacency relation with Voronoi tessellation. In *2009 Second International Symposium on Intelligent Information Technology and Security Informatics (IITSI)* (pp. 47–50). IEEE. https://doi.org/10.1109/IITSI.2009.17

Kumar, A., & Pushplata. (2015). Approach to formulate setback regulations for Indian hill towns. *International Journal of Sustainable Built Environment, 4*(1), 91–99. https://doi.org/10.1016/j.ijsbe.2015.03.001

Minai, N., Murota, A., & Fujisawa, M. (2016). Living environment of flag-shaped housing lot. *Journal of the Housing Research Foundation "Jusoken", 42*, 73–84. https://doi.org/10.20803/jusokenronbun.42.0_73

Osaragi, T., & Oki, T. (2012). Difficulty of evacuation in densely built-up area in the aftermath of a large-scale earthquake. *Journal of Architecture and Planning (Transactions of AIJ)., 77*(681), 2561–2567. https://doi.org/10.3130/aija.77.2561

Raji, A. U., & Attah, U. A. (2017). Enforcing building set backs as a viable strategy for an emerging city. Path of. *Science, 3*(6), 2.1–2.7. https://doi.org/10.22178/pos.23-2

Shaikh, S., Talpur, M. A. H., Baig, F., & Ogahi, M. A. (2019). Setback violation and residents' perception about living conditions in vertical buildings of Qasimabad, Hyderabad. *Journal of Architecture and Built Environment, 2*(1), 1–15. https://doi.org/10.32350/jaabe.21.01

Souza, M. T. (2009). *The effect of land use regulation on housing price and informality: A model applied to Curitiba, Brazil.* Lincoln Institute of Land Policy Working Paper No. WP09MS1. https://doi.org/10.2139/ssrn.1762658

Suppasri, A., Kitamura, M., Alexander, D., Seto, S., & Imamura, F. (2024). The 2024 Noto peninsula earthquake: Preliminary observations and lessons to be learned. *International Journal of Disaster Risk Reduction, 110*. https://doi.org/10.1016/j.ijdrr.2024.104611

Turok, I., & McGranahan, G. (2013). Urbanization and economic growth: The arguments and evidence for Africa and Asia. *Environment and Urbanization, 25*(2), 465–482. https://doi.org/10.1177/0956247813490908

Usui, H., & Asami, Y. (2010a). Comparison of accuracy of methods of judging adjacency of buildings to roads. *Theory and Applications of GIS, 18*(2), 161–170. https://doi.org/10.5638/thagis.18.161

Usui, H., & Asami, Y. (2010b). Method for determining buffer distance for judging adjacency of lots to roads. *Journal of Architecture and Planning (Transactions of AIJ), 75*(651), 1175–1180. https://doi.org/10.3130/AIJA.75.1175

Yoshihara, T., & Tanaka, T. (2020). Locational and physical categorizing of densely built-up areas. *Journal of the City Planning Institute of Japan, 55*(3), 274–281. https://doi.org/10.11361/journalcpij.55.274

Neighborhoods and Communities

Gentrification from the Sky: Using Remote Sensing and Machine Learning for Urban Change Detection

Javier Alfaro, Sanja Šćepanović, Stephen Law, and Daniele Quercia

Abstract Gentrification is an urban phenomenon marked by socioeconomic shifts that can displace long-term residents and increase inequality. Accurate measurement is essential for effective urban planning and equitable development. Traditional reliance on census data is costly, slow, and lacks the spatial and temporal resolution needed to detect neighborhood-level changes in real time. This study addresses these challenges by combining open satellite imagery with machine learning techniques to quantify gentrification more effectively. By analyzing high-resolution imagery, we detect physical changes—such as shifts in building density, rooftop materials, and green spaces—that are linked to gentrification but often overlooked by census-based approaches. Using the Greater London Area as a case study, our method improves measurement accuracy by up to 8%, achieving a balanced accuracy of 77% across 4,085 neighborhoods. Even a small improvement in accuracy can enable better identification of at-risk neighborhoods, helping policymakers intervene before displacement pressures become irreversible.

Keywords Open data · Satellite Imagery · Gentrification · Machine Learning

---

J. Alfaro (✉)
London School of Economics and Political Science, London, UK
e-mail: j.a.alfaro-garcia@lse.ac.uk

S. Šćepanović
Nokia Bell Labs, Cambridge, UK

University of Oxford, Oxford, UK
e-mail: sanja.scepanovic@nokia-bell-labs.com

S. Law
University College London, London, UK
e-mail: stephen.law@ucl.ac.uk

D. Quercia
Nokia Bell Labs, Cambridge, UK

King's College London, London, UK

Politecnico di Torino, Turin, Italy
e-mail: quercia@cantab.net

R. Goodspeed et al. (eds.), *Digital-Era Urban Transformations*, The Urban Book Series, https://doi.org/10.1007/978-3-031-98300-9_12

1 Introduction

In today's rapidly changing urban landscapes, the dynamics of gentrification have become increasingly critical to understand. Gentrification is characterized by socio-economic shifts that often lead to the displacement of long-term residents, alteration of neighborhood character, and rising social inequality. These changes pose significant challenges for urban planners, policymakers, and local communities. As cities evolve, accurately measuring gentrification is essential to mitigate its negative impacts and promote equitable development. Urban planning and policy decisions rely on timely and accurate data to understand these shifts, and without effective methods of quantification, interventions may fall short of addressing the needs of affected communities.

Traditionally, gentrification has been measured through census data, which focuses on socioeconomic indicators such as income, education, and housing characteristics. However, relying solely on census data presents several limitations. First, the process of collecting and analyzing census data is both time-consuming and costly, often resulting in delays that make it difficult to track the rapid evolution of neighbourhoods. Additionally, census data lacks the spatial and temporal resolution necessary to detect fine-grained changes at the neighbourhood level. As gentrification is often a gradual and localized process, subtle transformations—such as changes in building density, rooftop materials, urban greenery, or new infrastructure developments—can go unnoticed. These indicators, which reflect the physical aspects of urban change, are critical for a complete understanding of how neighbourhoods evolve over time. As a result, conventional methods are ill-suited for capturing the full picture of gentrification as it unfolds in real-time.

To address these challenges, this study introduces a novel approach that leverages open satellite imagery data and machine learning techniques to more effectively quantify the change of physical gentrification. By analyzing high-resolution satellite images, our method captures the physical changes in the urban environment that traditional datasets often overlook. These changes include modifications in building structures, rooftop materials, and green spaces, providing a more detailed and immediate picture of neighborhood transformations. Through the integration of satellite-derived features and machine learning models, we improve the accuracy of gentrification measurement by 8% over the baseline model, achieving a balanced accuracy of 77%. While this represents a measurable improvement, it also highlights the potential of linking physical changes observed in satellite imagery to socio-economic shifts, offering a complementary approach for studying gentrification dynamics.

The combination of physical and socio-economic data not only enhances the granularity and timeliness of gentrification measurement but also provides actionable insights for urban planners and policymakers. By applying this method to specific case studies within planning policy areas, this study demonstrates how satellite imagery and machine learning can contribute to a more equitable approach to managing urban development. We conclude with a discussion of the potential applications of this model, particularly in shaping policy and fostering community-driven interventions in rapidly gentrifying areas.

2 Literature Review and Related Work

2.1 Gentrification in London

First introduced by Glass (1960), the term *gentrification* describes the social and physical transformation of economically disadvantaged neighborhoods due to the influx of wealthier households. This process is marked by socioeconomic shifts, such as changes in demographic composition, alongside visible physical upgrades in local infrastructure, both of which signal rising economic status at the neighborhood level.

British studies on gentrification have traditionally favoured qualitative approaches—media analysis, interviews, and ethnography—over quantitative methods (Watt, 2008). This is partly due to concerns that secondary data may overlook the underlying class dynamics driving urban change. However, studies like Barton (2016), which used bivariate analysis, have uncovered gentrification in areas not typically recognized by qualitative assessments, suggesting that a broader, data-driven perspective can reveal more widespread transformations.

In London, research has often focused on emblematic neighbourhoods like Hackney, Hoxton, Stratford, and Brixton, where the social and cultural markers of gentrification are most visible (Harris, 2012; Butler et al., 2013; Benson and Jackson, 2017). This focus on well-known areas can obscure the subtler, yet significant changes occurring across the city. To gain a more comprehensive understanding of gentrification, it is essential to apply quantitative and multivariate analysis to a wider array of neighborhoods, identifying potential "gentrification frontiers".

This study employs machine learning (ML) techniques to analyze urban change across 4,835 neighbourhoods in London from 2011 to 2021. By leveraging deep learning to extract latent features from satellite imagery, we aim to uncover hidden aspects of urban dynamics and relate them to gentrification. Integrating these insights with London Planning Policy geospatial data provides a more precise understanding of neighbourhood changes and offers actionable guidance for urban planning and policy decisions.

2.2 Machine Learning for Gentrification

Machine learning (ML) has been increasingly applied to predict urban changes, including gentrification. One of the most notable studies by Reades et al. (2019) used a Random Forest (RF) model to predict gentrification in London, combining socio-economic indicators into a gentrification index using Principal Component Analysis (PCA). While this multidimensional approach helped capture the complexity of gentrification, it introduced challenges such as computational intensity and potential statistical errors.

Building on these advancements, researchers have explored alternative data sources to improve ML models. Jain et al. (2021) showed that Airbnb data, including listings and user reviews, can track neighborhood changes in real time. Similarly, Glaeser et al. (2018) used Yelp data on business activity to provide timely insights into urban transitions. Both studies address census limitations, offering real-time, and more granular information on gentrification.

Several studies have built upon this, exploring the advantages of ML over traditional regression models. Algorithms like RF and Gradient Boosted Machines (GBM) excel in their predictive accuracy and in prioritizing the most important features. For instance, Thackway et al. (2023a) used a wide range of census features and property data, including housing prices, Airbnb density, and proximity to points of interest like food and leisure establishments, to predict gentrification. These models offer flexibility in accounting for the varying predictive power of neighborhood characteristics across different contexts, such as proximity to transit or parks (Wardrip, 2011).

However, existing ML approaches still rely heavily on traditional demographic and socioeconomic data, similar to earlier studies, which often depend on infrequent tabular census data. These approaches overlook the potential of remote sensing, where open satellite data, available at weekly frequencies and continually improving in resolution, offer a more timely and detailed alternative for monitoring urban changes.

2.3 Urban Change Detection with Deep Learning

A key advancement in forecasting gentrification lies in the use of deep learning to analyze the built environment, often termed *deep mapping* or *machine mapping* (Ilic et al., 2019; Salesses et al., 2013). This approach, leveraging satellite, aerial, or Google Street View (GSV) imagery, automates the collection of detailed geospatial metrics, surpassing traditional land-use classification methods. By capturing subtle landscape changes, deep learning helps illuminate the evolving urban fabric crucial to understanding gentrification.

Computer vision techniques such as Convolutional Neural Networks (CNNs) are one of the core methodologies, which automatically learns image features—ranging from edges and textures to complex patterns—without predefined rules. This makes CNNs particularly powerful for analyzing urban imagery, as they can discern structural changes indicative of neighborhood transformations (Kang et al., 2018).

Recent studies, such as Ilic et al. (2019) and Thackway et al. (2023b), employed Siamese Convolutional Neural Networks (SCNNs) to detect gentrification by analyzing sequential GSV images. Ilic et al. (2019)'s study in Canada identified structural improvements in buildings over time, using these changes as proxies for gentrification. However, they noted a limitation: the absence of sociodemographic data could lead to overestimations, as not all property upgrades signify gentrification. Thackway et al. (2023b) extended this work in Sydney by integrating socioeconomic data into their predictions, creating a more robust model that accounted for both physical and demographic shifts.

Deep learning methodologies, particularly those using satellite imagery, have also been applied to urban vitality and housing price predictions (Law et al., 2019; Šćepanović et al., 2021a, b). These techniques excel at extracting nuanced details from high-dimensional data, offering new insights into urban dynamics. Building on a rich body of research in change detection (Bergamasco et al., 2022; Codegoni et al., 2022; Yin et al., 2023; Corley et al., 2024), our approach applies these advancements to enhance the spatial and temporal granularity for gentrification quantification, enhancing urban planning and decision-making through more precise, data-driven insights.

3 Data and Methods

This section describes the main data sources used for this study, the first dataset comprises census data from the UK and is used to profile neighborhoods in London. It comes from the Office for National Statistics (ONS, 2024). The second dataset entails remote-sensing imagery from the European Space Agency Copernicus Sentinel-2 mission (Copernicus Sentinel-2, 2021). A more detailed description of their processing is presented in the following sections.

3.1 Gentrification Score

To predict gentrification, we generate a score that captures changes in neighborhood socioeconomic conditions and create features from satellite data to predict such changes. We define neighborhoods as a Lower Layer Super Output Area (LSOA) in London because these represent granular administrative divisions for which governments provide socioeconomic data.

The LSOA contains between 1,000 and 3,000 inhabitants living in between 400 and 1,200 households: a geography small enough that even modest changes in the makeup of an area should show up, but large enough that the sample size of each is statistically robust and more privacy preserving. Whilst data are available on a finer scale (e.g., Output Areas) and a coarser scale (e.g., wards, Middle Layer Super Output Areas), LSOAs exemplify the characteristics of spatial proximity and social homogeneity which relate to *neighborhood effects* (Reades et al., 2019).

Inspired by Jain et al. (2021), we construct the gentrification score for each neighborhood (LSOA) based on changes in four socioeconomic measures: age,

Table 1 Socioeconomic measures for UK, London LSOAs

Measure	Definition	Data source 2011	Data source 2021
Age	Percent aged between 25	2011 ONS	2021 ONS
	and 34		
Education	Percent NVQ Level 4 or	2011 ONS	2021 ONS
Housing	Above Barriers to Housing and	IMD 2010	IMD 2019[a]
	Services Domain		
Income	Income Deprivation Domain	IMD 2010	IMD 2019

[a]The ONS collects data for the Indices of Multiple Deprivation (IMD) two years prior to release

education, housing, and income Table 1. These measures are grouped into two time periods: 2011 and 2021, then between these, the score is computed. First, socioeconomic measures are aggregated to create a neighborhood index for each LSOA. Values are standardized using percentile ranks, as raw Indices of Multiple Deprivation (IMD) values are not comparable across years (ONS, 2015). The neighborhood index is computed for the time windows:

$$\text{Neighborhood index}_{t1} = 1/4\text{x}\left(\text{age}_{t1}\ \text{education}_{t1}\ \text{housing}_{t1}\ \text{income}_{t1}\right) \tag{1}$$

Lower index values indicate more disadvantaged neighborhoods in terms of older, uneducated, poorer residents, or worse housing access.[1] Disadvantaged neighborhoods are defined as those in the bottom 50th percentile in 2011. Each neighborhood index is standardized using its percentile, and the gentrification score is aligned with literature:

$$\text{Gentrification score}_{t2} = \text{neighborhood index}_{t2} - \text{neighborhood Index}_{t1} \tag{2}$$

Here, t_2 is 2021, and t_1 is 2011. A higher gentrification score reflects a greater influx of younger, educated, wealthier residents, or improved housing access, whereas negative scores indicate disadvantaged neighborhoods with no signs of gentrification. These gentrification scores, derived from official census data, serve as the ground truth for evaluating our models. Although census data are collected at decadal intervals, our approach uses these labels to model gentrification trends, providing insights into long-term urban transformation. The spatial distribution of these scores is shown in Fig. 1.

In this study, we focus only on disadvantaged neighbourhoods because the concept of gentrification is usually discussed primarily in the context of less developed neighbourhoods. Even though affluent spaces may experience some change in

[1]According to the Office for National Statistics (ONS), 'Barriers to Housing and Services' refers to the physical and financial accessibility of housing and local services, captured through two subdomains: (1) 'geographical barriers', relating to the proximity of essential local services, and (2) 'wider barriers', which include factors such as affordability and homelessness.

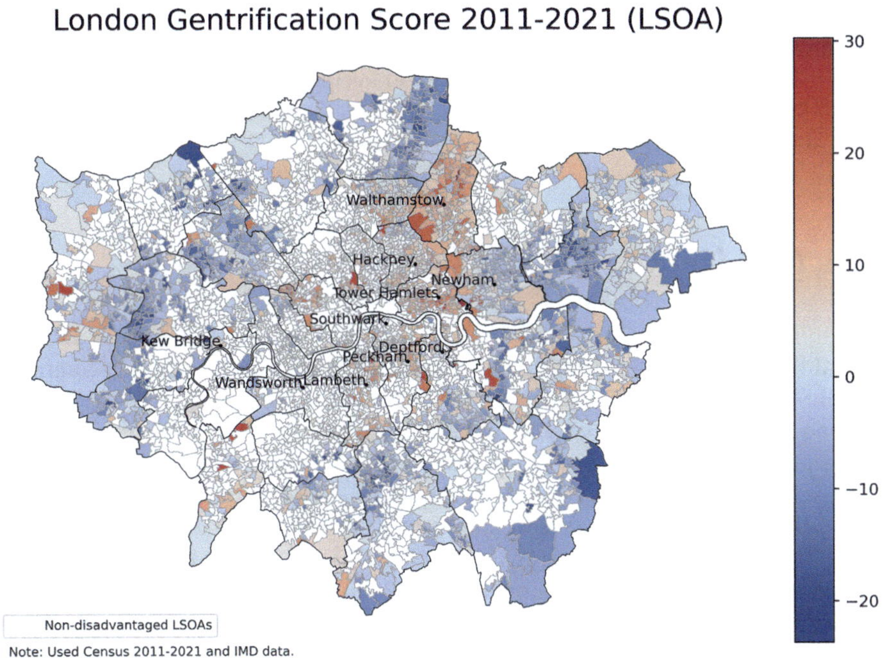

Fig. 1 Spatial distribution of gentrification scores, serving as the ground truth, calculated from changes in census data variables: age, education, housing, and income

socioeconomic measures, this is usually not considered to be gentrification (Zuk et al., 2018).

3.2 *Satellite Imagery*

In 2014, the European Space Agency (ESA) launched the first of the Sentinel satellites as part of the Copernicus Program (Torres et al., 2012), which aims at opening access to Earth Observation (EO) data to the public. Thanks to this investment, we can freely access accurate and timely data from the radar (Sentinel-1, Sentinel-3), altimeter (Sentinel-3), and optical (Sentinel-2) sensors (Šćepanović et al., 2021a, b).

For this study, we processed Sentinel-2 images for the years 2016 and 2021[2] through the (WASDI platform, 2023) by calculating average values during the summer season (June 1st to August 31st), as this period has fewer clouds over London, reducing obstructions in detecting physical changes. We focused on all 11 bands,

[2] The Sentinel-2 satellite was launched in 2015, limiting available data to June 2015 onwards. We focus on 2016 as the starting point, due to incomplete data for the 2015 composite.

capturing relevant environmental factors. To ensure data consistency and quality, we resampled the bands to a uniform resolution of 10 m, applied cloud masks to exclude affected pixels, and computed pixel-per-pixel averages over time under cloud-free conditions (Scepanovic et al., 2023). The resulting composite images represent the typical environmental characteristics for each season.

3.3 Deep Learning Feature Extractors

The full images comprise a dimension of (B, H, W) where B is the number of bands, H and W are the height and width of the image, the raw image size for the composite is (11, 4754, 6050), i.e., 11 bands each 4754×6050, each pixel represents parcels of land of 10×10 m^2. In order to process these images, we resized them to create imagelets of 256×256 pixels. This is also a standard practice in studies that process high resolution (HR) and very high resolution (VHR) satellite imagery (Corley et al., 2024; Fang et al., 2022; Šćepanović et al., 2021a, b; Yin et al., 2023).

To extract meaningful features from the satellite images, we applied a pipeline consisting of several deep learning techniques followed by image thresholding for change detection purposes. Given an image $i \in I$, where I is the set of imagelets, the goal was to extract a vector of visual components $v_{ib} \in V$ where i corresponds to each image, and b refers to each band. Some studies focus only on four bands (Papadomanolaki et al., 2019), but in order to fully leverage the whole dataset, the aforementioned strategy was employed.

In our experiments, we compare 6 methodologies, starting from a basic normalized image difference and then compiling their results. These are described as follows:

Simple-Diff (Sinergise, 2017) standardizes the image in t_2 with standard deviation and mean values from the image in t_1 to avoid detecting changes in brightness and contrast, and then computes the absolute difference between t_1 and the corrected t_2 image. Change is detected by using Multi-Otsu thresholding on this difference (Liao et al., 2001).

Res-Net demonstrates the use of Res-Net blocks (He et al., 2016) to extract deep image features for change detection with bi-temporal remote-sensed imagery. **FC-SiamDiff** (Daudt et al., 2018) is a fully convolutional (FC) Siamese approach, which, instead of concatenating both features from the encoding streams, calculates the absolute value of their difference.

CGNet (Han et al., 2023) is a change-guiding network that improves upon the U-Net architecture by incorporating an attention-based Change Guide Module, designed to capture long-range dependencies across pixels.

Bi-Temporal Siamese Model (Papadomanolaki et al., 2019) uses a simple Siamese U-Net model that first extracts spatial features from multi-date inputs, which are subsequently used for change detection. The LSTM block was not included in this study.

TinyCD (Codegoni et al., 2022) is another Siamese UNet change detection architecture, which uses an EfficientNet (Tan and Le, 2019) encoder backbone to extract convolutional features and feed them to a custom attention-based decoder network.

These methods aim to perform Change Detection (CD) in bitemporal images I_1 and I_2 acquired at times t_1 and t_2, where, in our case, t_1 is 2016 and t_2 is 2021. Among them, Simple Differencing follows a traditional differencing approach, ResNet serves as a feature extractor, and the remaining models operate as end-to-end change detection methods.

3.3.1 Bi-temporal Siamese Network

Assume that a set of N unlabeled samples $X = X_n$, $n = 1, \ldots, N$ extracted from I_1 is available. These methods use X to train from scratch the convolutional autoencoders (CAE) in an unsupervised way. Once the CAE is trained, we process I_1 and I_2, and extract bitemporal deep feature maps of the inputs from the models. These features are compared and fused to detect changed π_c and unchanged π_{nc} pixels, where π_c includes all the relevant changes that occurred in the image, while π_{nc} represents no change. The Siamese approach, which extracts features from bitemporal images, is shown in Fig. 2 as a demonstration of how the aforementioned methods process the data.

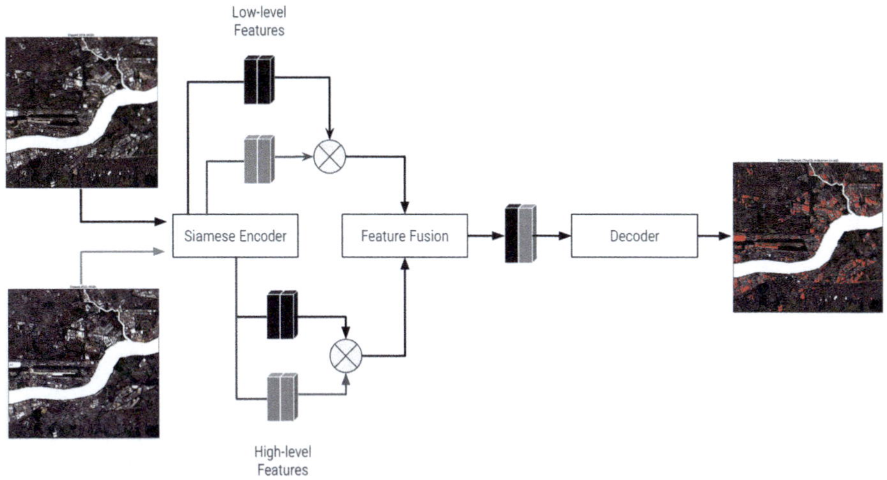

Fig. 2 Bitemporal Siamese Network Architecture for Urban Change Detection. The black and gray blocks represent satellite images captured at different times t_1 and t_2. Both high-level and low-level features are extracted from these images using a Siamese encoder. The feature maps from each image are fused and passed through a decoder, which detects urban changes, highlighted in red on the right. This process is applied independently to each spectral band of the satellite imagery

3.4 Implementation Details

The models are implemented in PyTorch (Paszke et al., 2017) and run on a single NVIDIA A100 GPU. We use the Adam optimizer (Kingma & Ba, 2014) with linear weight decay and a base learning rate of 0.001 for a single epoch across all models, except for the customized Res-Net, which runs for 100 epochs. The batch size is set to 8 for all experiments.[3]

For change detection, we employ a one-shot learning approach with Siamese neural networks, effective in learning object similarity (Nath, 2021). However, this was not applied to SimpleDiff, which relies on traditional differencing, or ResNet, which required additional training steps for meaningful feature extraction. The other models converged in a single training step, justifying the use of one-shot learning. This approach remains relatively unexplored in gentrification studies.

The models are trained using Mean Squared Error (MSE) loss following (Bergamasco et al., 2021, 2022). The goal is to minimize the reconstruction error between the original image and the predicted output over a batch $B \in X$ of patches:

$$MSE = \frac{1}{n}\sum_{n=1}^{B}\left(X_n - X_n'\right)^2 \tag{3}$$

To extract feature maps, the model processes I_1 and I_2 separately. The encoder's initial layers capture simple features like edges and lines, while the decoder layers extract more spatial context about the changes. We apply bilinear interpolation to the retrieved features to ensure the maps have the same spatial dimensions for each patch. The image is then reconstructed by applying a no-data mask from I_1.

The main idea is to associate the label π_c, π_{nc} with each pixel, and to do so at the most reliable level. To that end, we apply a threshold τ retrieved using the (Otsu, 1979) algorithm to the output features φ from each model. The pixel of the final change map CM in position i, j is assigned to π_c or π_{nc} according to the class detected in the pixel position i, j.[4]

$$CM\left(i,j\right) \in \{\pi_{nc}, \text{if}\, \theta\left(i,j\right) \leq \tau_h\, \pi_c, \text{if}\theta\left(i,j\right) > \tau^4{}_h \tag{4}$$

3.5 Changes at the LSOA Level

A raster image is generated for each change map and band. To contextualize this process into our neighbourhoods, we translate these outputs into vector features. To handle large datasets efficiently, this workflow includes raster-to-vector conversion, spatial joining, and parallel processing.

[3] The code implementation can be found in the following GitHub repository: https://github. com/jaroxciv/Gentrification-from-the-Sky.

[4] Scikit-image's implementation of the Multi-Otsu algorithm outputs three classes by default. The subscript h indicates the highest threshold level was selected for the change detection π.

The extracted change pixel counts are then aggregated for each LSOA. The total change counts for each band are normalized by the area of the respective neighborhood to calculate the percentage of change pixels, which provides a measure of the intensity of the change within each area. This results in a matrix Φ_{nb} where each row represents an LSOA neighborhood (n), and each column corresponds to a specific extracted satellite band feature (b), which is used as a predictor of neighborhood gentrification.

4 Modeling

4.1 Setup

We cast this as a binary classification task whose dependent variable is the gentrification score computed in Sect. 3.1.

The first step is the binarisation of the gentrification score based on percentiles used to classify neighborhoods as either gentrified or not gentrified. This process involves setting a threshold at the top and bottom 25th percentiles of the gentrification scores. The resulting binary classification assigns a value of 1 to neighborhoods with gentrification scores above the top 25th percentile, and a value of 0 to areas with scores below the bottom 25th percentile.

Let G_n represent the gentrification score for neighborhood n, $Q_{75}G$ be the 75th percentile (top 25%) of the gentrification scores, and $Q_{25}G$ be the 25th percentile (bottom 25%) of the gentrification scores. The binary gentrification label GB_n for neighborhood n is defined as:

$$GB_n \in \{1, if\ G_n > Q_{75}\left(G\right) 0, if\ G_n < Q_{25}\left(G\right) dropped, if Q_{25}\left(G\right) \leq G_n \leq Q_{75}\left(G\right) \quad (5)$$

The second step is to select a proper baseline model for these binary labels. To do so, we run four independent classification models using housing, population churn,[5] income, and Gini coefficient as predictors (The Centre for Labour and Social Studies, 2021; Suss, 2023). The results show that the Logistic Regression model and LinearSVC provides the best results to predict our binary gentrification target with a value of 0.72 for both balanced accuracy and F1-score. Additionally, for each of the models, since both the house and satellite feature vectors were skewed, we used the Yeo-Johnson power transformation Weisberg (2001), a method previously used to predict gentrification in London (Yee & Dennett, 2022).

[5] The CDRC Residential Mobility Index was used in a study to predict gentrification.

4.2 Classification Methods

Given the produced neighborhood-level feature vectors Φ_{nb} (calculated in sub-Sect. 3.5 and their corresponding labels GB_n (calculated in Equation 5), we aim to predict $\hat{y}$, using a logistic regression of the form:

$$\hat{y}_n^{\alpha} = w_0^{\alpha} + w_1^{\alpha} x_1 + \ldots + w_n^{\alpha} x_n, \quad n = 1, \ldots, N \tag{6}$$

$$\hat{y}_n^{\Omega} = w_0^{\Omega} + \sum\nolimits_{n=1}^{N} w_n^{\Omega} x_n + \sum\nolimits_{n=1}^{N} w_n^{\Omega} \Phi_{nb}, \quad n = 1, \ldots, N \tag{7}$$

where the coefficients w are learned by their respective model, y_n is equivalent to GB_n, and N is the total number of neighborhoods.

In addition to the **Logistic Regression**, which minimizes the following log-loss function:

$$\sum\nolimits_{(x,y)\in D} \left[-y log(\hat{y}) - (1-y)\log(1-\hat{y}) \right]$$

we also experimented with **Linear Support Vector Classification (SVC)**. The Linear SVC objective function is given by

$$\frac{1}{2}\|w\|^2 + C \sum\nolimits_n^N \max\left(0, 1 - y_n \bullet \hat{y}_n\right)$$

where $||w||$ is the norm of the weight vector (reflecting model complexity), and C is a regularization parameter that controls the penalty on misclassifications. We chose to experiment with these two models because Logistic Regression provides a well-calibrated, interpretable baseline for binary classification, while Linear SVC is a more flexible alternative that can optimize decision boundaries more effectively in some cases (Ghosh et al., 2019).

Finally, we also tested an ensemble of decision trees with extreme gradient boosting, equivalent to **XGBoost Classification**, which minimizes the log loss function. On small structured datasets, decision tree-based methods are considered to be the best-performing methods (Chen & Guestrin, 2016; Shwartz-Ziv & Armon, 2022).

5 Results

5.1 Evaluation Metrics

To evaluate the performance of each classification method in predicting our binary gentrification labels (Equation 7), we use standard classification metrics: accuracy (balanced), F1-score (weighted) (Chase Lipton et al., 2014), and ROC-AUC (Hand & Till, 2001). These metrics are calculated based on the counts of true positives tp, false positives *fp*, false negatives *fn*, and true negatives *tn*, typically represented in a confusion matrix. Under these terms, F1 can be expressed as a function of counts of true positives, false positives, and false negatives:

$$F1 = \frac{2tp}{2tp + fp + fn} \tag{8}$$

One last component to include, **specificity** $tn/(tn + fp)$, also known as true negative rate, helps us in defining balanced accuracy in a shorter way:

$$Balanced\ Accuracy = \frac{Sensitivity + Specificity}{2} \tag{9}$$

Balanced accuracy is crucial in this context to mitigate the bias towards the more frequent class (ungentrified areas), ensuring that both gentrified and ungentrified neighborhoods are given equal consideration during evaluation (Brodersen et al., 2010).

5.2 *Gentrification Classifier Results*

We compare the selected models from Sect. 4.2 using our evaluation metrics as a reference. For each model, we perform a Grid-Search ten-fold cross-validation to find the best possible baselines and optimal augmented models. After that, we apply a 30-fold cross-validation 5 times (repeated k-folds), and record the average scores and errors. The objective is to obtain results that are robust against a high variance derived from a low sample size (Šćepanović et al., 2021a, b).

The results are shown in Fig. 3. The first observation is that adding the satellite features yields improved scores in predicting gentrification. For the F1-score, this results in a 9% increase for the best-performing model (XGBoost), and an average improvement of 7% across all models. For ROC-AUC, the average improvement is 8%. Additionally, the scores between the Logistic Regression and the LinearSVC are comparable. These results suggest that a non-linear model is capable of capturing variation better. A plausible reason is that the Support Vector Machine algorithms are more geometrically motivated (Bennett & Bredensteiner, 2000), whilst the decision tree algorithm is better at discriminating the labels.

The table in Fig. 3 shows a numeric comparison of the findings, suggesting that the best model also has the lowest standard error in terms of ROC–AUC.

A heatmap analysis Fig. 4 highlights two key insights: **linear models' consistency** and **tree-based models' specialization**. Linear models show steady, reliable performance across different methods, while tree-based models yield more varied outcomes but offer flexibility through hyperparameter tuning. The result also shows that, among the different change detection methods, *TinyCD* and *Res-Net* appear to achieve the best overall performance (F1 = 0.78).

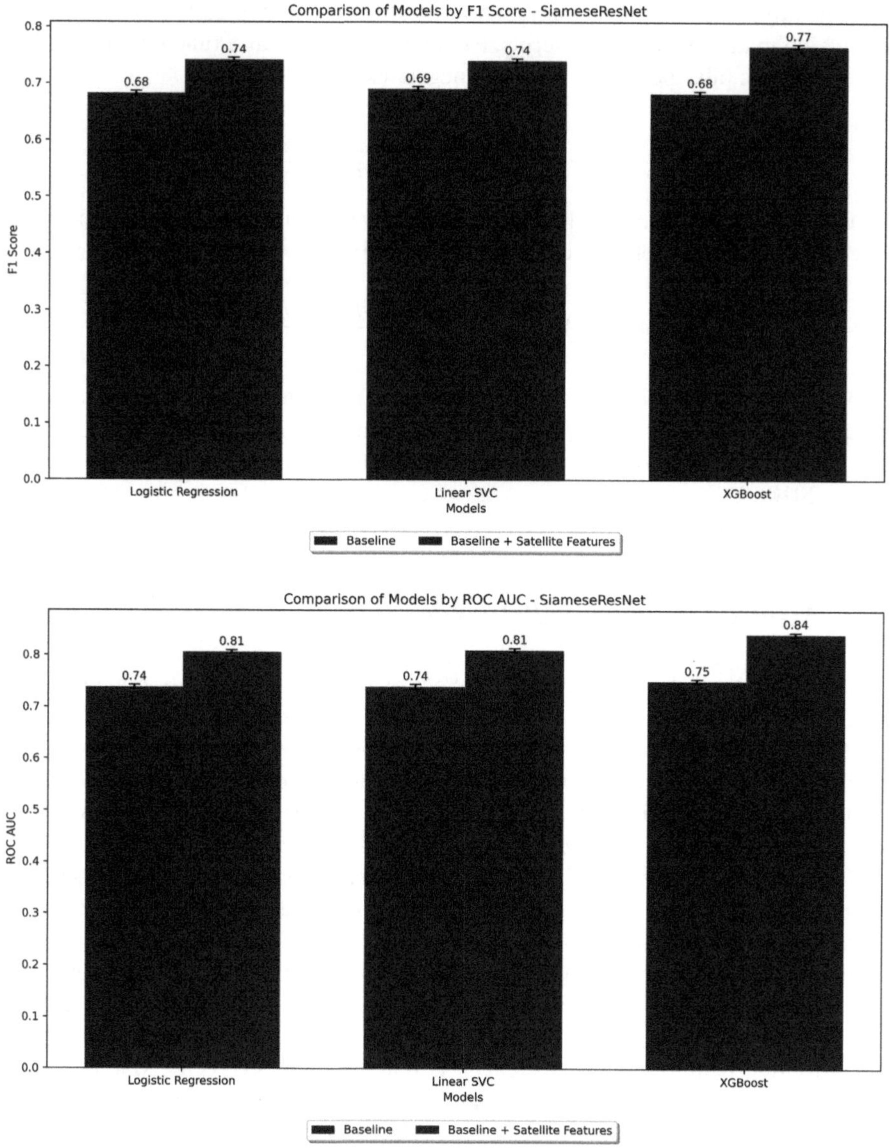

Method	Balanced Accuracy	F1 Score	ROC AUC
Logistic Regression	0.743 ± 0.051	0.742 ± 0.050	0.806 ± 0.047
Linear SVC	0.743 ± 0.050	0.740 ± 0.048	0.810 ± 0.046
XGBoost	**0.767 ± 0.049**	**0.766 ± 0.049**	**0.841 ± 0.043**

Fig. 3 Predicting gentrification from satellite features. The baseline model uses house median predictors, whilst the enhanced model adds an additional term $\sum_{n=1}^{N} w_n^{\Omega} \phi_{nb}$ from Equation 7. Balanced accuracy is omitted since it is equal to the F1-score results. Error bars represent the standard error of the mean across 30 folds

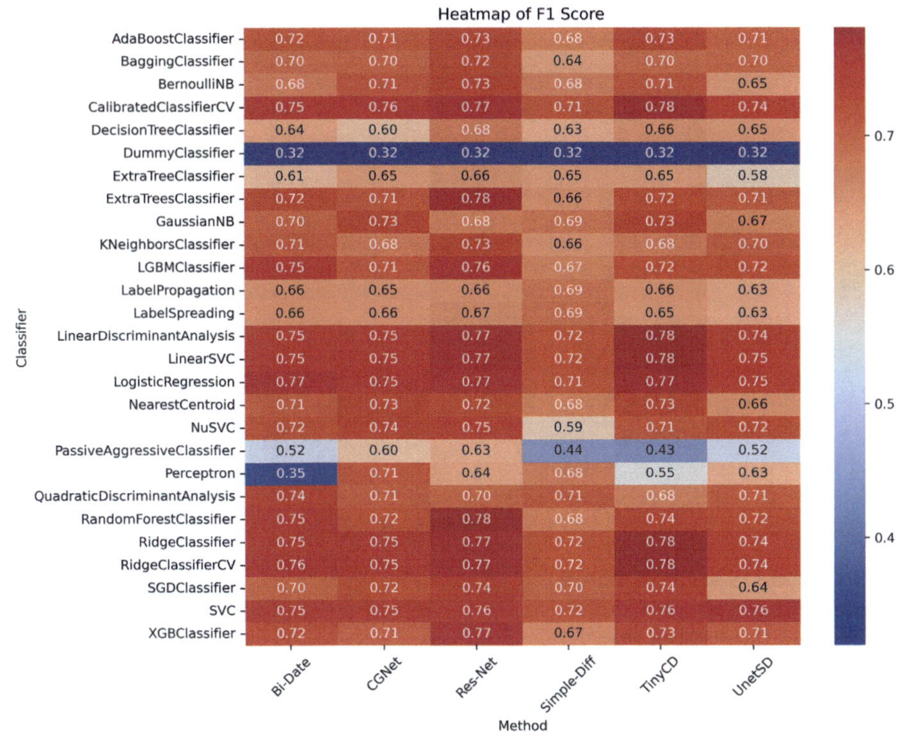

Classifier	Bi-Date	CGNet	Res-Net	Simple-Diff	TinyCD	UnetSD
AdaBoostClassifier	0.72	0.71	0.73	0.68	0.73	0.71
BaggingClassifier	0.70	0.70	0.72	0.64	0.70	0.70
BernoulliNB	0.68	0.71	0.73	0.68	0.71	0.65
CalibratedClassifierCV	0.75	0.76	0.77	0.71	0.78	0.74
DecisionTreeClassifier	0.64	0.60	0.68	0.63	0.66	0.65
DummyClassifier	0.32	0.32	0.32	0.32	0.32	0.32
ExtraTreeClassifier	0.61	0.65	0.66	0.65	0.65	0.58
ExtraTreesClassifier	0.72	0.71	0.78	0.66	0.72	0.71
GaussianNB	0.70	0.73	0.68	0.69	0.73	0.67
KNeighborsClassifier	0.71	0.68	0.73	0.66	0.68	0.70
LGBMClassifier	0.75	0.71	0.76	0.67	0.72	0.72
LabelPropagation	0.66	0.65	0.66	0.69	0.66	0.63
LabelSpreading	0.66	0.66	0.67	0.69	0.65	0.63
LinearDiscriminantAnalysis	0.75	0.75	0.77	0.72	0.78	0.74
LinearSVC	0.75	0.75	0.77	0.72	0.78	0.75
LogisticRegression	0.77	0.75	0.77	0.71	0.77	0.75
NearestCentroid	0.71	0.73	0.72	0.68	0.73	0.66
NuSVC	0.72	0.74	0.75	0.59	0.71	0.72
PassiveAggressiveClassifier	0.52	0.60	0.63	0.44	0.43	0.52
Perceptron	0.35	0.71	0.64	0.68	0.55	0.63
QuadraticDiscriminantAnalysis	0.74	0.71	0.70	0.71	0.68	0.71
RandomForestClassifier	0.75	0.72	0.78	0.68	0.74	0.72
RidgeClassifier	0.75	0.75	0.77	0.72	0.78	0.74
RidgeClassifierCV	0.76	0.75	0.77	0.72	0.78	0.74
SGDClassifier	0.70	0.72	0.74	0.70	0.74	0.64
SVC	0.75	0.75	0.76	0.72	0.76	0.76
XGBClassifier	0.72	0.71	0.77	0.67	0.73	0.71

Fig. 4 The heatmap visualizes model performance, where the darker red indicates higher F1 scores, and the darker blue represents lower scores. Both linear and tree-based methods perform strongly, with tree-based models showing slightly more variation. XGBoost is ultimately chosen due to its high scores and flexibility in hyperparameter tuning, offering potential for further optimization

Additionally, overlaying detected changes on the gentrification score map in Fig. 5 reveals that different bands capture distinct spatial features. For instance, Band 4 highlights urban areas and soil changes, while Band 8 is more sensitive to vegetation and biomass variations. Changes are detected in areas with both high and low gentrification scores, suggesting the models are identifying transformations even in stagnant or highly gentrified areas.

5.3 *Spatial Patterns*

We first examine the evaluation results across neighborhoods, focusing on town centers and the presence of underground stations.

Newham has four main centers; in two of them, the model correctly predicts gentrification, while in the other two, it does not (Fig. 6). The figure clearly shows that the gentrified area has better transport connectivity, with four underground stations, as well as access to both the Overground and the Docklands Light Railway

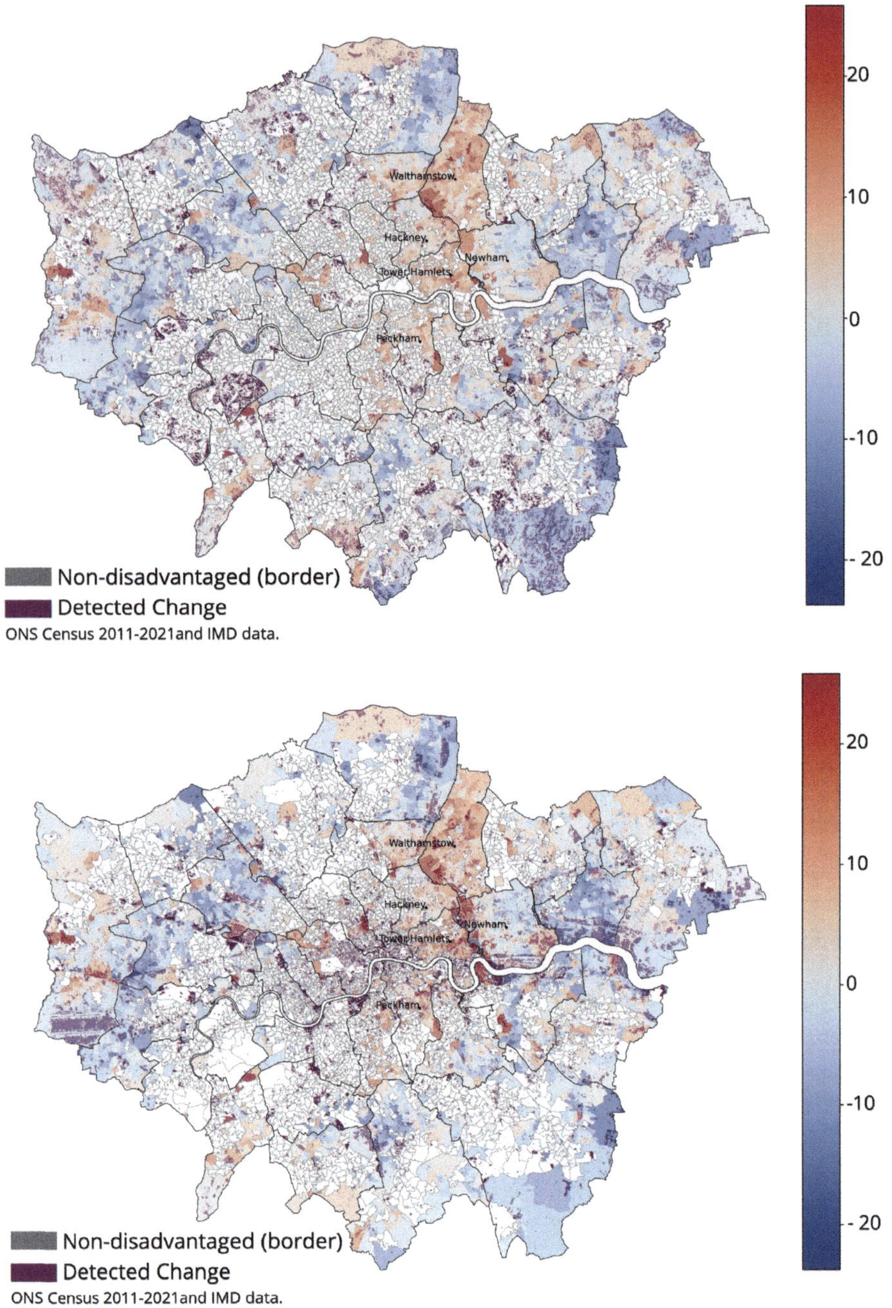

Fig. 5 London Gentrification Score with Detected Changes—LSOA. The top panel shows the detected urban changes from Band 8 (NIR), while the bottom illustrates changes from Band 4 (Red). Purple patches highlight areas of urban change as detected by our deep learning methods. Overlaying them reveals the spatial relationship between physical urban transformation and gentrification levels

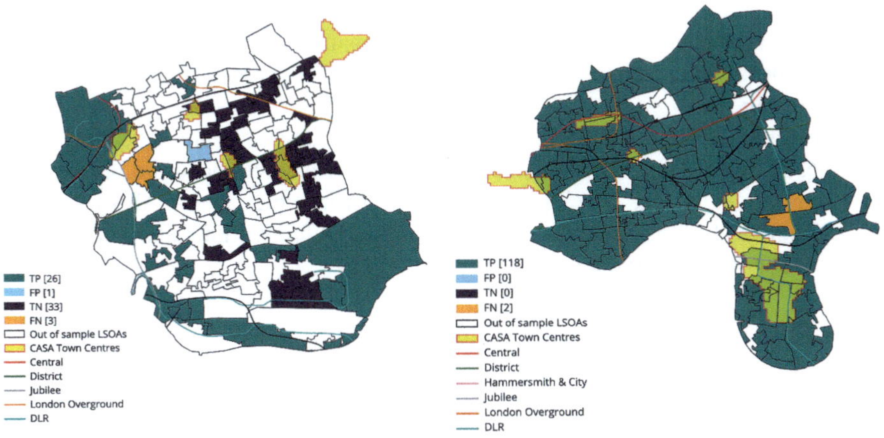

(a) (b)

Fig. 6 Classification metrics (TP, FP, TN, FN) by Borough: Newham and Tower Hamlets

(DLR). In contrast, the center in the false negative area has only a single connection to the District Line. These regions have an average population of 31,528 and relatively high house prices, around £400,000.

The situation is similar in Tower Hamlets, which has a high level of connectivity with five underground stations, as well as access to both the Overground and the Docklands Light Railway (DLR). This borough contains six town centers, with average house prices ranging from £300,000 to £885,000. These centers have a lower average population of 22,358. Our model largely predicts gentrification correctly in this borough, aligning with previous academic studies and gray literature (Reades et al., 2019; The Centre for Labour and Social Studies, 2021).

5.4 Urban Change Detection

The previous maps offer a clear visual representation of the model's performance, showcasing areas where it successfully detects urban changes and areas where its predictions need improvement. However, understanding these changes goes beyond identifying where they are likely to occur—it also involves validating the accuracy and broader implications of these predictions.

The literature on urban change detection often uses well-labeled datasets like the Onera Satellite Change Detection (OSCD) dataset (Daudt et al., 2018), with models trained on them (Chen & Shi, 2020; Ji et al., 2019; Liu et al., 2022). Lacking such data, we validate our model using Conservation Areas—zones with strict regulations limiting physical and structural modifications to the built environment (City of London Corporation, 2024). By comparing satellite-derived change maps from

Equation 4 with Conservation Areas as a proxy for non-change, we can assess the accuracy of the change detection model. False positives may arise within these protected zones, while changes outside likely represent real urban transformations, as shown in Fig. 7.

The overlaid map validates our deep learning approach for detecting urban changes in Tower Hamlets, showing that 99% of the changes occurred outside conservation zones, where development is more permissible. Across all of London, 41,230 pixels (1.72%) of detected changes in Band 4 fell within conservation areas, further supporting the model's accuracy.

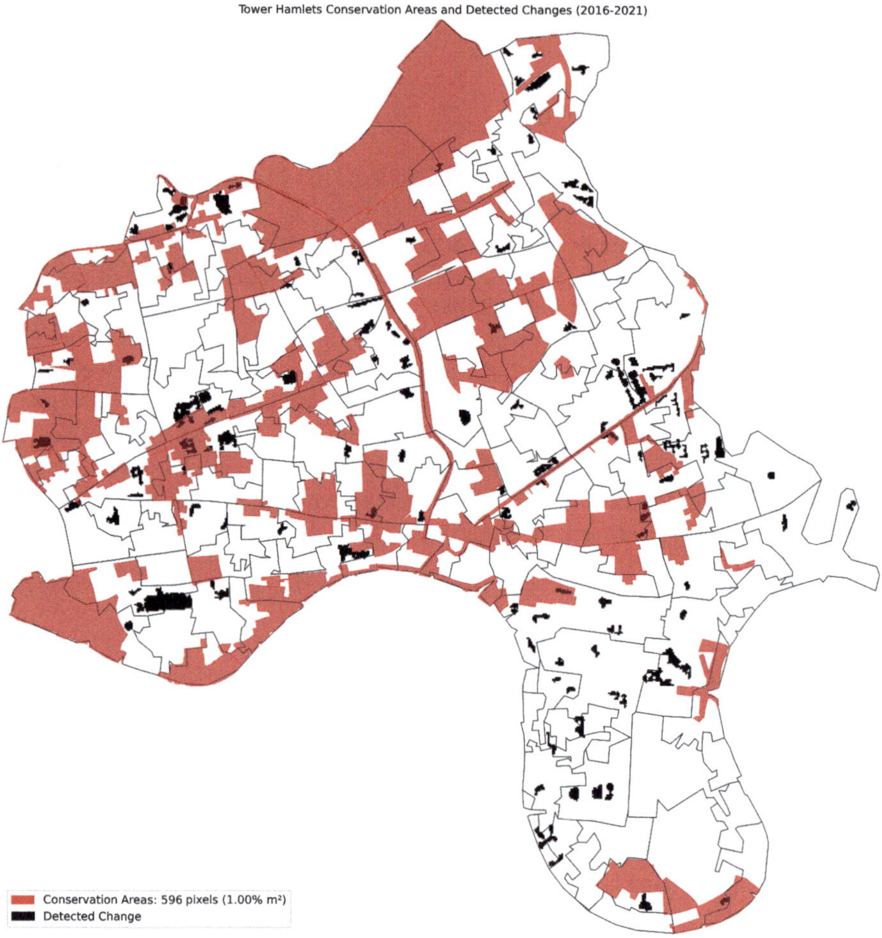

Fig. 7 Changes detected in Tower Hamlets within conservation areas (in red), using Band 4 of the TinyCD algorithm. Detected changes (in black) mostly occur outside these protected zones, where development is more likely. Changes within conservation areas may represent false positives due to strict regulations, while those outside likely reflect genuine urban transformations. This comparison provides insight into how well conservation policies prevent significant developments

Coverage across individual bands varies, from 0.22% (Band 2) to 42.39% (Band 8), with Bands 6 to 9 showing over 35% coverage within conservation zones, indicating sensitivity to certain urban changes. In contrast, Bands 2, 3, and 4 show more selective detection. On average, 19.55% of changes occur within conservation areas, reflecting the variability in how different bands capture urban transformations.

5.5 *Ablation Study*

The methods outlined in 3.3 define change by setting a threshold on the satellite imagery data. The choice of threshold can significantly impact the evaluation metrics as it influences the sensitivity of the model to detect true changes.

To investigate the impact of thresholding, we conducted an ablation study by defining a range of thresholds based on the minimum and maximum values identified for each spectral band. Changes were then computed at multiple points within this range.[6] Finally, we regressed only the satellite features on gentrification, accumulating accuracy and F1 scores at each threshold.

The ablation study reveals distinct curves for different models, suggesting variability not only at the threshold level but also in model specification (Fig. 8). The

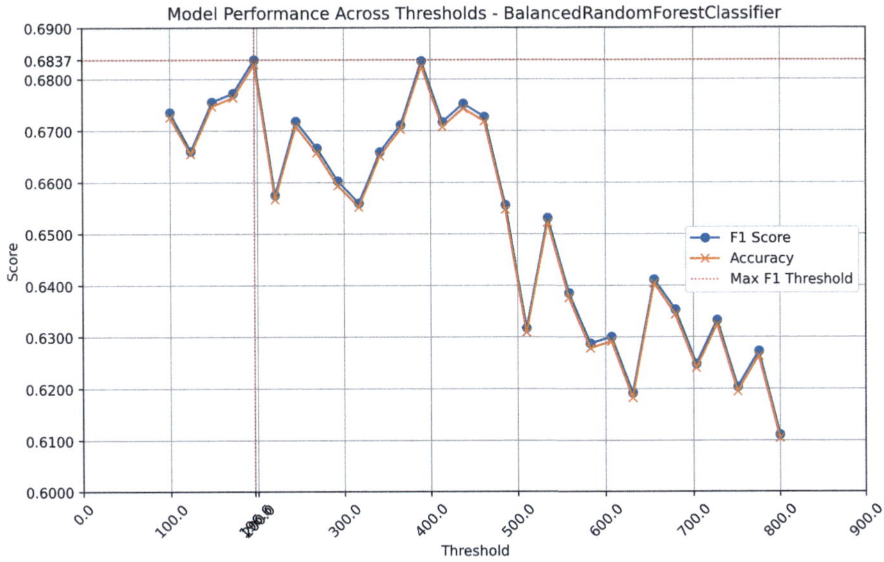

Fig. 8 Performance metrics across thresholds. Predictive model performance is sensitive to the choice of threshold, with an optimal range required to balance accuracy and F1 score effectively

[6]We use the simple-diff approach described in Sect. 3.3, which is both efficient in detecting changes and effectively demonstrates the robustness check.

study highlights two key peaks in performance: one around a threshold value of 200, which temporarily declines, and another around 400, where accuracy and F1 score reach a similar level. However, beyond 400, performance declines significantly. It is important to note that these threshold values may vary depending on the method used, further emphasizing the need for method-specific calibration to maintain predictive accuracy and reliability.

6 Discussion

6.1 *Planning Policy*

The observed spatial patterns in gentrification probabilities reveal associations with factors like transport connectivity and proximity to town centers, suggesting that accessibility and urban density are closely related to areas experiencing gentrification. However, these spatial patterns raise further questions about how policy decisions, infrastructure investments, and urban planning initiatives might correlate with these urban changes.

To explore these relationships, we analyze the UK government's official Planning Policy layers, which provide insights into land use designations, conservation areas, and transport infrastructure. By overlaying these policy layers with the detected urban changes from deep learning techniques, we aim to understand the associations between policy frameworks, infrastructure developments, and socio-economic shifts. Visualizing these interactions, as shown in Fig. 9, offers insights into how planning policies coincide with urban transformations, supporting arguments like those from Meltzer (2016), who noted that gentrifying neighbourhoods tend to experience employment growth and industrial restructuring.

Several key observations arise from analyzing the map. Local Industrial Locations account for the highest percentage of detected changes, with 12.48% of the total area, suggesting significant activity and potential redevelopment in these zones. The London Legacy Development Corporation (LLDC) within Newham follows with 9.90%, likely driven by ongoing large-scale projects, particularly in the context of post-Olympic regeneration (The Centre for Labour and Social Studies, 2021). Strategic Sites and Employment Hubs collectively represent around 17% of the detected changes.

Interestingly, a weaker relationship appears between detected changes and predicted gentrification probability in Local Industrial Locations. These areas, often smaller and concentrated, may lack the space for large-scale redevelopment, which could explain their low aggregated gentrification probability of 0.42 (Hubscher, 2021). The limited room for transformative projects likely constrains significant socio-economic shifts.

In contrast, larger zones like the Legacy Development Corporation LLDC within Newham and Strategic Sites exhibit higher gentrification probabilities, likely due to

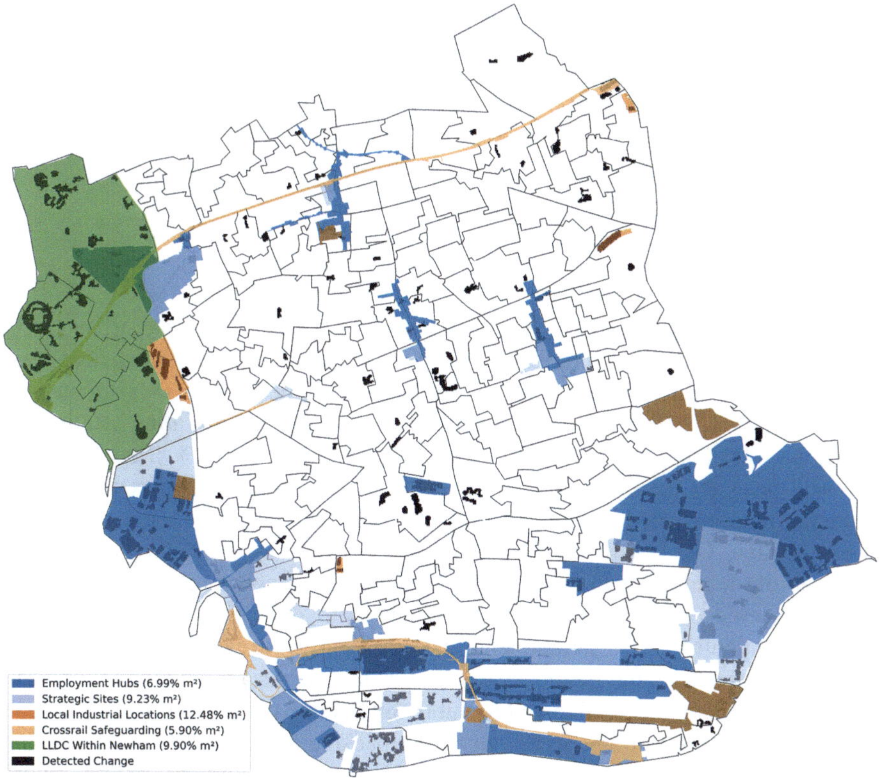

Policy Layer	Avg Probability
LLDC Within Newham	0.707
Strategic Sites	0.554
Employment Hubs	0.502
Crossrail Safeguarding	0.484
Local Industrial Locations	0.417

These probabilities are calculated using the XGBoost model across all neighborhoods.

Fig. 9 Detected changes within policy areas in Newham (top) and average probability of gentrification by policy layer (bottom). The legend shows the percentages of area in square meters corresponding to detected changes in the specified policy areas

their capacity for extensive infrastructure projects (Hubscher, 2021). These larger areas can more easily absorb spillover effects from development, stimulating gentrification across broader regions. This highlights how the scale and concentration of policy areas are associated with both physical changes and their socio-economic impacts.

While these findings offer valuable insights into the interaction between policy areas and detected changes, it is important to approach these results with caution. These results should be interpreted as indicative rather than definitive, highlighting potential areas of change but requiring further validation against ground truths and additional data sources.

7 Conclusion

Based on our analysis and results, we propose a method to evaluate urban change detection and gentrification probabilities using Sentinel-2 imagery and a deep learning framework. Our approach began with a baseline model, having the median house prices for 2011, 2016, and 2021 as predictors, and by incorporating satellite features, we were able to increase the predictive power by up to 8%. These enhanced probabilities were then used to uncover spatial patterns that, when paired with policy planning data, revealed deeper insights into the interplay of economic drivers and urban change. Our results are significant for two reasons. First, they demonstrate that satellite features, both alone (with 64% predictive power) and combined with baseline socioeconomic features like median housing price (77%), enable faster and large-scale gentrification assessments using globally available data. Second, the high temporal and spatial frequency of satellite data allows not only for detecting but also for tracking gentrification and its subtle interactions with the urban landscape.

Larger policy-supported areas, such as expansive policy zones or development sites, tend to have a higher potential for changes and, consequently, higher gentrification probabilities. The spread of new developments within these large zones often leads to spillover effects, where the impact of gentrification extends beyond the immediate vicinity of the changes.

Conversely, smaller, more concentrated development areas, like compact industrial zones or specialized policy areas, may exhibit a high percentage of detected changes due to their focused nature. However, their limited surface area and compactness constrain the scope of redevelopment and the potential for widespread gentrification. As a result, these areas may show lower aggregated probabilities of gentrification despite significant physical changes. This dynamic illustrates how the scale and density of a development zone (region) influence both the extent of urban change detected and the socio-economic outcomes that follow.

7.1 Practical Implications

The findings of this study offer several practical applications that can significantly influence urban planning and policy-making, especially in areas undergoing rapid change or facing potential gentrification pressures.

Enhanced Decision-Making through Probability Maps: The probability maps generated in this study can serve as powerful tools for urban planners and policymakers. By visualizing the likelihood of gentrification across different policy areas, these maps can help prioritize interventions in neighbourhoods at higher risk of socio-economic shifts. This targeted approach allows for more effective allocation of resources, ensuring that areas most in need of support receive timely attention.

Open Public Access Platforms for Community Engagement: Making these findings accessible to the public through open platforms and dashboards can empower communities by providing them with valuable information about their neighbourhoods. Residents can use this data to engage in informed discussions with policymakers, advocate for their needs, and participate more actively in the urban planning process. This transparency fosters a collaborative environment where both community members and decision-makers can work together towards equitable urban development.

Near Real-Time Monitoring of Urban Change: The use of open satellite data to detect changes within policy areas, rather than relying on private platform data such as Airbnb Jain et al. (2021) or Yelp Glaeser et al. (2018), offers an opportunity to develop a near real-time monitoring system for urban environments. By continuously updating the detected changes and probabilities of gentrification, cities can respond more swiftly to emerging trends, adapting policies as needed to mitigate negative impacts or capitalize on positive developments.

Guiding Policy Interventions in Vulnerable Areas: The variability in gentrification probabilities across different policy areas suggests that tailored interventions are necessary. Policymakers can use these insights to design specific strategies that address the unique challenges and opportunities within each area, ensuring that policies are both context-sensitive and effective in promoting sustainable urban growth.

7.2 Limitations and Future Work

Our approach for predicting and validating the gentrification prediction from satellite data comes with three main limitations:

Lack of Ground Truth Labels: Since ground-truth data on gentrification are unavailable, we follow previous work Jain et al. (2021) in assuming that socio-economic changes serve as a reliable proxy. Additionally, we infer the accuracy of our unsupervised gentrification detection method by examining its relationship with conservation areas. However, without explicit validation data, these relationships cannot be definitively confirmed.

Need for Human Annotations: To enhance the robustness of our model, integrating human annotations could be highly beneficial. Expert reviewers could validate a subset of the detected changes, providing ground truth labels that the model can then use for retraining. This process would improve the accuracy and reliability of the model, ensuring it is better at distinguishing between true changes and noise.

Band Overfitting: While we have analyzed all available spectral bands, some such as Band 10 (used for cirrus cloud detection) might lead to "overdetection", potentially introducing noise and bias into the classification. Being wary of this issue can help refine the model's overall performance.

While our approach has demonstrated the potential of using Sentinel-2 imagery and deep learning to analyze urban change and gentrification probabilities, there are several avenues for future work that could further enhance the accuracy and utility of our method. First, improving the deep learning feature extraction process could lead to more precise detection of urban changes. This could be coupled with refined validation techniques that incorporate expert manual labelling, ensuring that the detected changes align closely with on-the-ground realities. Additionally, addressing potential multicollinearity issues within the spectral bands through the use of indices could further refine the models' predictions, making them more robust.

One limitation of our study is the use of medium-resolution imagery (10 m), which, while effective, could be improved with Synthetic Aperture Radar (SAR) data, such as Sentinel-1, which is unaffected by cloud cover, or very high-resolution satellite imagery, providing a more detailed view of urban dynamics. Additionally, integrating explainable AI techniques (Höhl et al., 2024) could enhance transparency in the feature extraction process, offering clearer insights into how specific urban features influence gentrification predictions. These advancements would not only refine the technical capabilities of our model but also create a more precise and reliable tool for urban planners and policymakers.

Acknowledgements The content of this chapter is primarily based on the first author's Masters Thesis, supported by the Consumer Data Research Centre (CDRC) and Nokia Bell Labs. The funding support from these institutions is gratefully acknowledged.

References

Barton, M. (2016). An exploration of the importance of the strategy used to identify gentrification. *Urban Studies, 53*(1), 92–111.

Bennett, K. P. & Bredensteiner, E. J. (2000). Duality and geometry in svm classifiers. In *ICML* (Vol. 2000, pp. 57–64). Citeseer.

Benson, M., & Jackson, E. (2017). Making the middle classes on shifting ground? r esidential status, performativity and middle-class subjectivities in contemporary l ondon. *The British Journal of Sociology, 68*(2), 215–233.

Bergamasco, L., Martinatti, L., Bovolo, F., & Bruzzone, L. (2021). An unsupervised change detection technique based on a super-resolution convolutional autoencoder. In *2021 IEEE International Geoscience and Remote Sensing Symposium IGARSS* (pp. 3337–3340). https://doi.org/10.1109/IGARSS47720.2021.9553859

Bergamasco, L., Saha, S., Bovolo, F., & Bruzzone, L. (2022). Unsupervised change detection using convolutional-autoencoder multiresolution features. *IEEE Transactions on Geoscience and Remote Sensing, 60*, 1–19. https://doi.org/10.1109/TGRS.2022.3140404

Brodersen, K. H., Ong, C. S., Stephan, K. L., & Buhmann, J. M. (2010). The balanced accuracy and its posterior distribution. In *2010 20th International Conference on Pattern Recognition* (pp. 3121–3124). https://doi.org/10.1109/ICPR.2010.764

Butler, T., Hamnett, C., & Ramsden, M. J. (2013). Gentrification, education and exclusionary displacement in e ast l ondon. *International Journal of Urban and Regional Research, 37*(2), 556–575.

The Centre for Labour and Social Studies. Pushed to the margins: A quantitative analysis of gentrification in london in the 2010s. Technical report, The Centre for Labour and Social Studies, United Kingdom, 2021. Retrieved from https://policycommons.net/artifacts/2650208/pushed-to-the-margins/3673081/ on 08 Aug 2024. CID: 20.500.12592/1wdtb3.

Chen, T., & Guestrin, C. (2016). Xgboost: A scalable tree boosting system. In *Proceedings of the 22nd ACM SIGKDD International Conference on Knowledge Discovery and Data Mining, KDD '16* (pp. 785–794). Association for Computing Machinery. ISBN 9781450342322. https://doi.org/10.1145/2939672.2939785

Chen, H., & Shi, Z. (2020). A spatial-temporal attention-based method and a new dataset for remote sensing image change detection. *Remote Sensing, 12*(10). https://doi.org/10.3390/rs12101662. ISSN 2072-4292.

City of London Corporation. (2024). *Conservation areas*. https://www.cityoflondon.gov.uk/services/planning/historic-environment/conservation-areas. Accessed: 2024-08-11.

Codegoni, A., Lombardi, G., & Ferrari, A. (2022. ISSN 0941-0643). Tinycd: a (not so) deep learning model for change detection. *Neural Comput. Appl., 35*(11), 8471–8486. https://doi.org/10.1007/s00521-022-08122-3

Copernicus Sentinel-2. (2021). Msi level-2a boa reflectance product. collection 1. *European Space Agency*. https://sentinels.copernicus.eu/web/sentinel/sentinel-data-access/sentinel-products/sentinel-2-data-products/collection-1-level-2a. Processed by ESA

Isaac Corley, Caleb Robinson, and Anthony Ortiz. A change detection reality check. *arXiv preprint arXiv:2402.06994*, 2024. https://arxiv.org/abs/2402. 06994.

Daudt, R. C., Le Saux, B., & Boulch, A. (2018). Fully convolutional siamese networks for change detection. In *2018 25th IEEE International Conference on Image Processing (ICIP)* (pp. 4063–4067). IEEE.

Fang, S., Li, K., Shao, J., & Li, Z. (2022). Snunet-cd: A densely connected siamese network for change detection of vhr images. *IEEE Geoscience and Remote Sensing Letters, 19*, 1–5. https://doi.org/10.1109/LGRS.2021.3056416

Ghosh, S., Dasgupta, A., & Swetapadma, A. (2019). A study on support vector machine based linear and non-linear pattern classification. In *2019 International Conference on Intelligent Sustainable Systems (ICISS)* (pp. 24–28). https://doi.org/10.1109/ISS1.2019.8908018

Glaeser, E. L., Kim, H., & Luca, M. (2018). Nowcasting gentrification: Using yelp data to quantify neighborhood change. *AEA Papers and Proceedings, 108*, 77–82. https://doi.org/10.1257/pandp.20181034

Han, C., Chen, W., Guo, H., Meiqi, H., Li, J., & Chen, H. (2023). Change guiding network: Incorporating change prior to guide change detection in remote sensing imagery. *IEEE Journal of Selected Topics in Applied Earth Observations and Remote Sensing, 16*, 8395–8407. https://doi.org/10.1109/JSTARS.2023.3310208

Hand, D. J., & Till, R. J. (2001). A simple generalisation of the area under the roc curve for multiple class classification problems. *Machine Learning, 45*, 171–186. https://api.semanticscholar.org/CorpusID:43144161

Harris, A. (2012). Art and gentrification: pursuing the urban pastoral in hoxton, london. *Transactions of the Institute of British Geographers, 37*(2), 226–241.

He, K., Zhang, X., Ren, S., & Sun, J. (2016). Deep residual learning for image recognition. In *Proceedings of the IEEE Conference on Computer Vision and Pattern Recognition (CVPR)* (pp. 770–778). https://doi.org/10.1109/CVPR.2016.90

Höhl, A., Obadic, I., Torres, M. Á. F., Najjar, H., Oliveira, D., Akata, Z., Dengel, A., & Zhu, X. X. (2024). Opening the black-box: A systematic review on explainable ai in remote sensing. *arXiv preprint arXiv:2402.13791.*

Hubscher, M. (2021. ISSN 2071-1050). Megaprojects, gentrification, and tourism. a systematic review on intertwined phenomena. *Sustainability, 13*(22). https://doi.org/10.3390/su132212827

Ilic, L., Sawada, M., & Zarzelli, A. (2019). Deep mapping gentrification in a large canadian city using deep learning and google street view. *PLoS One, 14*(3), 1–21. https://doi.org/10.1371/journal.pone.0212814

Jain, S., Proserpio, D., Quattrone, G., & Quercia, D. (2021). Nowcasting gentrification using airbnb data. *Proceedings of the ACM on Human-Computer Interaction, 5*(CSCW1), 1–21.

Ji, S., Wei, S., & Meng, L. (2019). Fully convolutional networks for multisource building extraction from an open aerial and satellite imagery data set. *IEEE Transactions on Geoscience and Remote Sensing, 57*(1), 574–586. https://doi.org/10.1109/TGRS.2018.2858817

Kang, J., Körner, M., Wang, Y., Taubenböck, H., & Zhu, X. X. (2018). Building instance classification using street view images. *ISPRS Journal of Photogrammetry and Remote Sensing, 145*, 44–59. https://doi.org/10.1016/j.isprsjprs.2018.02.006. ISSN 0924-2716. Deep Learning RS Data.

Diederik P Kingma and Jimmy Ba. Adam: A method for stochastic optimization. *arXiv preprint arXiv:1412.6980*, 2014

Law, S., Paige, B., & Russell, C. (2019. ISSN 2157-6904). Take a look around: Using street view and satellite images to estimate house prices. *ACM Transactions on Intelligent Systems and Technology, 10*(5). https://doi.org/10.1145/3342240

Liao, P.-S., Chen, T.-S., Chung, P.-C., et al. (2001). A fast algorithm for multilevel thresholding. *Journal of Information Science and Engineering, 17*(5), 713–727.

Lipton, Z. C., Elkan, C., & Narayanaswamy, B. (2014). Thresholding classifiers to maximize f1 score. *arXiv e-prints*, arXiv–1402.

Liu, M., Chai, Z., Deng, H., & Liu, R. (2022). A cnn-transformer network with multiscale context aggregation for fine-grained cropland change detection. *IEEE Journal of Selected Topics in Applied Earth Observations and Remote Sensing, 15*, 4297–4306. https://doi.org/10.1109/JSTARS.2022.3177235

Meltzer, R. (2016). Gentrification and small business: Threat or opportunity? *Cityscape, 18*(3), 57–86. ISSN 1936007X. http://www.jstor.org/stable/26328273

Nath, S. (2021). Style change detection using siamese neural networks. In *CLEF (Working Notes)* (pp. 2073–2082).

ONS. (2015). *The english indices of deprivation technical report.* Technical report, Office of National Statistics. https://assets.publishing.service.gov.uk/government/uploads/system/uploads/attachment_data/file/464485/English_Indices_of_Deprivation_2015_-_Technical-Report.pdf. Accessed: 2024-08-07

ONS. *Census statistics*. https://www.ons.gov.uk, 2024. Retrieved January 2024.

Otsu, N. (1979). A threshold selection method from gray-level histograms. *IEEE Transactions on Systems, Man, and Cybernetics, 9*(1), 62–66. https://doi.org/10.1109/TSMC.1979.4310076

Papadomanolaki, M., Verma, S., Vakalopoulou, M., Gupta, S., & Karantzalos, K. (2019). Detecting urban changes with recurrent neural networks from multitemporal sentinel-2 data. In *IGARSS 2019–2019 IEEE International Geoscience and Remote Sensing Symposium* (pp. 214–217). IEEE.

Paszke, A., Gross, S., Chintala, S., Chanan, G., Yang, E., DeVito, Z., Lin, Z., Desmaison, A., Antiga, L., & Lerer, A. (2017). Automatic differentiation in pytorch. In *NIPS 2017 Workshop on Autodiff.* https://openreview.net/forum?id=BJJsrmfCZ

Reades, J., De Souza, J., & Hubbard, P. (2019). Understanding urban gentrification through machine learning. *Urban Studies, 56*(5), 922–942. https://doi.org/10.1177/0042098018789054

Ruth Lazarus Glass. (1960). *London: aspects of change* (Vol. 3). MacGibbon & Kee.

Salesses, P., Schechtner, K., & Hidalgo, C. A. (2013). The collaborative image of the city: mapping the inequality of urban perception. *PLoS One, 8*(7), e68400.

Šćepanović, S., Joglekar, S., Law, S., & Quercia, D. (2021a). Jane jacobs in the sky: Predicting urban vitality with open satellite data. *Proc. ACM Hum.- Comput. Interact., 5*(CSCW1). https://doi.org/10.1145/3449257

Šćepanović, S., Antropov, O., Laurila, P., Rauste, Y., Ignatenko, V., & Praks, J. (2021b). Wide-area land cover mapping with sentinel-1 imagery using deep learning semantic segmentation models. *IEEE Journal of Selected Topics in Applied Earth Observations and Remote Sensing, 14*, 10357–10374. https://doi.org/10.1109/JSTARS.2021.3116094

Scepanovic, S., Obadic, I., Joglekar, S., Giustarini, L., Nattero, C., Quercia, D., & Zhu, X. (2023). Medsat: A public health dataset for england featuring medical prescriptions and satellite imagery. In A. Oh, T. Naumann, A. Globerson, K. Saenko, M. Hardt, & S. Levine (Eds.), *Advances in Neural Information Processing Systems* (Vol. 36, pp. 77810–77851) Curran Associates, Inc.

Shwartz-Ziv, R., & Armon, A. (2022). Tabular data: Deep learning is not all you need. *Information Fusion, 81*, 84–90.

Sinergise. (2017. https://medium.com/sentinel-hub/next-mission-automated-detection-of-land-changes-8e988dce55ff). *Next mission: Automated detection of land changes*. Sentinel Hub Blog.

Suss, J. H. (2023). Measuring local, salient economic inequality in the UK. *Environment and Planning A: Economy and Space, 55*(7), 1714–1737.

Tan, M., & Le, Q. (2019). Efficientnet: Rethinking model scaling for convolutional neural networks. In *International conference on machine learning* (pp. 6105–6114). PMLR.

Thackway, W., Ng, M., Lee, C.-L., & Pettit, C. (2023a. ISSN 0264-2751). Building a predictive machine learning model of gentrification in sydney. *Cities, 134*, 104192. https://doi.org/10.1016/j.cities.2023.104192

Thackway, W., Ng, M., Lee, C.-L., & Pettit, C. (2023b. ISSN 0198-9715). Implementing a deep-learning model using google street view to combine social and physical indicators of gentrification. *Computers, Environment and Urban Systems, 102*, 101970. https://doi.org/10.1016/j.compenvurbsys.2023.101970

Torres, R., Snoeij, P., Geudtner, D., Bibby, D., Davidson, M., Attema, E., Potin, P., Rommen, B. Ö., Floury, N., Brown, M., Traver, I. N., Deghaye, P., Duesmann, B., Rosich, B., Miranda, N., Bruno, C., L'Abbate, M., Croci, R., Pietropaolo, A., Huchler, M., & Rostan, F. (2012). Gmes sentinel-1 mission. *Remote Sensing of Environment, 120*, 9–24. https://doi.org/10.1016/j.rse.2011.05.028. ISSN 0034-4257. The Sentinel Missions—New Opportunities for Science.

Wardrip, K. (2011). *Public transit's impact on housing costs: A review of literature*. Center for Housing Policy. https://scholar.google.com/scholar_lookup?title=Public%20Transit%E2%80%99s%20impact%20on%20housing%20costs&author=K.%20Wardrip&publication_year=2011

WASDI platform. *Earth observation tech for everyone*. https://www.wasdi.cloud, 2023. Online; accessed 3-June-2023.

Watt, P. (2008). The only class in town? gentrification and the middle-class colonization of the city and the urban imagination. *International Journal of Urban and Regional Research, 32*(1), 206–211.

Weisberg, S. (2001). *Yeo-johnson power transformations*. Department of Applied Statistics, University of Minnesota. Retrieved 1 June 2003.

Yee, J., & Dennett, A. (2022). Stratifying and predicting patterns of neighbourhood change and gentrification: An urban analytics approach. *Transactions of the Institute of British Geographers, 47*(3), 770–790.

Yin, H., Ma, C., Weng, L., Xia, M., & Lin, H. (2023. ISSN 2072-4292). Bitemporal remote sensing image change detection network based on siamese-attention feedback architecture. *Remote Sensing, 15*(17). https://doi.org/10.3390/rs15174186

Zuk, M., Bierbaum, A. H., Chapple, K., Gorska, K., & Loukaitou-Sideris, A.- s. (2018). Gentrification, displacement, and the role of public investment. *Journal of Planning Literature, 33*(1), 31–44. https://doi.org/10.1177/0885412217716439

Increasing Urban Greenery Through Reconfiguring Street Spaces to Mitigate Urban Heat Island Effects in Metropolitan Adelaide, Australia

Jinhui Yu, Raziyeh Teimouri, and Andrew Allan

Abstract Urban Heat Island (UHI) is a significant variable in climate and environmental studies. Urban greening and the development of green spaces, through modifications of roads and their ancillaries, can mitigate UHI effects. This study aims to develop a Planning Support Tool to evaluate opportunities for retrofitting roads and ancillaries in urban areas by enhancing urban greening to reduce UHI effects. To achieve this goal, the potential for developing green spaces was investigated across four case study suburbs in the Adelaide Metropolitan Area, Australia. In pursuit of this goal, traffic volumes, UHI hotspots, and Normalised Difference Vegetation Index (NDVI) were analysed in ArcGIS 10.2 to prioritise the case studies for reconfiguring street spaces. The results revealed the potential benefits and applications of this technique in identifying opportunities to modify the urban road environment through increased urban greenery, supporting urban planners in creating more sustainable, climate-resilient environments.

Keywords Urban Heat Island (UHI) · Normalised Difference Vegetation Index (NDVI) · Mitigation · Green space development · Roads · Ancillary spaces

J. Yu · R. Teimouri · A. Allan (✉)
Australian Research Centre for Interactive and Virtual Environments, University of South Australia, Adelaide, South Australia, Australia
e-mail: raziyeh.teimouri@mymail.unisa.edu.au; Andrew.Allan@unisa.edu.au

R. Goodspeed et al. (eds.), *Digital-Era Urban Transformations*, The Urban Book Series, https://doi.org/10.1007/978-3-031-98300-9_13

1 Introduction

Rapid urbanisation has contributed to major environmental challenges for cities worldwide, particularly through the phenomenon of Urban Heat Islands (UHI) (Singh et al., 2020; Li et al., 2018; Karimi et al., 2023). In recent decades, large cities have been subject to increasing global warming, with worsening UHI impacts and associated environmental consequences (Santamouris, 2020; He, 2019; Elliott et al. 2020; Rizwan et al. 2008). This has intensified the pressure to develop solutions that mitigate climate change impacts (Kataoka et al., 2009; Streutker, 2003). The UHI phenomenon has led to increased energy consumption, caused thermal discomfort, and heightened risks to human health (Fadhil et al., 2023; Venhari et al., 2017; Sturiale & Scuderi, 2019; Zhang et al., 2024). The synergy between climate change impacts and UHI effects poses a significant risk to urban climate adaptation and sustainability capacity (Ananyeva & Emmanuel, 2023).

Therefore, practical measures to mitigate UHI and reduce urban air temperatures need to be developed. In this regard, developing green spaces in urban areas has generally been recognised as a key nature-based solution (He, 2022; Marando et al., 2022). According to Deilami et al. (2018), drawing on 75 studies globally, the most common factors affecting the UHI include: vegetation cover (44%), the season (33%), built-up areas (28%), diurnal variations (25%), population density (14%), and water bodies (12%).

Urban greening appears to play a crucial role in UHI adaptation and mitigation (Irfeey et al. 2023, Razzaghmanesh et al., 2016). For example, in addition to increasing solar energy, tree shade reduces the temperature of surfaces and increases latent heat exchange through the evapotranspiration process (Wong et al., 2021; Wang & Akbari, 2016). Vegetation cover also increases wind flow throughout urban areas due to temperature differences which can reduce UHI effects and improve urban microclimates (Al-Obaidi et al., 2021; Wang & Zacharias, 2015; Xi et al., 2023).

Urban roads, streets, and ancillary surfaces are key urban elements contributing to the creation of UHI. Urban green cover, such as street trees, is a practical and effective strategy for mitigating overheating in cities (Xu et al., 2024).

The focus on roads and ancillary facilities, such as public parking areas, is integral to this study's aim of mitigating Urban Heat Island (UHI) effects in car-dependent urban environments like metropolitan Adelaide. These infrastructural elements, being in the public domain, are more easily subject to planning controls and regulation. To a large extent, these surfaces are excessive and most likely to yield significant impacts on UHI with changes (i.e., reductions) through improved greening.

The primary goal of this study is to investigate how *UHI can be mitigated through urban road greenery*. To achieve this, the following objectives are considered:

- Identifying UHI hotspots throughout the study's suburbs in the Adelaide Metropolitan Area.
- Investigating vegetation cover across the case study suburbs.

– Exploring traffic volumes in the main roads of the case studies to estimate the potential for converting road spaces into greenery.
– Identifying public parking areas within the case study areas that could be adapted to increase greenery.
– Prioritising road spaces and ancillary areas for motor vehicles within the case studies for the development of green spaces, such as along road corridors and in public parking areas.

Identifying areas that need to be prioritised for optimisation requires the use of spatial classification methods (Hsu et al., 2022). Urban spatial classification is a classic task, and there are various spatial classification methods based on statistical features. *K*-means algorithm is a common algorithm for urban spatial classification, and many popular algorithms currently in use are proposed based on the *K*-means algorithm (Liu et al., 2021). Many studies have shown that *K*-means can achieve better results and is easier to implement in urban spatial classification tasks such as urban vegetation classification (Lassiter & Darbari, 2020) and the identification of urban–rural fringes in polycentric cities (Ji et al., 2024). Therefore, this work applied the *K*-means for clustering and selecting the optimal number of clusters using methods that calculate initial cluster centres for *K*-means clustering.

2 Methodology

2.1 Study Area

The metropolitan area of Adelaide, situated at 34.55° South and 138.35° East on Australia's southern coast, experiences a hot Mediterranean climate (based on the Köppen–Geiger climate classification), with cool, wet winters and extremely hot and dry summers. Adelaide's hot and dry summers tend to exacerbate the UHI effect because the city experiences hotter temperatures and more hours of sunshine compared to other cities at similar latitudes, such as Sydney or Auckland.

Adelaide has earned various accolades, including being named the third most liveable city in the world and the sixth most sustainable destination in 2021 (Kellett, 2010). With a population of 1.4 million, Adelaide has 30% open green space, and 95% of its residents live within 400 m of public green spaces. Adelaide is the world's second globally recognised National Park City (Green Adelaide, 2024). However, despite Adelaide's liveability, it experiences high UHI effects highlighted by Soltani and Sharifi's (2017) study finding of Adelaide's CBD near-surface UHI peak, reaching 5.9 °C.

The selection of the four metropolitan Adelaide suburban case studies of Burton, Mawson Lakes, Tonsley, and West Lakes was guided by environmental, infrastructural, and land-use criteria to comprehensively address Urban Heat Island (UHI) effects in Adelaide. These suburbs represent a range of urban settings across metropolitan Adelaide, ensuring a holistic and representative analysis. Key factors for their selection include their status as UHI hotspots, with varying intensity levels from mild to severe, providing diverse scenarios for testing mitigation strategies. The suburbs also feature mixed land uses such as residential, commercial, and industrial—allowing evaluation of UHI effects across different urban forms and activities. Significant arterial roads with varying traffic volumes in these areas offer opportunities for urban greening through road space redesign for both high-traffic and low-traffic conditions. Additionally, the suburbs vary in vegetation cover and tree canopy, as shown by NDVI analysis, enabling assessments of greening interventions in areas with limited greenery. This selection not only ensures locally applicable findings but also enhances the relevance of the research to other car-dependent cities facing similar urban heat challenges globally.

The selection criteria highlighted in Table 1 formed the basis for selecting the sample suburbs with these characteristics to determine UHI impacts and how these can be ameliorated. The selected case study suburbs (Fig. 1) are in different locations throughout the metropolitan area and possess various features, are representative of a range of car-dependent suburban environments with various UHI, NDVI, and Traffic volume thresholds across Adelaide.

2.2 *Data Sources*

The UHI data was downloaded from the South Australian Government Data Directory. This dataset is based on daytime thermal data from metropolitan Adelaide in March/April 2022, where surface temperatures were mapped during both day and night using thermal imaging cameras under warm to hot conditions. The map resolution is 125 m × 125 m, and the map represents the difference between the surface temperature of each pixel and the average surface temperature of all pixels at different spatial scales. Urban heat islands are categorised into three levels: high (2 °C above the average but <3 °C), severe (3 °C above the average but <4 °C), and extreme (4 °C above the average).

The NDVI, tree canopy, and green space cover datasets for Adelaide were also downloaded from the South Australia Data Directory (Department for Environment and Water, 2022). These datasets were derived from LiDAR data and high-resolution

Table 1 The case studies selection criteria

	Criteria	Requirement (current features)
1	UHI hotspots	Various UHI intensities, from mild to severe
2	Land use diversity	Diverse land uses
3	Arterial roads and traffic volume	Arterial roads with varying traffic volumes in the suburbs
4	Urban green space (UGS)	Differing levels of existing vegetation cover and tree canopies
5	Geographic representation	Geographically diverse suburbs across Adelaide

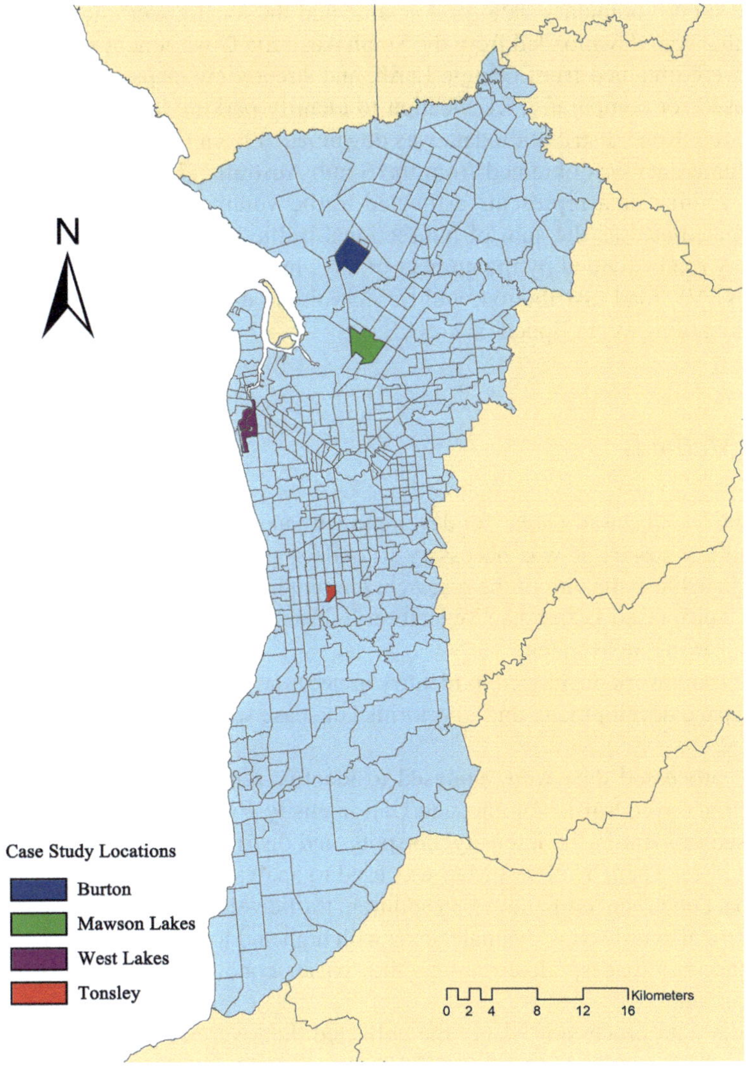

Fig. 1 The sample suburbs selected as case studies in Adelaide metropolitan area

satellite multispectral imagery. Most areas of Adelaide were captured in 2018 and 2019, and additional aerial surveys were conducted in January and February 2022 to update the previous data. NDVI values range from −1 to 1, with higher values indicating denser vegetation. Tree canopy data represents the coverage of tree canopies ≥3 m in height, creating a 100-m resolution raster grid (100 m × 100 m) that indicates the percentage of tree canopy coverage per 100-m unit area. The green space cover dataset maps all actively growing vegetation (e.g., trees, shrubs, and grass), where vegetation ≥2 m is identified through LiDAR point cloud classification, and vegetation <2 m is identified through NDVI classification.

The suburb boundary data for Adelaide and the local government area data for Australia were downloaded from the South Australia Government website. Satellite maps were obtained from Google Earth, and street view maps from Google Maps were used for comparative verification to identify parking lot locations within the study area. Road distribution data was downloaded from OpenStreetMap, and traffic volume data was obtained from the South Australia Government website. The traffic volume data represents estimated traffic volumes from July 2022 to June 2023, calculated as the sum of bidirectional traffic passing observation points on two-way roads divided by the number of days in a year. After data acquisition, the UHI, NDVI, road distribution, and traffic volume data corresponding to our four study locations were clipped in ArcGIS.

2.3 Methods

To identify and classify areas for developing urban greenery alongside the roads and in ancillary spaces, it was necessary to collect and analyse relevant data. In this study, to achieve the aim of the research, the required data to analyse included: UHI maps; Normalised Difference Vegetation Index (NDVI) maps; traffic volumes and public parking areas maps.

The framework developed to identify unnecessary roads and ancillary spaces for green space development, and to prioritise the case studies for greenery, is detailed in Fig. 2.

The processed data were analysed to identify relationships between land use, vegetation cover, traffic volume, and UHI intensity. Firstly, Adelaide UHI map data was used to identify the intensity, hotspots, and distribution of UHI throughout the case studies. Then, an NDVI Map was used to analyse vegetation cover and assess its impact on urban temperature. In addition, traffic volumes and public parking area data were also utilised to evaluate areas with high and low traffic volumes and identify parking spaces, determining the basic conditions for urban greenery development.

In the data processing stage, the collected data were refined and prepared for analysis. This involved identifying UHI hotspots and vegetation cover, reclassifying NDVI raster data, and addressing any missing values. Parking lots were identified using Google satellite maps and street view images. The NDVI and UHI data were then projected, and their type were converted as needed. Outliers were detected using box plot analysis to ensure that extreme values do not skew results. Finally, the data were standardised and normalised to improve the accuracy and reliability of subsequent analyses.

The next stage focused on identifying the demand level for optimisation for each road and car park area. This involved evaluating NDVI, UHI effects, and traffic volume within every car park and a 5-m radius of each road. To determine the

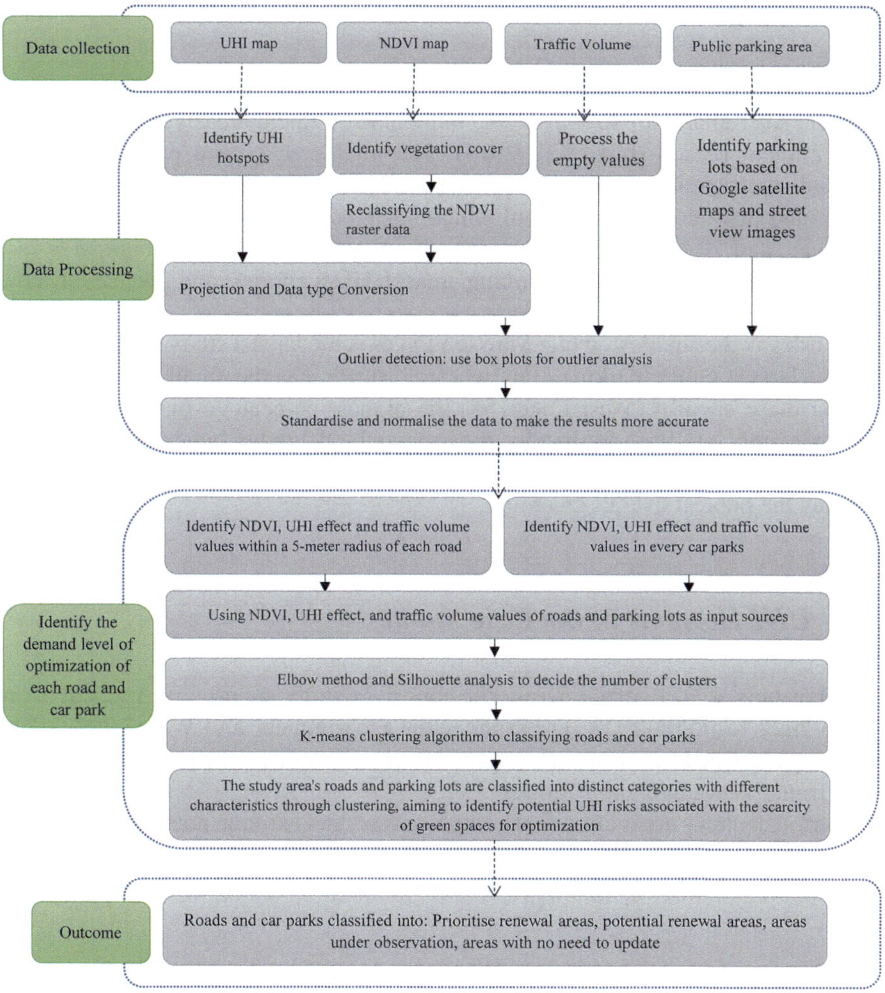

Fig. 2 The research methodology and analysis applied to each case study

appropriate number of clusters for analysis, the Elbow Method and Silhouette Analysis were applied. A *K*-means clustering algorithm (Zalik, 2008; Santoso Haw, 2023) was then used to classify roads and car parks based on their unique characteristics and identified needs.

The final stage is classification and categorisation of roads and car parks into four distinct groups: prioritised renewal areas, potential renewal areas, areas under observation, and areas with no need for updates. This classification is a tool that assists urban planners and policymakers to make informed decisions regarding infrastructure priorities.

3 Findings

According to a previous study, Adelaide has the least equitable distribution of green space in Australia, with 20% available in the most affluent areas versus 12% in the least affluent (Astell-Burt et al. 2014). The total green space per capita in the Adelaide metropolitan area is 272 m^2/person (386 km^2 in total), encompassing both public and private green spaces, of which 4.8% belongs to inner urban, 24.6% to suburban, and 70.6% to peri-urban areas (Teimouri, 2023). Table 2 compares case studies in relation to public car parking areas, UHI categories, and tree canopy coverage (NDVI).

Tonsley is the smallest, while Mawson Lakes is the largest study suburb. Tonsley has the largest public parking area, whereas Burton has the smallest parking area among these suburbs. West Lakes has the highest percentage of green cover, at approximately 14.2%. Additionally, the average NDVI index indicates that the area with the best overall greening is West Lakes, followed by Mawson Lakes, with Tonsley the poorest overall greening.

3.1 *UHI Hotspots in the Case Studies*

UHI hotspots were mapped across the four case study suburbs in Adelaide from 18-03-2022 to 06-01-2023 (Department for Environment and Water, 2022). The mapping classified heat islands into four categories based on the temperature difference from the average temperature.

Burton contains areas classified as high, severe, and extreme hotspots. The industrial and commercial activities, along with heat-absorbing infrastructure, are likely the main reasons for the extreme extent of UHI in this suburb. The UHI map of

Table 2 UHI hotspots by severity and NDVI for the case studies (Metropolitan Adelaide)

			UHI hotspot areas categories (%)			NDVI	
Case studies	Suburb area (km^2)	Public parking area (%)	High	Severe	Extreme	Greening area	Average NDVI index
Burton	5.8	0.7%	25.1%	15.6%	9.3%	7.3%	−0.29
Mawson Lakes	60.4	0.1%	2.5%	1.2%	0.6%	1.08%	−0.15
West Lakes	3.5	2.9%	4.3%	1.7%	0.2%	14.21%	−0.03
Tons-ley	1.0	10.9%	26.7%	15.3%	4.4%	4.69%	−0.3

Note: Method for calculating green space area: Using to the natural breaks method, the NDVI values are divided into five categories: −0.99 to 0.53, −0.53 to 0.18, −0.18 to 0.04, 0.04 to 0.29, and 0.29 to 0.99. The category with the highest NDVI values (0.29 to 0.99) is considered the green space area and is then quantified; The Average NDVI is calculated by determining the mean NDVI value within a region, with values ranging from −1 to 1.

Mawson Lakes primarily shows modest UHIs, with some high hotspots and a few severe and extreme heat spots throughout the suburb. This can be attributed to the presence of lakes and sufficient green cover. Similarly, most parts of West Lakes exhibit modest UHIs due to the surrounding West Lakes rivers and proximity to St Vincent Gulf. However, some areas experience severe and extreme heat island effects, particularly concentrated in the eastern portions, which are influenced by proximity to major roads and a high-density built environment.

UHI mapping of Tonsley showed varied UHIs, particularly around educational zones. In this suburb, industrial and commercial activities, along with heat-absorbing materials in the built environments, contribute to the high and severe heat spots.

3.2 Normalised Difference Vegetation Index (NDVI) in the Case Studies

Normalised Difference Vegetation Index (NDVI) is a commonly used measure for evaluating vegetation cover, obtained from satellite images that measure the reflectance of red and near-infrared spectra (Zenouzi et al., 2022). In this study, NDVI values range from −0.99 to 0.99, with colours indicating different levels of green cover density. The greener the shades, the closer the value is to 1, indicating a higher vegetation cover. Reddish shades represent areas with little to no vegetation, such as water bodies or surfaces made of heat-absorbing materials like asphalt, stone, and concrete.

NDVI mapping demonstrated that Mawson Lakes and West Lakes have relatively higher vegetation density, likely influenced by natural elements such as green cover or water bodies. By contrast, Burton and Tonsley had lower NDVI values, reflecting more urbanised areas with limited green cover, particularly in Tonsley. These variations in vegetation cover are due to differences in types and densities of urban development and vegetation, highlighting opportunities to integrate more green spaces to enhance urban sustainability.

3.3 Public Parking and Traffic Volumes

The distribution of public parking spaces within the study areas was mapped to determine whether these were associated with UHI hotspots. The resulting maps revealed that parking lots in Burton, covering 0.7% of the total area, are widely distributed across the suburb. Parking lots in Mawson Lakes, covering 0.1% of the suburb, follow a similar pattern, being mainly clustered around educational and commercial facilities, indicating a concentration of access points to serve institutions and local businesses. West Lakes presented a different pattern, with 2.9% of the suburb allocated to public parking lots, which are primarily located near water bodies and recreational areas, such as shopping centres. In addition, Tonsley, with 10.9% of its areas dedicated to parking, had the largest concentration of public parking spaces compared to other suburbs. Most of these parking areas are located around large commercial and industrial complexes.

Traffic volumes were mapped across arterial roads in the case study suburbs. The data included estimated daily traffic volume information for South Australian roads (Department for Infrastructure and Transport, 2024). Traffic Volume is the sum of traffic travelling in both directions on a two-way road, passing a roadside observation point over the period of a full year, divided by the number of days in the year. The data is automatically updated weekly.

From the mapping analysis of key arterial roads, Burton experiences very light traffic volumes, typically ranging from 3 to 100 vehicles per day (v/d) across its street network. Mawson Lakes also had light traffic volumes but a slightly higher range of 101 to 500 v/d. West Lakes had moderate traffic volumes, ranging between 501 and 3,000 v/d. By contrast, Tonsley experienced heavy traffic, with volume ranging from 20,001 to 50,000 v/d.

3.4 Data Processing

In this study, ArcGIS 10.2 was used to analyse the collected data. Due to the large volume of NDVI data, the software could not process it directly. Therefore, the Natural Breakpoint Method was first applied to reclassify the NDVI raster data. The data were reclassified into 5 types to facilitate processing of the large volume of data.

Projection and data type conversion were performed on the NDVI and Urban Heat Island (UHI) data, after which the NDVI and UHI effect values within the road buffer were obtained using 'Clip' tool. The NDVI and UHI values were assigned to each road using 'Identity' tools to calculate the values within a 5-m radius of each road. The road network in this study includes all types of roads. The nature strips/sidewalks (or 'buffer' zones) on side roads are narrow, while those on main roads tend to be wider. For the sake of calculation convenience and to reflect standardised road design practice in metropolitan Adelaide, the road buffer zones were set to 5 m.

3.5 Classification Process of Roads

Due to the focus of this work on addressing the heat island effect of roads and ancillary facilities, with a particular emphasis on the UHI effect of land parcels, the categorisation of update type for greening potential was determined based on the existing level of 'greenness' and the most recent mapped vehicle traffic demand. Q-type clustering was applied to divide roads and parking spaces into independent types in order to identify potential renewal areas for greening. This study adopted *K*-means clustering method for data classification and analysis using the Elbow Method (Gove, 2017), with mean deviations used to determine the appropriate number of clusters.

Cluster analysis was conducted using $K = 6$ and $K = 7$, where the Silhouette coefficient is at its maximum, and the comparison results indicated that $K = 6$ provided better clustering performance. Therefore, $K = 6$ was chosen as the final value.

The *k*-means algorithm was used to classify the various roads at the four case sites in Adelaide, and the clustering results are shown in Table 3. Six classifications were developed: Class 1 (priority renewal areas); Class 2 (potential renewal areas); Classes 3 and 4 (marginal areas); and classes 5 and 6 (deemed satisfactory). The roads were divided into six categories, considering relative values rather than absolute values, with lower levels of greenery in categories 1, 4, and 5, significant heat island effects in categories 1, 2, and 3, and lower traffic flow in categories 3, 5, and 6. Based on the NDVI level, UHI level, and traffic flow size of different categories, the clustering results were distributed according to these classifications (Table 3).

3.6 *Classification Process of Parking Lots*

Similarly, performing the same operation on the parking lot area resulted in the following outcome. Cluster analysis was conducted with $K = 3$ and $K = 4$, respectively. The comparison results showed that $K = 4$ provided better clustering performance. Therefore, the final value of K was determined to be 4.

The clustering results determined that parking lots were classified as being priority renewal areas if they exhibited an extreme heat island effect and extremely poor greening level. The category with a severe heat island effect but a marginal greening characteristic was classified as a potential renewal area. Areas where the heat island effect is not significant were classified as not requiring improvement. No parking lot was classified as marginal. As shown in Table 4 and based on the above

Table 3 Road cluster outcomes

Class	NDVI	UHI	Traffic_volume
1	0.232247	0.5402047	0.015605
2	0.445153	0.8457121	0.018699
3	0.545519	0.5034504	0.003107
4	0.307454	0.06896552	0.447641
5	0.317252	−9.99E-16	0.00167
6	0.578271	−8.60E-16	0.001252

Table 4 Car park cluster outcome

Class	NDVI	UHI
1	0	1
2	0.705093	0.545454545
3	0.503781	0
4	0.780893	$-2.78E^{-17}$

classification criteria, Class 1 belongs to the priority renewal area, Class 2 belongs to the potential renewal area, and Classes 3 and 4 belong to the satisfactory category.

3.7 Results of the Optimisation Priority

The distribution of clustering results is shown in Fig. 3. As depicted in the maps of the studied suburbs, the Tonsley area has the highest number of roads that require optimisation. The presence of industrial and commercial activities in this area creates a denser network of traffic-heavy roads. In addition, the lack of vegetation cover (such as tree canopies) along the main roads in Tonsley presents opportunities for greening interventions.

The analysis results for Burton also indicate that much of the suburb potentially requires renewal and greening initiatives. Burton is dominated by residential and commercial areas, with parts of the suburb, particularly the northern section, still in the development phase. This area would benefit from greenery interventions as part of the ongoing development process.

The road cluster results for the West Lakes and Mawson Lakes areas have the lowest number of roads requiring renewal. In these suburbs, most roads do not need

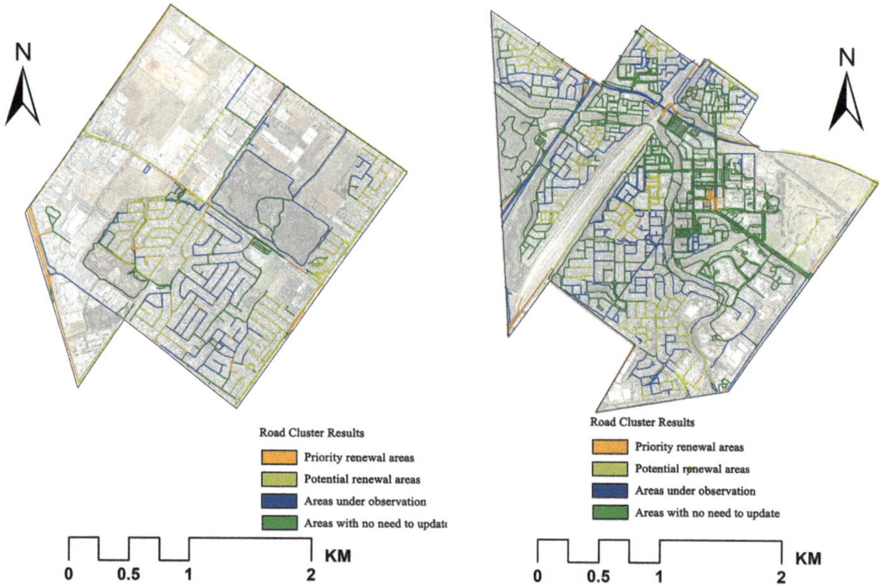

Fig. 3 Roads cluster results of optimisation priority

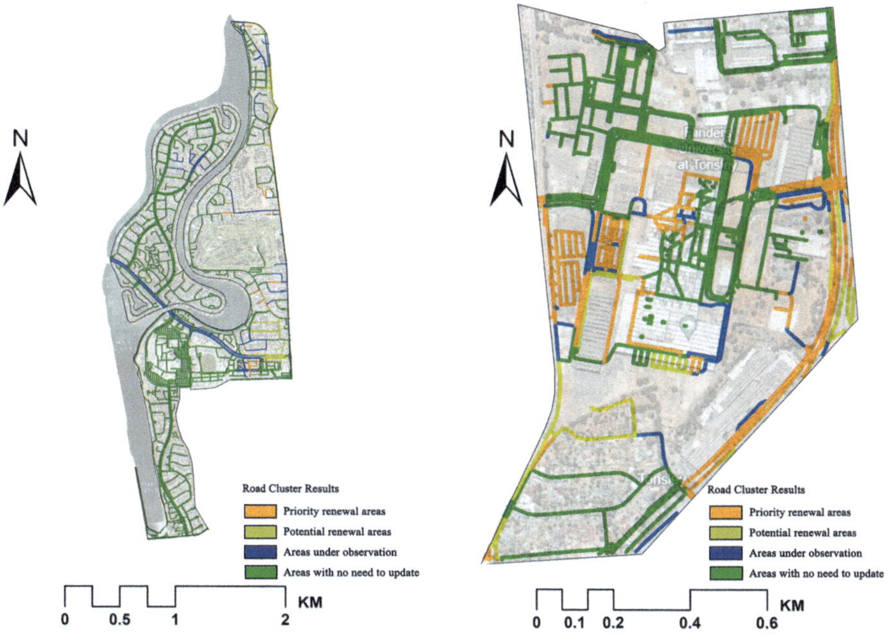

Fig. 13.3 (continued)

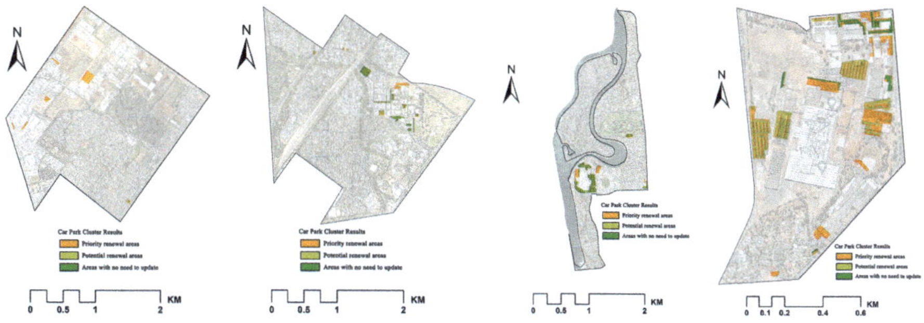

Fig. 4 Public car parking lots cluster results for prioritisation of greening actions

additional greenery, as the presence of a water body has had a significant positive impact on the travel environment.

Fig. 4 represents the priorities for developing greenery in public car parking spaces across the four studied suburbs in the Adelaide metropolitan area. Almost all of Burton's parking lots require priority optimisation, green space development, and improvements, while Tonsley's parking lots are primarily categorised as areas requiring priority or potential renewal. Although relatively few parking lots in the West Lakes and Mawson Lakes areas require priority renewal, there are still several areas identified as priority for renewal that need urgent optimisation.

4 Discussion

This study emphasised the impact of urban road greenery in mitigating UHI effects. To this end, four suburbs from different parts of the Adelaide metropolitan area were investigated in terms of their need for green space development along main roads and ancillaries (i.e. parking lots). The findings of this study align with broader trends that link reduced vegetation cover to increased UHI effects.

The UHI spatial distributions showed that West Lakes outperformed the other suburbs, having the fewest UHI hotspots. Based on the road cluster analysis, this suburb was assessed as having the lowest priority for reconfiguration of roads. The large parking areas in West Lakes, particularly around shopping centres, contributed significantly to UHI due to their heat-absorbing surfaces. Redeveloping these parking areas to incorporate greenery and permeable surfaces presents a key opportunity to mitigate UHI in the area. This finding supports previous studies (Sen et al., 2020; Sharifi et al., 2021), which advocate for urban sustainability goals by addressing localised heat stress without requiring extensive road reconfiguration.

The results of the analysis for Burton demonstrated that, with a high UHI impact and low vegetation cover, there is considerable potential for modifying roads and parking areas to increase greening. This potential arises from relatively light traffic volumes, the recent residential development that lacks mature vegetation, and the reduced need for arterial road space and large parking lots.

Factors such as water bodies and vegetation in the Mawson Lakes area help maintain relatively mild temperatures and create a more comfortable environment with fewer UHI hotspots. However, there are a few extreme hotspots, mainly due to high-density residential areas in certain parts of the suburb. As a result, Mawson Lakes has fewer critical areas in need of road reconfiguration. Nonetheless, the clustering of parking spaces around commercial and educational centres in this suburb still requires improvements to their parking lots and roadways to further reduce UHI impacts.

Tonsley, a more industrialised suburb among the studied areas in Adelaide, which is undergoing regeneration, transitioning from industrial to residential uses, shows some UHI hotspots, indicating a need for improvements in green infrastructure. Based on optimisation priority analysis for both roads and ancillary areas, this suburb has the highest number of roads and public parking lots that urgently need optimisation. Therefore, mitigation strategies, such as increasing green spaces and tree cover, could significantly alleviate UHI impacts in Tonsley.

The cluster analysis in this study identified areas requiring priority optimisation, particularly in Tonsley and Burton, where the need for urban renewal is most pressing due to severe UHI effects and limited vegetation cover. The presence of priority renewal areas in these suburbs indicates that strategic urban interventions, such as increasing tree canopy and transforming parking lots into greener spaces, could significantly improve local environmental conditions. While Mawson Lakes and West Lakes exhibit fewer critical renewal areas due to their proximity to water bodies and greater green cover, there is still room for improvement in their parking lots and roadways to mitigate UHI effects.

5 Conclusion

This study highlights the importance of road and ancillary greenery in mitigating UHI effects. The findings from the analysis of four suburbs in the Adelaide metropolitan area demonstrate a need for increased green infrastructure development, particularly in industrial areas like Tonsley and Burton, to enhance urban sustainability and resilience.

This study has developed a tool to identify the potential for retrofitting increased greenery in car-dependent urban environments where there is an overabundance of hard surfaces serving motor vehicles, which exacerbate the UHI effect and contribute to climate change. Thus, the results and approach of this study can be considered an important Planning Support Tool for planners and other built environment professionals to mitigate UHI effects through roads and ancillary areas by enhancing urban greenery.

Therefore, based on the prioritisation of the urban areas for developing green spaces, urban planners and stakeholders can tailor urban planning strategies to address both environmental and urban challenges related to the UHI effects of roads and ancillary areas. Increased integration of green spaces into urban environments can create more liveable and environmentally sustainable urban areas through cooler microclimates, improved visual aesthetics, and increased community amenities.

This study recommends that future studies apply the tool developed in this work to investigate the types and massing of vegetation within road corridors and in ancillary vehicle areas best suited to mitigate UHI effects. Furthermore, studying the standards for road widths, including the design of vegetated buffers, tree canopies, sidewalks, and road medians, for different types of greening development based on traffic volume is essential for improving urban heat management, with a view to developing planning policies to better manage UHI effects.

Acknowledgements This research was supported through the 2024 Virtual Internship Program through the Australian Research Centre for Interactive and Virtual Environments (IVE), University of South Australia.

Appendix: Data Sources and Copyright Statement

UHI Data: Department for Environment and Water, South Australia, Metropolitan Adelaide urban heat mapping 2022, Sourced on 20 March 2024, https://data.sa.gov.au/data/dataset/8e9f3a9e-c365-404d-be97-4b611e140384.

NDVI, Tree Canopy, and Green Space Data: Department for Environment and Water, South Australia, Metropolitan Adelaide tree canopy, green spaces and built environment 2022, Sourced on 2024-03-19, https://data.sa.gov.au/data/dataset/0f7ab193-326b-4894-aec4-298b5e6ea0ba.

Suburb Boundary Data: Department for Housing and Urban Development, South Australia, Suburbs, Sourced on 14 March 2013, https://data.sa.gov.au/data/dataset/e4d3a355-29d7-4bdc-a81d-13fd5ed09ef9.

Local Government Areas Data: Department for Housing and Urban Development, South Australia, Local Government Areas, Sourced on 14 May 2013, https://data.sa.gov.au/data/dataset/968c8768-802a-48ed-92d9-eb927fe50c46.

Traffic Volume Data: Department for Infrastructure and Transport, South Australia, Traffic Volumes, Sourced on 04 July 2014, https://data.sa.gov.au/data/dataset/e5d6588a-f163-4f6a-bc57-25e95c87b5bd.

Road Distribution Data: This data can be accessed via the OpenStreetMap (OSM) platform at the following URL: https://www.openstreetmap.org/. This dataset is made available under the Open Database License (ODbL). For detailed information regarding the licensing terms and copyright policies, please refer to the official OpenStreetMap copyright page: https://www.openstreetmap.org/copyright.

In this study, satellite maps provided by Google Earth (https://earth.google.com) and street view data provided by Google Maps (https://maps.google.com) were utilized. The satellite imagery was processed by reducing both opacity and fill to serve as the background for certain figures, while the street view data was used as a reference source to validate the accuracy of our maps. All data were obtained through Google's publicly accessible platforms and adhere to its terms of service as well as the licensing agreements of third-party data providers. As the imagery from Google Maps and Google Earth is protected by copyright, this study strictly limited its use to academic analysis, with no commercial purposes or large-scale data downloads conducted.

References

Al-Obaidi, I., Rayburg, S., Półrolniczak, M., & Neave, M. (2021). Assessing the impact of wind conditions on urban heat islands in large Australian cities. *Journal of Ecological Engineering, 22*(11), 1–15. https://doi.org/10.12911/22998993/142967

Ananyeva, O., & Emmanuel, R. (2023). Street trees and urban heat island in Glasgow: Mitigation through the 'Avenues Programme'. *Urban Forestry & Urban Greening, 86*, 128041. https://doi.org/10.1016/j.ufug.2023.128041

Astell-Burt, T., Feng, X., Mavoa, S. et al. (2014) Do low-income neighbourhoods have the least green space? A crosssectional study of Australia's most populous cities. *BMC Public Health 14*, 292. https://doi.org/10.1186/1471-2458-14-292

Deilami, K., Kamruzzaman, M., & Liu, Y. (2018). Urban heat island effect: A systematic review of spatio-temporal factors, data, methods, and mitigation measures. *International Journal of Applied Earth Observation and Geoinformation, 67*, 30–42. https://doi.org/10.1016/j.jag.2017.12.009

Department for Environment and Water (2022). Retrieved June 2024, from https://data.sa.gov.au/data/dataset/metropolitan-adelaide-urban-heat-mapping-2022

Department for Infrastructure and Transport (2024). https://data.sa.gov.au/data/dataset/traffic-volumes Retrieved June 2024.

Elliot T., Almenar J.B. and Rugani, B. (2020). Modelling the relationships between urban land cover change and local climate regulation to estimate urban heat island effect. *Urban Forestry & Urban Greening, 50*, 126650, ISSN 1618-8667. https://doi.org/10.1016/j.ufug.2020.126650. (https://www.sciencedirect.com/science/article/pii/S161886671930812X).

Fadhil, M., Hamoodi, M. N., & Ziboon, A. R. T. (2023). Mitigating urban heat island effects in urban environments: Strategies and tools. In *IOP conference series: Earth and environmental science (Vol. 1129, no. 1)* (p. 012025). IOP Publishing.

Gove, R. (2017). Using the elbow method to determine the optimal number of clusters for k-means clustering. URL: https://blocks.org/rpgove/0060ff3b656618e9136b, pp. 17–19.

Green Adelaide (2024). viewed 2 Jun 2024, https://www.greenadelaide.sa.gov.au/ access date May 2024

He, B. J. (2019). Towards the next generation of green building for urban heat island mitigation: Zero UHI impact building. *Sustainable Cities and Society, 50*, 101647. https://doi.org/10.1016/j.scs.2019.101647

He, B. J. (2022). Green building: A comprehensive solution to urban heat. *Energy and Buildings, 271*, 112306. https://doi.org/10.1016/j.enbuild.2022.112306

Hsu, Y. Y., Hawken, S., Sepasgozar, S., & Lin, Z. H. (2022). Beyond the backyard: GIS analysis of public green space accessibility in Australian metropolitan areas. *Sustainability, 14*(8), 4694. https://doi.org/10.3390/su14084694

Irfeey, A. M. M., Chau, H. W., Sumaiya, M. M. F., Wai, C. Y., Muttil, N., & Jamei, E. (2023). Sustainable mitigation strategies for urban heat island effects in urban areas. *Sustainability, 15*(14), 10767. https://doi.org/10.3390/su151410767

Ji, D., Tian, J., Zhang, J., Zeng, J., & Namaiti, A. (2024). Identification and spatiotemporal evolution analysis of the urban–rural fringe in polycentric cities based on K-means clustering and multi-source data: A case study of Chengdu City. Land. https://doi.org/10.3390/land13111727.

Karimi, A., Mohammad, P., García-Martínez, A., Moreno-Rangel, D., Gachkar, D., & Gachkar, S. (2023). New developments and future challenges in reducing and controlling heat island effect in urban areas. *Environment, Development and Sustainability, 25*(10), 10485–10531. https://doi.org/10.1007/s10668-022-02530-0

Kataoka, K., Matsumoto, F., Ichinose, T., & Taniguchi, M. (2009). Urban warming trends in several large Asian cities over the last 100 years. *Science of the Total Environment, 407*(9), 3112–3119. https://doi.org/10.1016/j.scitotenv.2008.09.015

Kellett, J. E. (2010). Adelaide: the garden city on steroids (Doctoral dissertation, Istanbul Technical University).

Lassiter, A., & Darbari, M. (2020). Assessing alternative methods for unsupervised segmentation of urban vegetation in very high-resolution multispectral aerial imagery. *PLoS One, 15*(5), e0230856.

Li, G., Zhang, X., Mirzaei, P. A., Zhang, J., & Zhao, Z. (2018). Urban heat island effect of a typical valley city in China: Responds to the global warming and rapid urbanization. *Sustainable Cities and Society, 38*, 736–745. https://doi.org/10.1016/j.scs.2018.01.033

Liu, B., Deng, Y., Li, M., Yang, J., & Liu, T. (2021). Classification schemes and identification methods for urban functional zone: A review of recent papers. *Applied Sciences, 11*(21), 9968. https://doi.org/10.3390/app11219968

Marando, F., Heris, M. P., Zulian, G., Udías, A., Mentaschi, L., Chrysoulakis, N., Parastatidis, D., & Maes, J. (2022). Urban heat island mitigation by green infrastructure in European functional urban areas. *Sustainable Cities and Society, 77*, 103564. https://doi.org/10.1016/j.scs.2021.103564

Razzaghmanesh, M., Beecham, S., & Salemi, T. (2016). The role of green roofs in mitigating urban Heat Island effects in the metropolitan area of Adelaide, South Australia. *Urban Forestry & Urban Greening, 15*, 89–102. https://doi.org/10.1016/j.ufug.2015.11.013

Rizwan, A. M., Dennis, L. Y., & Chunho, L. I. U. (2008). A review on the generation, determination and mitigation of urban Heat Island. *Journal of Environmental Sciences, 20*(1), 120–128. https://doi.org/10.1016/S1001-0742(08)60019-4

Santamouris, M. (2020). Recent progress on urban overheating and heat island research. Integrated assessment of the energy, environmental, vulnerability and health impact. Synergies with the global climate change. *Energy and Buildings, 207*, 109482. https://doi.org/10.1016/j.enbuild.2019.109482

Santoso, H. A., & Haw, S. C. (2023). Improvement of k-means clustering performance on disease clustering using Gaussian mixture model. *Journal of System and Management Sciences, 13*(5), 169–179.

Sen, S., Fernandèz, J. P. R. M. R., & Roesler, J. (2020). Reflective parking lots for microscale urban heat island mitigation. *Transportation Research Record, 2674*(8), 663–671. https://doi.org/10.1177/0361198120919401

Sharifi, E., Zawarus, P., & Lehmann, S. (2021). Optimization of urban cooling strategies for parking lots in hot and dry climates: Case study of Las Vegas and Adelaide. Urban Heat Island (UHI) Mitigation: Hot and Humid Regions, pp. 169–182. https://doi.org/10.1007/978-981-33-4050-3_8

Singh, N., Singh, S., & Mall, R. K. (2020). Urban ecology and human health: Implications of urban heat island, air pollution and climate change nexus. In *Urban ecology* (pp. 317–334). Elsevier. https://doi.org/10.1016/B978-0-12-820730-7.00017-3

Soltani, A., & Sharifi, E. (2017). Daily variation of urban heat island effect and its correlations to urban greenery: A case study of Adelaide. *Frontiers of Architectural Research, 6*(4), 529–538. https://doi.org/10.1016/j.foar.2017.08.001

Streutker, D. R. (2003). Satellite-measured growth of the urban heat island of Houston, Texas. *Remote Sensing of Environment, 85*(3), 282–289. https://doi.org/10.1016/S0034-4257(03)00007-5

Sturiale, L., & Scuderi, A. (2019). The role of green infrastructures in urban planning for climate change adaptation. *Climate, 7*(10), 119. https://doi.org/10.3390/cli7100119

Teimouri, R. (2023). An approach to planning and designing urban green space to enhance social and ecological sustainability of residential areas, PhD Thesis in Urban and Regional Planning, University of South Australia.

Venhari, A. A., Tenpierik, M., & Hakak, A. M. (2017). Heat mitigation by greening the cities, a review study. *Environment, Earth and Ecology, 1*, 1.

Wang, Y., & Akbari, H. (2016). The effects of street tree planting on urban Heat Island mitigation in Montreal. *Sustainable Cities and Society, 27*, 122–128. https://doi.org/10.1016/j.scs.2016.04.013

Wang, Y., & Zacharias, J. (2015). Landscape modification for ambient environmental improvement in central business districts–a case from Beijing. *Urban Forestry & Urban Greening, 14*(1), 8–18. https://doi.org/10.1016/j.ufug.2014.11.005

Wong, N. H., Tan, C. L., Kolokotsa, D. D., & Takebayashi, H. (2021). Greenery as a mitigation and adaptation strategy to urban heat. *Nature Reviews Earth & Environment, 2*(3), 166–181. https://doi.org/10.1038/s43017-020-00129-5

Xi, C., Han, L., Wang, J., Feng, Z., Kumar, P., & Cao, S. J. (2023). How can greenery space mitigate urban heat island? An analysis of cooling effect, carbon sequestration, and nurturing cost at the street scale. *Journal of Cleaner Production, 419*, 138230. https://doi.org/10.1016/j.jclepro.2023.138230

Xu, C., Huang, Q., Haase, D., Dong, Q., Teng, Y., Su, M., & Yang, Z. (2024). Cooling effect of green spaces on urban Heat Island in a Chinese megacity: Increasing coverage versus optimizing spatial distribution. *Environmental Science & Technology, 58*(13), 5811–5820. https://doi.org/10.1021/acs.est.3c11048

Zalik, K. (2008). An efficient k'-means clustering algorithm. *Pattern Recognition Letters, 29*, 1385–1391. https://doi.org/10.1016/j.patrec.2008.02.014

Zenouzi, A. S., Yenneti, K., Teimouri, R., Abbasiyan, F., & Palme, M. (2022). Analysis of changes in vegetation index during the rapid urban spatial development period (1990–2020) in Tehran metropolis, Iran. *Atmosphere, 13*(12), 2010. https://doi.org/10.3390/atmos13122010

Zhang, H., Kang, M. Y., Guan, Z. R., Zhou, R., Zhao, A. L., Wu, W. J., & Yang, H. R. (2024). Assessing the role of urban green infrastructure in mitigating summertime urban heat island (UHI) effect in Metropolitan Shanghai, China. *Sustainable Cities and Society*, 105605. https://doi.org/10.1016/j.scs.2024.105605

Sensing and Analysing Urban Heat Islands Using Internet of Things: A Real Time Microclimate Network for London

Dongyi Ma, Andrew Hudson-Smith, Martin De Jode, and Leah Lovett

Abstract Urban Heat Islands (UHIs) refer to the phenomenon where cities become significantly hotter than their rural surroundings, posing one of the most pressing concerns in urban climate studies and effective urban management (Oke, T.R., Urban climates. Cambridge University Press, Cambridge, United Kingdom, 2017). This chapter explores the concept of improved planning and urban management for UHI mitigation by integrating Internet of Things (IoT) technology, specifically real-time monitoring through bespoke Long Range Wide Area Network (LoRaWAN) based heat sensors. We present the development and deployment of three UHI sensor designs through ongoing deployments in London, with 43 devices currently collecting data throughout the city, and the ultimate aim of establishing a citizen science-based microclimate network in London. Case studies are documented with the Greater London Authority (GLA) and the London Legacy Development Corporation (LLDC), providing preliminary results which suggest the feasibility of using low-cost networked devices to sense and communicate Urban Heat Islands.

Keywords Internet of Things · Urban Climate Mitigation · London · Citizen-Science · Urban Heat Island

1 Understanding Urban Heat Islands and Heatwaves in Urban Management

There is a growing need for improved urban planning and management in cities due to the effects of climate change. According to the United Kingdom Climate Projection (UKCP18) from the UK Meteorological Office, the UK is expected to experience notable climate changes in the coming decades. Their projections

D. Ma (✉) · A. Hudson-Smith · M. De Jode · L. Lovett
University College London, The Bartlett Centre for Advanced Spatial Analysis, London, UK
e-mail: dongyi.ma.21@ucl.ac.uk; a.hudson-smith@ucl.ac.uk; m.dejode@ucl.ac.uk; l.lovett@ucl.ac.uk

R. Goodspeed et al. (eds.), *Digital-Era Urban Transformations*, The Urban Book Series, https://doi.org/10.1007/978-3-031-98300-9_14

indicate a higher likelihood of milder, wetter winters and hotter, drier summers, with an increased frequency and intensity of extreme weather events such as heatwaves (Met Office, 2018). Urban areas, particularly cities like London, are increasingly vulnerable to the combined effects of heatwaves and what is known as the Urban Heat Island (UHI) phenomenon, where urban regions become significantly warmer than their rural surroundings due to factors such as reduced vegetation and the prevalence of impervious surfaces (Oke et al., 2017). The UHI effect makes urban populations more vulnerable to heat-related illnesses and even death (Zhao et al., 2018). Recent statistics from the Office for National Statistics (ONS) show an average of 1272 deaths annually in the UK due to extreme weather, with an additional 12,086 hospitalisations linked to hot days (GOV.UK, U, 2020). The impact of extreme heat is not limited to health. The existing literature highlights the relationship between heat stress and various socioeconomic and environmental factors, revealing a clear link between extreme heat and increased mortality rates, particularly among vulnerable populations such as the elderly and those with pre-existing health conditions (Michelozzi et al., 2005, 2009; Taylor et al., 2015; Mavrogianni et al., 2011). Moreover, heat stress disproportionately impacts lower-income communities due to inadequate housing and lack of access to cooling infrastructure, exacerbating social inequalities (Alizadeh et al., 2022). Elevated ambient temperatures lead to increased energy consumption, particularly in the building sector, where both peak and total electricity demand rise dramatically during extreme heat events (Santamouris et al., 2015). In addition, UHIs contribute to environmental degradation by altering urban ecosystems, affecting material flows, hydrological systems, and wildlife behaviours (Yang et al., 2016).

The increasing impacts of extreme heat, particularly in urban areas, underscore the critical need to measure, monitor, and mitigate UHIs and heatwaves to safeguard public health, ensure social equity, and promote sustainable urban development. Broadly, UHIs manifest in two forms: the air UHI and the surface UHI. While the former pertains to atmospheric temperature variations in the canopy and boundary layers, the latter represents temperature disparities captured through remote sensing data between urban and non-urban surfaces (Zhou et al., 2019). Of these, canopy-level UHI is of particular interest, as it directly affects residents' daily experiences with heat. Current approaches to measuring canopy-level UHI include both direct and indirect methods. Remote sensing provides an indirect measure of canopy-level UHI by capturing surface temperatures and adopting models like the UMEP (Urban Multi-scale Environmental Predictor) for the estimation of mean radiant temperature using open-source data. In contrast, direct measurement techniques, such as mobile transects and fixed weather stations, provide more precise data but are limited in their ability to cover large areas (Sun et al., 2019). Existing crowdsourced personal weather station networks in London, such as Netatmo and Weather Underground, leave gaps in coverage, particularly in less-monitored areas. As a result, they fall short in providing the high spatial and temporal resolution, real-time data needed to study London's Urban Heat Island (UHI) effect and heatwaves. Given the urgency of the climate adaptation measures and the scarcity of field data, it is crucial to further explore the potential of low-cost sensing and citizen science for open-source UHI dataset. Such data could complement current datasets to enhance understanding and inform urban planning and urban management for London to mitigate the impacts of UHI.

The Internet of Things (IoT), defined as a network of interconnected devices that collect and exchange data in real-time, is becoming a key tool for addressing urban environmental challenges in urban management. As IoT technologies become cheaper and more compact, they offer significant opportunities for mass data collection and analysis, fundamentally changing how we manage and plan our cities. The opportunities for mass data collection through urban IoT devices are notable, providing valuable insights for urban modelling and policy making. The value, volume, velocity, and variety of data collected from IoT devices offer opportunities for improving urban management strategies (Hudson-Smith et al., 2019, 2020, 2021). One effective example of using IoT to address urban temperature is the Birmingham Urban Climate, which deployed a network of 250 low-cost air temperature sensors and 30 automatic weather stations, utilising Wi-Fi networks for data transmission (Warren et al., 2016). This network provided a heat map and interpolated map of the city to the public, addressing several key challenges: supplementing sparse weather station data, providing more accurate air temperature data than surface temperature proxies, and overcoming the accessibility issues of local and global weather station datasets. However, Wi-Fi-based sensor networks face limitations due to their high power consumption, which is unsuitable for long-term, battery-powered deployments. Long Range Wide Area Network (LoRaWAN) based sensor network offers broad coverage and low power consumption, and therefore presents a promising alternative for a long-range, wide-area UHI sensor network (Karunarathne et al., 2018; Sadek & Elbadawy, 2022; Mekki et al., 2019). With the impacts of climate change and the need for improved urban planning and management noted, this study introduces the development of a low-cost and scalable sensor network capable of real-time, localised air temperature and humidity measurement, utilising low-power, long-range radio networks (LoRaWAN) to overcome the limitations of Wi-Fi-based systems. Central to this is the ability to provide data for monitoring UHI intensity, identifying hot spots and cool zones across the city, and tracking microclimate variations in near real-time. The following sections of this chapter discuss the detailed design, deployment, and findings from these geo-located sensor networks, providing early insights into the spatial distribution of UHIs in London. Furthermore, we explore the aim to provide an open-source hardware and software system design dataset, which researchers, urban planners, and policymakers are able to access and use freely, fostering collaboration and encouraging further innovation.

2 Sensor Deployment and Iteration

Given this context, we have developed three iterations of sensors tailored to distinct urban deployment scenarios: the Pilot, Solar-Powered, and Citizen Scientists Versions, as shown in Figs. 1, 2 and 3. The design and capabilities of these sensors build on prior work that demonstrated their effectiveness in monitoring

Fig. 1 The deployed pilot version sensor in St. Vincent School

Fig. 2 The solar-powered version sensors deployed in the Queen Elizabeth Olympic Park

microclimate effects within East London (Ma et al., 2024). The Pilot Version, designed for small-scale, cost-effective (under £80) deployments, was utilized in collaboration with the Greater London Authority to monitor temperatures of playgrounds in London schools (Fig. 4). The Solar-Powered Version, optimized for

Fig. 3 The designed PCB for citizen scientists version sensor

long-term deployments for years, was deployed across 15 sites in Queen Elizabeth Olympic Park to capture urban heat island effects in varied landscapes, including impervious surface, waterways, vegetation, or a combination. Figure 5 shows the live air temperature map generated from the sensor network data, providing a visual representation of the spatial distribution of air temperature across QEOP with live data overlayed. Lastly, the Citizen Scientists Version features a custom PCB design that integrates the microcontroller and sensor modules, as well as the switch from LiPo batteries to standard AA batteries, making the sensors safer and easier to maintain. This iteration has been specifically designed for large-scale, community-driven projects, enabling non-experts to easily assemble and deploy sensors, enhancing the scalability for a wider monitoring network.

The sensor used in all three of these versions has a typical temperature accuracy of ±0.2 °C, meeting ISO Standard 7726 (ISO 1985), ensuring measurements within a range of ±0.5 °C, with a preferred accuracy of ±0.2 °C. To guarantee reliable data, sensors were installed in Stevenson screens, which shield against direct sunlight and moisture while allowing adequate airflow. Besides, placement at 2–3 m high minimized ground-level heat effects while capturing canopy-level variations on existing infrastructure like lamp posts and trees, eliminating extra costs for installation. Data quality checks involved excluding outliers (values exceeding three standard deviations) and periodic maintenance of devices. All three iterations employ the same network architecture, which enables real-time data transmission and processing using LoRaWAN. The sensors for our network were put into sleeping mode to conserve energy and wake up every 5 min to relay air temperature and humidity data to LoRa Gateways, which then forward it to The Things Network, a community-driven platform for handling LoRaWAN messages. Once transmitted through The Things Network, the data is routed to an MQTT broker, which allows for efficient, lightweight messaging. From there, the sensor data is processed and stored in a time-series database (InfluxDB), with Telegraf managing the continuous capture of new readings. End users can visualize and interact with the data using the Grafana dashboard, which provides access to real-time temperature trends and downloadable data for further analysis. The preliminary results from deployment will be discussed in detail in the following section, utilizing the data collected from London‘s schools.

Fig. 4 GLA funded pilot version sensors deployment

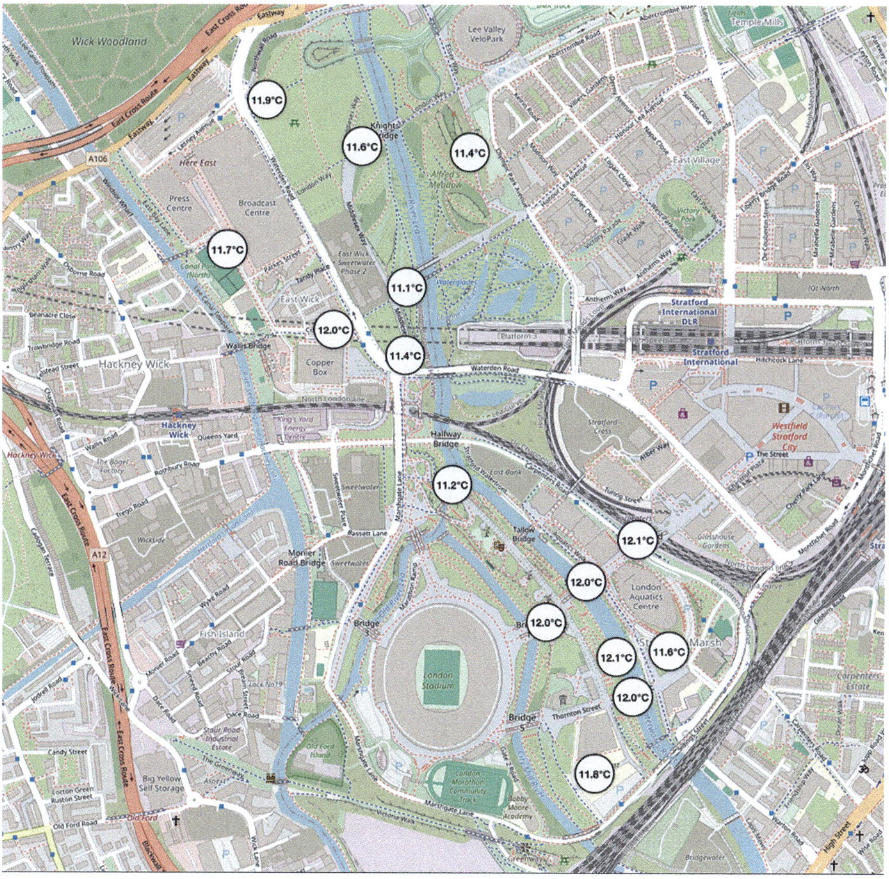

Fig. 5 Live air temperature map of Queen Elizabeth Olympic Park at 12:36 PM, October 9, 2024

3 Data and Findings: Preliminary Findings in Schools and Park in London

As noted, the first iteration of the sensor network was deployed in a collaborative effort with the Greater London Authority (GLA), facilitating contact with ten schools across London. The aim was to monitor local microclimates and provide real-time data on temperature variations, contributing to a better understanding of the UHI effect in diverse urban settings. This project also sought to raise awareness about the significance of real-time temperature monitoring and to address disparities in access to local temperature data. The deployment targeted a diverse range of schools, including mainstream schools, special educational needs schools, and an alternative provision school, as shown in Table 1. The selection criteria included socioeconomic factors such as levels of deprivation, the presence of minority communities, and areas with poor coverage of personal weather stations. Additionally,

Table 1 Summary of GLA project deployment details

School name	School location	Topology	Sensor location
William Davis Primary School	East London	Open sky view, significant number of trees	Playground
School 360	East London	Open sky view	Playground
Griffin Primary School	South London	Close to vegetation	Playground
Mayflower Primary School	East London	Open sky view, close to vegetation	Playground
Abingdon School	Central London	No vegetation or playground outdoor space limited	School Gate
St. Vincent's Catholic Primary School	Central London	Playground on roof no trees	Courtyard at ground level with limited sky view
ArtsXchange School Hackney	East London	No outdoor space and close proximity to busy road	Pavement
Harrington Hill Primary School	North London	Close to vegetation	Playground
Harmood School Camden	North London	Proximity to vegetation	Playground
Ian Mikardo Tower Hamlets	East London	Close to tube station but also proximity to vegetation	School car park

the proximity to the existing LoRaWAN network in London was considered to ensure reliable data transmission. The sensors were strategically installed in school playgrounds and outdoor spaces, mounted at a height of 2–3 meters to capture air temperature at the canopy level.

The air temperature measurements collected from the ten London schools between first February and 31st August 2024 reveal significant differences when compared to data from the Met Office station at St James's Park, a more open, vegetated area within London. This location is deliberately chosen to minimise the UHI effect, representing baseline or background climate conditions without the heat retention typical of dense urban areas. Outliers, identified as values beyond 3 standard deviations, were excluded for validation. Table 2 provides a statistical summary of the temperature differences between schools and the weather station, illustrating overall average, median, standard deviation, averaged monthly variations, and averaged hot/non-hot day differences. Figure 6 shows the spread of UHI across schools, indicating that central London schools generally experience higher temperatures than others. As Fig. 7 highlights, UHI intensifies during summer months, and Fig. 8 shows that UHI is exacerbated on hot days (defined as days when Met Office temperatures exceed 30 °C). Besides, it is worth mentioning that the minimum UHI intensity dropped to −2.15 and −2.09 °C at William Davis School and School 360, respectively. These negative values suggest moments when urban schools were cooler than St James's Park, possibly due to variations in the local weather, as suggested by the deployment map in Fig. 4. It is also important to

Table 2 Summary of UHI patterns for schools

School	MeanUHI	MedianUHI	Max_UHI	Min_UHI	SD_UHI	Feb	Mar	Apr	May	Jun	Jul	Aug	Hot_Days_UHI	Non_Hot_Days_UHI
Abingdon	1.04	0.82	4.14	−1.23	0.91	0.39	0.64	0.85	1.14	1.53	1.51	1.45	1.51	1.03
ArtsXchange	0.50	0.37	2.91	−1.76	0.69	0.41	0.45	0.32	0.66	0.62	0.56	0.50	0.69	0.50
Griffin	0.60	0.43	3.00	−1.61	0.70	0.13	0.24	0.47	0.72	1.00	0.85	0.79	1.05	0.59
Harmood	0.34	0.20	2.70	−1.85	0.70	−0.10	0.00	0.19	0.48	0.63	0.58	0.58	0.79	0.33
HarringtonHill	0.48	0.40	2.99	−1.97	0.76	0.10	0.19	0.34	0.42	0.79	0.74	0.74	0.56	0.48
IanMikardo	0.40	0.29	2.79	−1.86	0.68	0.30	0.33	0.32	0.26	0.49	0.46	0.55	0.48	0.40
Mayflower	0.65	0.50	3.14	−1.72	0.73	0.26	0.36	0.46	0.72	1.16	1.11	NA	1.72	0.65
School360	0.99	0.69	4.67	−2.09	1.13	0.19	0.28	0.72	1.13	1.61	1.46	1.73	1.65	0.98
StVincent	0.97	0.78	3.60	−1.38	0.79	0.54	0.67	0.76	1.01	1.42	1.22	1.16	1.22	0.96
WilliamDavis	0.29	0.21	2.89	−2.15	0.75	0.14	0.06	0.07	0.28	0.54	0.48	0.49	0.30	0.29

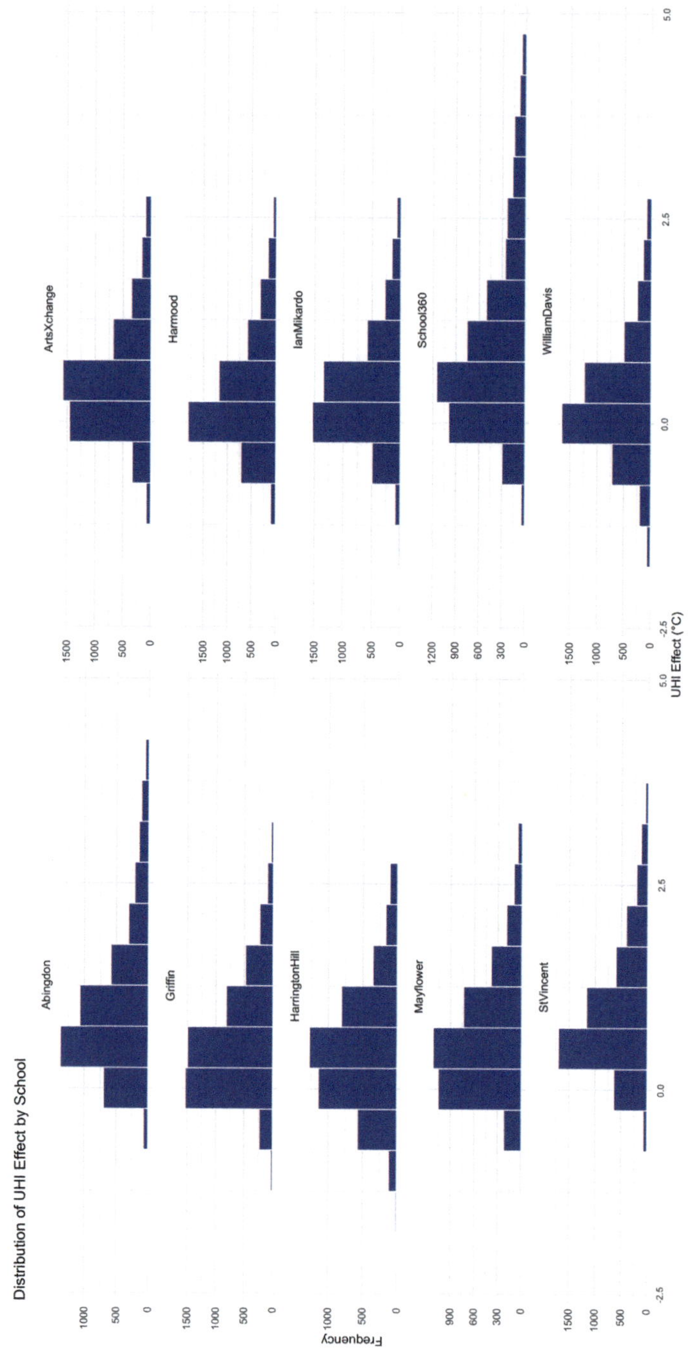

Fig. 6 UHI distribution overview

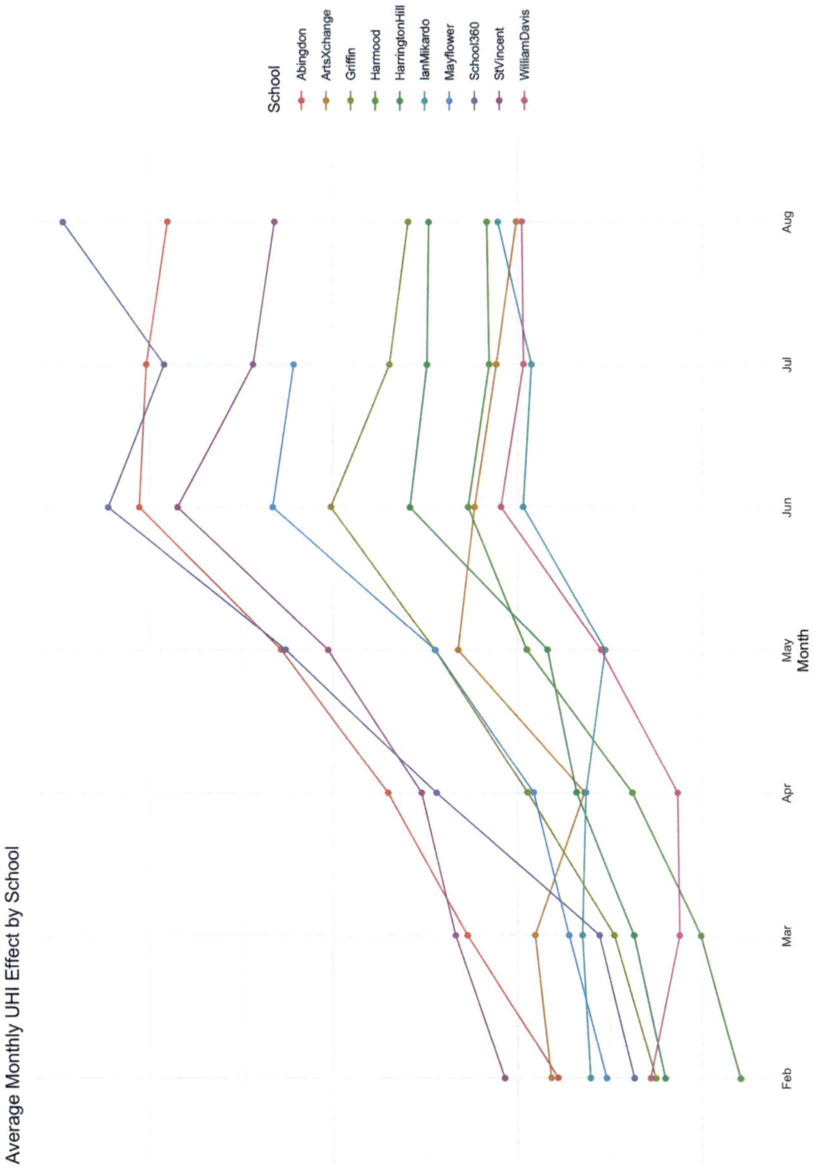

Fig. 7 Average UHI effect February to August for schools

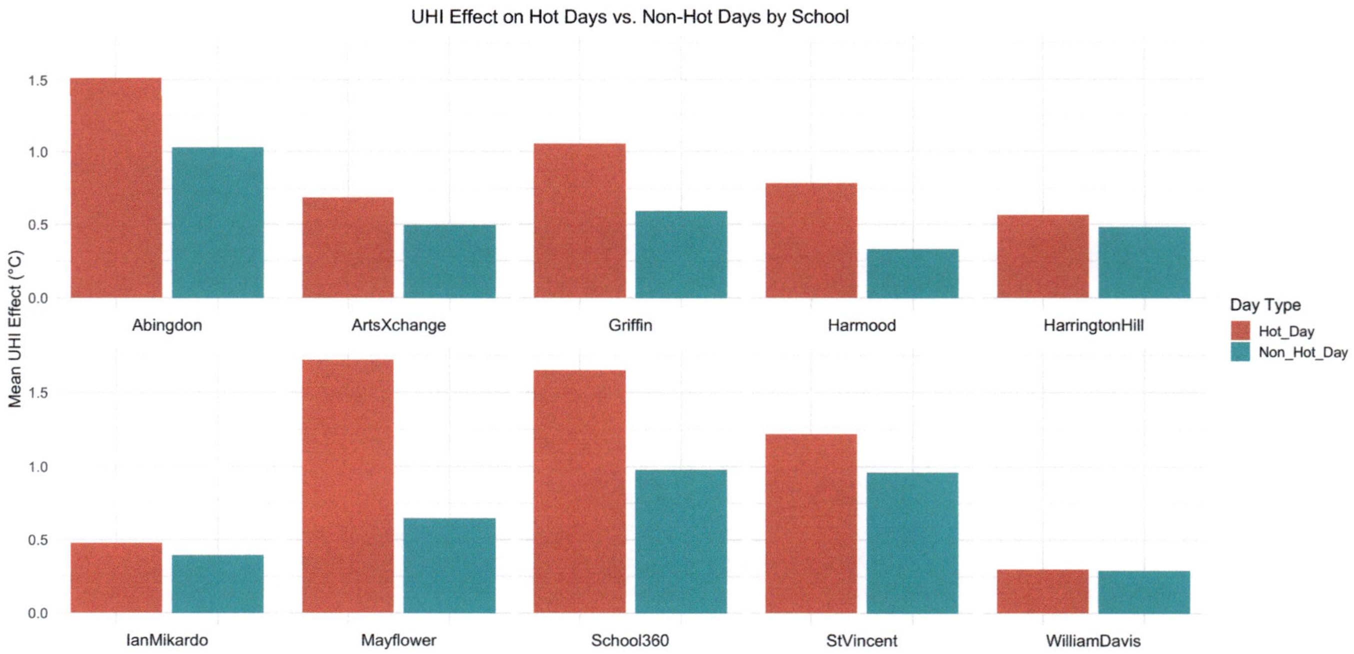

Fig. 8 UHI effect comparison for hot days and non-hot days

acknowledge certain limitations in the dataset. During the monitoring period, some data gaps emerged due to sensor battery failures. The first version of the sensor system relied on 2000 mAh LiPo batteries that required regular maintenance and replacement. Delays in replacing dead batteries caused interruptions in data collection. This limitation emphasized the need for power sustainability in sensor networks, driving the development of the second version, which incorporated solar-powered systems to reduce maintenance demands and ensure data continuity.

4 Next Stages: Citizen Scientists

The deployments of our sensor network in Greater London Authority (GLA) schools and the Queen Elizabeth Olympic Park (QEOP) demonstrate the potential of IoT technology for real-time Urban Heat Island (UHI) monitoring and microclimate analysis. In total, we have 43 devices now operational across various settings and have established a foundation for building a comprehensive urban climate monitoring system for London, potentially applicable to other cities around the world. Although this study focused on establishing an independent monitoring network, there is significant potential for integrating data from other existing networks. This would enhance spatial-temporal coverage and facilitate cross-validation of datasets to improve overall data accuracy and robustness for a more comprehensive understanding of UHI dynamics across London. Future work exploring the microclimate effect with other sensing networks and personal weather stations is currently being carried out in East London.

Building on the work carried out thus far, we aim to expand our network through citizen science involvement. By inviting community participation, we hope to create a more extensive and detailed microclimate monitoring system that spans London. To facilitate this engagement, we have made the designs of the Printed Circuit Board (PCB) available to the public and the Stevenson screen to be downloaded and 3D-printed. In acknowledgment of the technical barriers to engagement of the public with these resources, we also aim to offer workshops to raise awareness of UHIs and guide participants on how to produce, assemble, and deploy the sensors effectively. One potential venue for these workshops is the Eastbank Summer School, an annual program at Queen Elizabeth Olympic Park that engages young people in hands-on educational and creative activities. This collaborative approach serves multiple purposes. It will broaden the spatial coverage of our network, allowing for a more nuanced understanding of UHI patterns and the effect of variations in land use/land cover on local microclimate. Moreover, the engagement of citizen scientists aims to foster a collective sense of responsibility regarding urban climate challenges, bridging the gap between scientific research, public awareness, and policy. The insights gained from this expanded IoT-based sensor network have the potential to increase our understanding of the spatial distribution of UHIs in London, reinforcing the importance of data-driven urban management strategies. By providing open access to system designs and datasets, we aim to foster collaboration and innovation among researchers, urban planners, and policymakers, potentially, worldwide.

The introduction of IoT sensors into urban planning is particularly innovative because these devices provide real-time, high-resolution data on various urban parameters. IoT sensors offer continuous monitoring, delivering dynamic insights into how cities function at a granular level. For UHI studies, IoT sensors are especially valuable as they capture localized temperature variations caused by different land cover types, building materials, and human activities. Real-time monitoring of UHI hotspots can guide interventions such as strategically locating green spaces, optimizing urban design to improve airflow, and implementing zoning regulations to reduce heat exposure. Additionally, real-time data can support emergency response systems during heat waves, enabling rapid action to protect vulnerable populations. In summary, the integration of IoT sensor networks and citizen science initiatives offers a scalable and participatory approach to urban climate monitoring. This strategy not only enhances data collection but also empowers communities to engage in the development of targeted, data-driven climate adaptation strategies, contributing to the creation of more adaptive, resilient, and sustainable cities.

Acknowledgments Thanks go to the Greater London Authority and the London Legacy Development Corporation for their help and cooperation in deploying sensors. Thanks also go to the hosting Schools—notably William Davis Primary School, School 360, Griffin Primary School, Mayflower Primary School, Abingdon House School and College, St. Vincent's Catholic Primary School, ArtsXchange School Hackney, Harrington Hill Primary School, Harmood Secondary School, and Ian Mikardo High School. Finally, the authors are grateful to The Met Office for sharing the St James Park weather station's hourly temperature data.

References

Alizadeh, M. R., Abatzoglou, J. T., Adamowski, J. F., Prestemon, J. P., Chittoori, B., Akbari Asanjan, A., & Sadegh, M. (2022). Increasing heat-stress inequality in a warming climate. Earth's. *Future, 10*, 2. https://doi.org/10.1029/2021ef002488

GOV.UK, U. (2020). PHE heatwave mortality monitoring. GOV.UK. Available at: https://www.gov.uk/government/publications/phe-heatwave-mortality-monitoring (Accessed 16 October 2023).

Hudson-Smith, A., Hay, D., Wilson, D., & Gray, S. (2019). *Little book of connected environments and the internet of things* (pp. 4–5). Imagination Lancaster, Lancaster University, UK.

Hudson-Smith, A., Hügel, S., & Roumpani, F. (2020). Self-monitoring, analysis and reporting technologies: Smart cities and real-time data. In K. S. Willis & A. Aurigi (Eds.), *The Routledge companion to smart cities* (pp. 383–394). Routledge.

Hudson-Smith, A., Wilson, D., Gray, S., & Dawkins, O. (2021). Urban IoT: Advances, challenges, and opportunities for mass data collection, analysis, and visualization. In W. Shi et al. (Eds.), *Urban informatics. The urban book series*. Springer. https://doi.org/10.1007/978-981-15-8983-6_38

Karunarathne, G. G. K. W. M. S. I. R., Kulawansa, K. A. D. T., & Firdhous, M. F. M. (2018). Wireless communication technologies in internet of things: A critical evaluation. In *2018 international conference on intelligent and innovative computing applications (ICONIC). 2018 international conference on intelligent and innovative computing applications (ICONIC)* (pp. 1–5). https://doi.org/10.1109/ICONIC.2018.8601226

Ma, D., Brousse, O., De Jode, M., & Hudson-Smith, A (2024). 'Using Bespoke LoRaWAN heat sensors to explore microclimate effects within the London Urban Heat Islands–A Pilot Study in East London'. 32nd Annual GIS Research UK Conference (GISRUK), Zenodo. doi: https://doi.org/10.5281/zenodo.10897767.

Mavrogianni, A., et al. (2011). The comfort, energy and health implications of London's urban heat Island. *Building Services Engineering Research and Technology, 32*(1), 35–52. https://doi.org/10.1177/0143624410394530

Mekki, K., Bajic, E., Chaxel, F., & Meyer, F. (2019). A comparative study of LPWAN technologies for large-scale IoT deployment. *ICT Express, 5*(1), 1–7. https://doi.org/10.1016/j.icte.2017.12.005

Met Office. (2018) UKCP18 overview slidepack notes. Retrieved March 7, 2024, from https://www.metoffice.gov.uk/binaries/content/assets/metofficegovuk/pdf/research/ukcp/ukcp18-overview-slidepack-notes.ff.pdf.

Michelozzi, P., de Donato, F., Bisanti, L., Russo, A., Cadum, E., DeMaria, M., D'Ovidio, M., Costa, G., & Perucci, C. A. (2005). The impact of the summer 2003 heat waves on mortality in four Italian cities. *Euro Surveillance, 10*(7), 161–165.

Michelozzi, P., Accetta, G., De Sario, M., D'Ippoliti, D., Marino, C., Baccini, M., Biggeri, A., Anderson, H. R., Katsouyanni, K., Ballester, F., Bisanti, L., Cadum, E., Forsberg, B., Forastiere, F., Goodman, P. G., Hojs, A., Kirchmayer, U., Medina, S., Paldy, A., Schindler, C., Sunyer, J., Perucci, C. A., & PHEWE Collaborative Group. (2009). High temperature and hospitalizations for cardiovascular and respiratory causes in 12 European cities. *American Journal of Respiratory and Critical Care Medicine, 179*(5), 383–389. https://doi.org/10.1164/rccm.200802-217OC. Epub 2008 Dec 5.

Oke, T. R., et al. (2017). *Urban climates* (pp. 197–236). Cambridge University Press.

Sadek, R. A. and Elbadawy, H. M. (2022). 'Towards IoT Era with current and future wireless communication technologies: An overview'. in 2022 39th National Radio Science Conference (NRSC). 2022 39th National Radio Science Conference (NRSC), pp. 343–353. doi: https://doi.org/10.1109/NRSC57219.2022.9971196.

Santamouris, M., Cartalis, C., Synnefa, A., & Kolokotsa, D. M. (2015). On the impact of urban heat Island and global warming on the power demand and electricity consumption of buildings—A review. *Energy and Buildings, 98*, 119–124. https://doi.org/10.1016/j.enbuild.2014.09.052

Sun, C.-Y., Kato, S., & Gou, Z. (2019). Application of low-cost sensors for urban Heat Island assessment: A case study in Taiwan. *Sustainability, 11*(10), 2759. https://doi.org/10.3390/su11102759

Taylor, J., et al. (2015). Mapping the effects of urban heat Island, housing, and age on excess heat-related mortality in London. *Urban Climate, 14*, 517–528. https://doi.org/10.1016/j.uclim.2015.08.001

Warren, E. L., Young, D. T., Chapman, L., Muller, C., Grimmond, C. S. B., & Cai, X.-M. (2016). The Birmingham urban climate laboratory—A high density, urban meteorological dataset, from 2012–2014. *Scientific Data*. Nature Publishing Group, *3*(1), 160038. https://doi.org/10.1038/sdata.2016.38

Yang, L., Qian, F., Song, D.-X., & Zheng, K.-J. (2016). Research on urban Heat-Island effect. In *Procedia engineering. (fourth international conference on countermeasures to urban Heat Island, 30–31 may and 1 June 2016), 169* (pp. 11–18). https://doi.org/10.1016/j.proeng.2016.10.002

Zhao, L., Oppenheimer, M., Zhu, Q., Baldwin, J. W., Ebi, K. L., Bou-Zeid, E., Guan, K., & Liu, X. (2018). Interactions between urban heat islands and heat waves. *Environmental Research Letters, 13*(3), 034003. https://doi.org/10.1088/1748-9326/aa9f73

Zhou, D., Xiao, J., Bonafoni, S., Berger, C., Deilami, K., Zhou, Y., Frolking, S., Yao, R., Qiao, Z. and Sobrino, J. (2019). 'Satellite remote sensing of surface urban Heat Islands: Progress, challenges, and perspectives'. Remote sensing (Basel, Switzerland) : MDPI AG, 11 (1), 48 doi: https://doi.org/10.3390/rs11010048.

Daytime to Nighttime Street View Image Generation for 24-h Safety Perception Mapping

Jiajing Dai, Zhiyi Liu, Tingting Li, Tianyi Ren, Waishan Qiu, Da Chen, and Wenjing Li

Abstract Urban safety perception is crucial for quality of life, yet research on nighttime safety perception remains limited due to the lack of nighttime street view images (SVIs). This study uses generative models to translate daytime SVIs into nighttime ones (D2N), addressing this data gap. However, challenges exist in understanding how micro-level streetscape features affect nighttime perceived safety. To

Jiajing Dai, Zhiyi Liu and Tingting Li contributed equally with all other contributors.

J. Dai (✉)
Weitzman School of Design, University of Pennsylvania, Philadelphia, PA, USA
e-mail: daijiajing@alumni.upenn.edu

Z. Liu
School of Architecture and Urban Planning, Beijing University of Civil Engineering and Architecture, Beijing, China
e-mail: 202006010225@bucea.edu.cn

T. Li
School of Architecture, South Minzu University, Chengdu, China
e-mail: 202031707075@stu.swun.edu.cn

T. Ren
Department of Product Research and Development, Smart Gwei Tech, Shanghai, China

W. Qiu
Department of Urban Planning and Design, The University of Hong Kong, Pokfulam, HK SAR, China
e-mail: waishanq@hku.hk

D. Chen
Department of Computer Science, University of Bath, Bath, UK
e-mail: da.chen@bath.edu

W. Li
Center for Spatial Information Science, The University of Tokyo, Tokyo, Japan
e-mail: liwenjing@csis.u-tokyo.ac.jp

R. Goodspeed et al. (eds.), *Digital-Era Urban Transformations*, The Urban Book Series, https://doi.org/10.1007/978-3-031-98300-9_15

fill these gaps, this study collected 1,461 paired SVIs from high-density areas and 222 from low-density areas across four Chinese cities and Boston. We validated the high-density D2N (HD2N) and low-density D2N (LD2N) models using standard metrics and human-machine adversarial scoring. Our findings show that: (1) Urban density affects D2N translation accuracy, with higher densities posing challenges; (2) HD2N requires over 760 samples to converge; (3) Roads improve safety perception, while trees reduce it. These results offer insights for urban planning and safety improvements.

Keywords Street view image · Nighttime safety perception · Day-to-night translation · Day and night discrepancy · Urban environment auditing

1 Introduction

Urban safety is a critical concern in large cities like Chicago, Boston, Sydney, and New York City (Chalfin et al., 2020, 2022; Lee et al., 2021; Phan et al., 2010; Wadds, 2015) where it significantly influences quality of life and economic values (Brands & Schwanen, 2014; Cui et al., 2023). Notably, nighttime activities, which are vital for economic prosperity (Cui et al., 2021), residents' physical health (Saidi et al., 2020), and well-being (McCallum et al., 2022), often coincide with times of higher rates of crime compared to daytime (Liu et al., 2020), impacting the sense of safety, particularly among vulnerable groups (Keane, 1998; Phan et al., 2010; Truntsevsky et al., 2018).

Safety perception, distinct from actual safety metrics and crime rates, is largely subjective and typically assessed through self-reports and questionnaires (Li et al., 2013; Lin & Moudon, 2010; Syropoulos et al., 2024). With advancements in crowd-sourced surveys (Salesses et al., 2013) based on geo-tagged datasets like Street View Images (SVIs), researchers now have unprecedented tools to analyze urban elements and safety perceptions over a broad area (Kang et al., 2023). Tools like the TrueSkill rating system have been used to predict safety perception scores based on survey data from online volunteers who chose the safer-looking SVI in pairs (Naik et al., 2014). Furthermore, studies have shown that convolutional neural networks can learn human scoring patterns from limited datasets and classify images based on perceived safety (Dubey et al., 2016).

Despite these advancements, the lack of nighttime SVI data remains a significant gap, hindering our understanding of urban environments after dark. This gap is critical as the same urban settings can evoke vastly different perceptions between day and night, for example, green spaces that appear inviting during the day may seem intimidating at night. This study explores the potential of Generative Adversarial Networks (GANs) to generate nighttime SVIs from daytime ones, raising the possibility for subsequent research on nighttime safety perception (Liu et al., 2024). By extending Liu et al.'s training dataset, this research aims to enhance the accuracy of day-to-night (D2N) image conversions for urban areas of varying densities.

2 Literature Review

2.1 Safety Perception on Nighttime Streetscape

The importance of nighttime environmental quality is underscored by its impact on commercial vibrancy, crime rates, and public health (Chalfin et al., 2022; Foster & Giles-Corti, 2008). Nighttime environments, when well-designed, not only foster active public engagement through increased outdoor activities (Basu et al., 2023; Chen et al., 2024; Ngesan et al., 2013) but also mitigate psychological stress (Wang et al., 2019) and enhance public space utilization (Rahm et al., 2021; Wang & Shi, 2023). Furthermore, effective nighttime design can significantly reduce crime (Chalfin et al., 2022) and promote safety through measures like improved lighting and increased surveillance (Lis et al., 2023; Murray & Feng, 2016).

Gender differences in safety perception are notable, with women often feeling more vulnerable, particularly in transport scenarios (Carr, 2001; Cui et al., 2023). This perception affects their use of public transit at night, contributing to inequities in mobility (Plyushteva & Boussauw, 2020). For example, compared to men, women perceive a 15-min waiting time at unsafe transit stops as about 20% longer than the actual waiting time (Fan et al., 2016).

Research also highlights that well-lit public spaces are crucial for enhancing nighttime safety perceptions (Gargiulo et al., 2020; Park & Garcia, 2020; Svechkina et al., 2020). However, the challenge remains to balance the benefits of lighting with its environmental and economic costs, as excessive lighting can lead to increased energy consumption and light pollution, affecting both human health and natural ecosystems (Alzubaidi & Soori, 2012; Gallaway et al., 2010).

2.2 Measurements for Nighttime Environments

Recent advancements in urban studies have leveraged nighttime satellite imagery to assess aspects such as nighttime brightness (Barentine et al., 2020; Levin, 2017) and the extent of illumination (Bennett & Smith, 2017). However, this approach is limited by its resolution (Barentine et al., 2021), missing the micro-level streetscape details and making it challenging to connect directly with human subjective perceptions due to its lack of human eye-level perspective.

An alternative approach involves utilizing daytime SVI for analyzing built environments and safety perceptions (Biljecki & Ito, 2021; Li et al., 2022a, b; Sevtsuk et al., 2021; Wu, 2023). Techniques from computer vision (CV) and machine learning (ML) are applied to extract urban indices, such as the sky, buildings, roads, etc. (Gong et al., 2018, 2019; Yin et al., 2015). Human perceptions and urban indices have been applied to predict various urban dynamics, including house prices (Qiu et al., 2022), on-street crime (Su et al., 2023), etc. However, human perception of nocturnal SVI has not been studied due to the lack of images.

2.3 D2N Image-to-Image Translation

Advanced rendering techniques are employed to simulate nighttime SVI, yet computational challenges persist (Chandler et al., 2015; Conte, 2019; Ershov et al., 2022; Li et al., 2022a, b). Image translation models such as ToDayGAN (Anoosheh et al., 2019), ForkGAN (Zheng et al., 2020), Daydriex (Lee & Kang, 2021), and PatchGAN (Adhikari et al., 2022), predominantly employed within autonomous driving, focus on transforming nighttime SVIs into daytime to enhance the nocturnal driving experience. However, CycleGAN (Zhu et al., 2017) stands out for its suitability for D2N translation due to its capability to learn bidirectional mappings between distinct domains. Recent enhancements to CycleGAN involve incorporating transfer learning to better preserve semantic information during the translation process (Lee et al., 2023). Liu et al. (2024) have introduced a medium-density dataset of D&N SVIs from four cities in China and tested various generative models, including CycleGAN, Pix2Pix, and StableDiffusion. However, the high- and low-density urban areas have yet to be explored.

3 Data and Methods

3.1 Analysis Framework

To predict the safety perception of nighttime SVIs and the discrepancy between daytime and nighttime, this study implements the framework outlined in Fig. 1. First, the daytime SVIs from New York City (NYC) are processed using the HD2N model to generate corresponding nighttime SVIs. Second, subjective safety perception measures are collected through an online crowdsourcing survey, where the TrueSkill algorithm is applied to the nighttime SVIs to obtain safety perception scores. Simultaneously, the corresponding daytime SVIs are analyzed to extract view indices for various streetscape features. Third, the view indices, along with their associated safety perception scores, are fed into an ML model—Random Forest (RF) algorithm—designed to predict nighttime safety perception. Finally, to further analyze the discrepancy between daytime and nighttime safety perceptions, a multiple regression analysis is performed using Ordinary Least Squares (OLS) regression. This analysis investigates the relationships and variances between daytime and nighttime safety perceptions based on objectively measured streetscape features and the TrueSkill-derived safety scores.

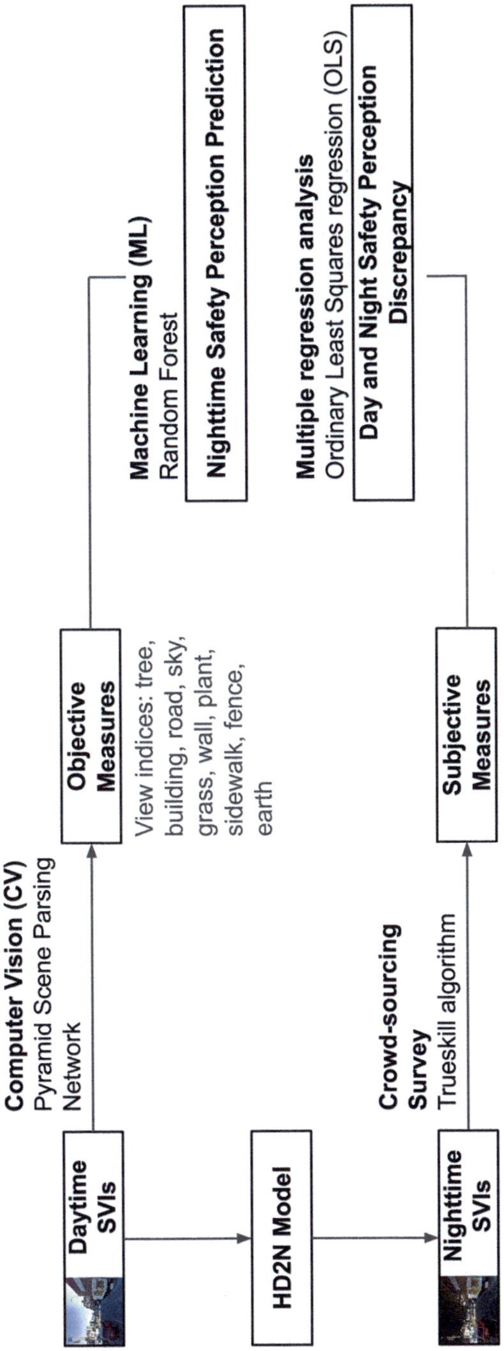

Fig. 1 Analysis framework

3.2 *Model Architecture*

3.2.1 Dataset Collection

The training data is sourced from high- and low-density urban areas in Chinese cities, including Shanghai, Wuhan, Chengdu, and Hong Kong, with a smaller portion representing street scenes from the central business district (CBD) of Boston, USA. The high-density training set size is 1,461, and the low-density training set size is 222. Figure 2 illustrates paired D&N SVI samples collected in Boston as an example. The testing set consists of 48 paired D&N samples from the CBD of Boston to conduct ground-truth validation of the HD2N and LD2N models.

3.2.2 Model Training and Validation

Based on previous research, CycleGAN was selected as the generative model to perform D2N image translation in high- and low-density urban areas (Fig. 3). Several quantitative assessment metrics were introduced to evaluate the quality of the generated images. These metrics include L1 distance, L2 distance, Structural Similarity Index (SSIM), Inception Score (IS), Inception Score standard deviation (IS_std), and Frechet Inception Distance (FID).

3.3 *Model Sensitivity Test*

In image generation tasks, expanding the dataset size often enhances the model's generalization ability while mitigating the risk of overfitting, indicating that the model's parameters have stabilized. Much of the focus has been on evaluating the quality of the generated D2N images. Research on the relationship between dataset

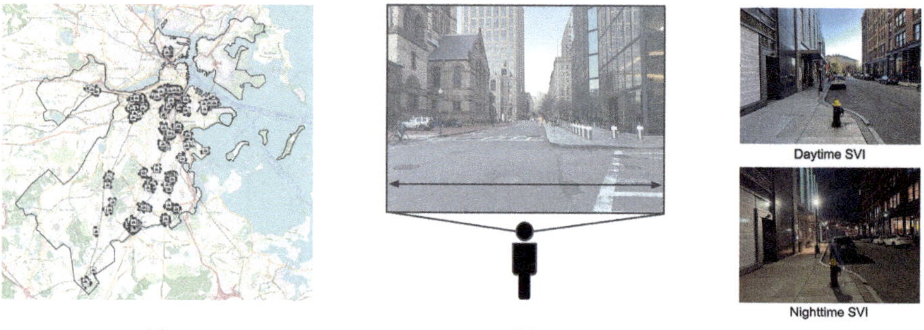

Fig. 2 SVI dataset collection process. (**a**) sampling area in Boston; (**b**) viewpoints of all SVIs; (**c**) daytime and nighttime SVI counterparts

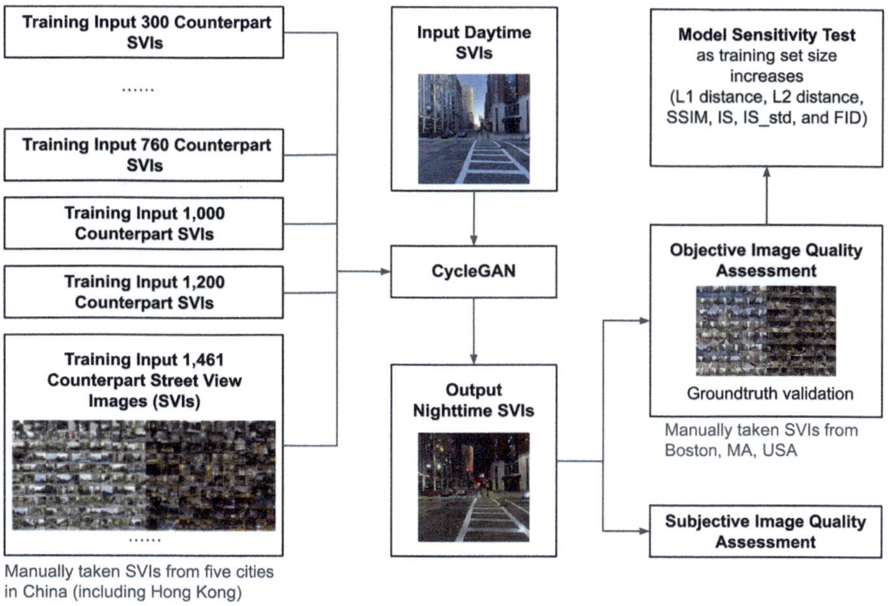

Fig. 3 High-density D2N model pipeline

size and model convergence, i.e., the model's sensitivity to variations in the dataset, remains limited. This gap limits our understanding of the models'capabilities and the optimal dataset size. Therefore, to quantitatively assess the impact of training set size on the convergence of the HD2N model, we evaluated performance metrics at key intervals as the training size increased (300, 400, 500, 654, 760, 1,000, 1,200, and 1,461 paired SVIs).

3.4 *Streetscape Features and Safety Perceptual Scores*

3.4.1 Safety Perception Consistency of Generated Nighttime SVIs

Before leveraging the model for broader urban safety applications, it is essential to establish the perceptual consistency between real nighttime SVIs and generated nighttime SVIs. Figure 4 shows the survey interface comprising two images: one ground truth nighttime SVI and one corresponding image generated by the HD2N model. Four participants were recruited for this study, each from a different professional background to encompass a variety of perspectives on safety perception. Participants were presented with pairs of images in a randomized order. For each pair, they were asked to judge which one felt safer between the real and generated night SVI. Each participant took a survey once a day for 2 days to check the consistency of the same participants in the two measurements. Intra-rater reliability (IRR)

Fig. 4 Generated and ground truth SVIs consistency survey

was analyzed to assess the consistency of each rater's safety perception over multiple sessions. This measure ensures that each participant's perception remains stable over time, lending credibility to the observed Intraclass Correlation Coefficient (ICC) results.

3.4.2 Collecting Subjective Safety Perception Scores from Online Survey

To gather safety perception scores for both daytime and nighttime SVIs, we developed a survey website where participants were asked to select the safer SVI (Fig. 5) from two randomly paired options. A total of 217 participants were recruited, all from diverse backgrounds. Notably, the spatial locations of the SVIs were not disclosed to the participants, and they were required to choose a winner in each pairwise comparison.

One hundred eight four SVIs from the high-density urban area of NYC were collected to generate nighttime SVIs utilizing the HD2N model. QGIS and Google Street View API were employed to collect SVIs. To ensure a consistent viewing perspective, identical camera settings, and image resolution were applied using three key API parameters. The 'heading' parameter was set to align with the street centerline, the 'FOV' (field of view) was fixed at 90°, and the 'pitch' (camera tilt angle) was kept at 0°. The image resolution was uniformly standardized at 640×400 pixels, and only the view parallel to the street centerline's tangent was captured.

These pairwise preferences from the surveys were converted into ranked scores using the TrueSkill algorithm (Microsoft, 2005; Minka et al., 2018; Qiu et al., 2022). Statistical studies indicate that a training sample size of at least 75 to 100 (Figueroa et al., 2012), or about ten times the number of parameters (Haldar, 2015), is necessary to train robust models. Each SVI needed to be compared more than 10 times to reach convergence (Herbrich et al., 2006). These 184 SVIs, labeled with

Fig. 5 Collecting safety perception scores for SVIs with an online survey

safety perception scores, then formed the training set for the RF model aimed at predicting subjective perception scores.

3.4.3 Classification of the Physical Features

Semantic scene parsing is a key technology for feature recognition, segmentation, and analysis in SVI (B. Zhou et al., 2017). PSPNet is a cutting-edge semantic segmentation method with high pixel-level accuracy (Zhao et al., 2017), employed in this study. Previous studies have explored how street features influence both urban design quality (Chen et al., 2023; Ewing & Handy, 2009; Ma et al., 2021) and perception quality (Verma et al., 2020; Zhang et al., 2018). The view index, the percentage of feature pixels relative to the total number of pixels in SVIs (Fu et al., 2019), is a key indicator of quantifying various physical features in SVIs, shown in formula (1).

$$VI_{obj} = \frac{\sum_{i=1}^{n} PIXEI_{obj}}{\sum_{i=1}^{n} PIXEL_{obinl}}, obj \in \{tree, building, sky, etc\} \quad (1)$$

3.4.4 Predicting Subjective Safety Perception Scores

Ten extracted view indices were utilized as explanatory variables. The dataset, comprising 184 images, was partitioned into training and testing sets with an allocation of 80% and 20%, respectively. RF has demonstrated efficacy in predicting safety perception scores in previous urban scene perception studies (Qiu et al., 2022; Su et al., 2023; Wang et al., 2019; Yao et al., 2019). The model's quality was assessed using the coefficient of determination (R^2) and the Mean Absolute Error (MAE).

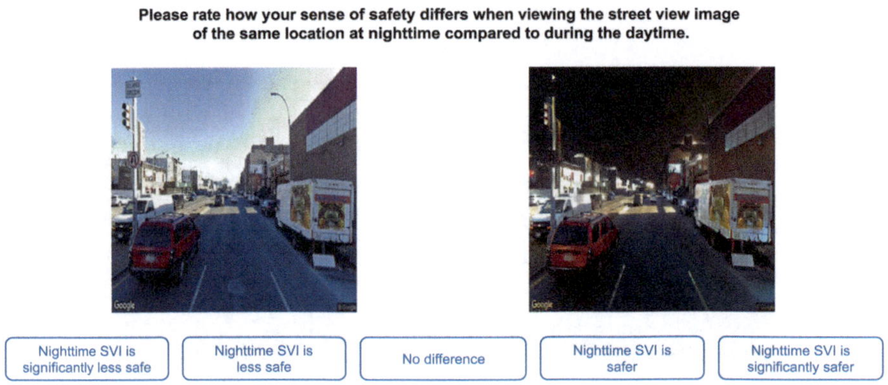

Fig. 6 A pairwise voting system to measure the safety perception discrepancy between D&N

Furthermore, the feature importance of ten types of streetscape objective features was calculated, which quantifies the impact of each variable on the model's predictive accuracy.

3.4.5 Quantification of Safety Perception Discrepancy between D&N

Multivariate regression analysis is conducted to assess the impact of various streetscape features on nighttime safety perceptions. As depicted in Fig. 6, the dependent variable in Model 0 was assessed using a Likert scale that ranged from −2 (significantly less safe) to +2 (significantly safer). Model 1 calculated the difference between the nighttime and daytime TrueSkill scores of perceived safety. The streetscape features analyzed encompassed roads, trees, plants, buildings, grass, walls, earth, and sidewalks, which were quantified using streetscape feature view indices.

4 Results and Findings

4.1 Objective Image Quality Assessment

4.1.1 Model Sensitivity Test Results

Six objective measures are used to evaluate the quality of images generated by the HD2N as the training size increases. Figure 7 suggests that the HD2N model benefits significantly from increased training data in terms of both error reduction and improved quality of generated images. However, the diminishing returns beyond certain points (e.g., after 760 or 1000 samples for some metrics) indicate the model might be reaching its learning capacity under the current configuration.

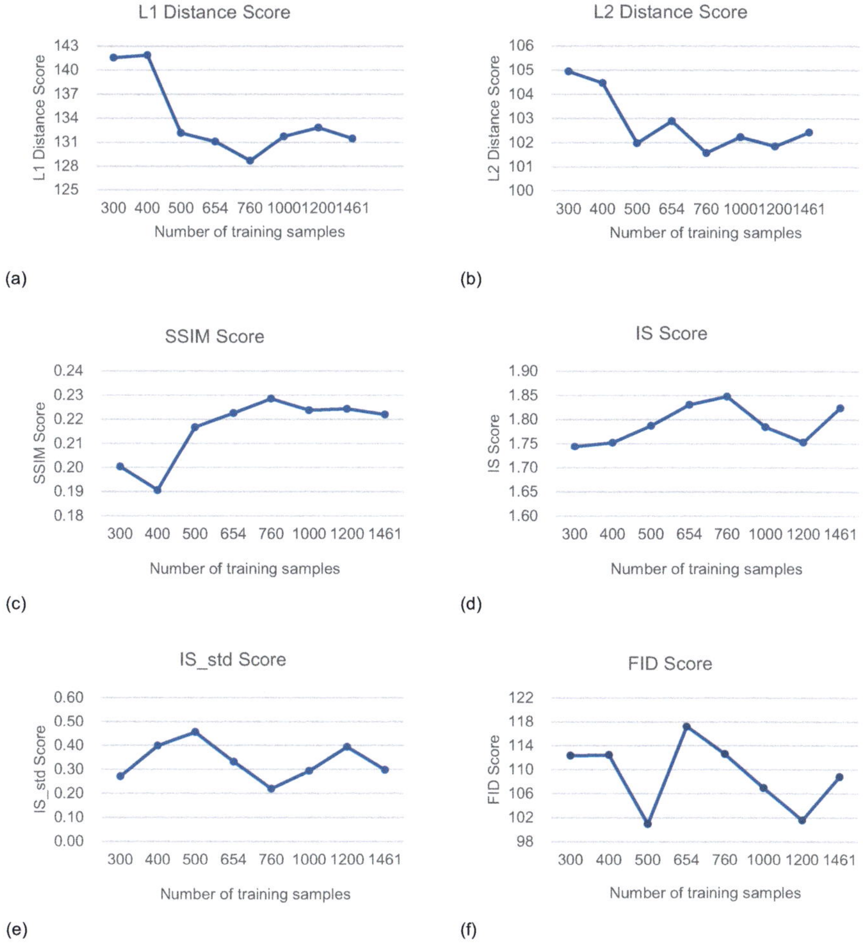

Fig. 7 Metrics convergence as the training size increases. Parts (**a**) to (**f**)

4.1.2 Comparison between Various Densities

Table 1 presents the performance comparison across several models—CycleGAN (D2N), D2N, LD2N, and HD2N—trained on datasets of varying urban densities and training sizes. Liu et al. made some human-guided improvements to the output of CycleGAN (D2N), making the D2N model yield slightly better. In summary, while the CycleGAN (D2N) and D2N models serve as reliable baselines for D2N image translation in medium-density urban environments, the LD2N model demonstrated better pixel-level accuracy but struggled with feature distribution. The HD2N models, despite increasing the training set size, failed to surpass the baseline models in most metrics, particularly in terms of pixel accuracy and structural preservation. However, some improvement in FID was observed with larger datasets.

Table 1 Performance comparison of D2N models across various urban density

Model	Training size	L1 distance	L2 distance	SSIM	IS score	FID
CycleGAN (D2N)	638	123.03	100.53	0.24	2.54	115.23
D2N	638	122.89	100.54	0.24	2.48	115.17
LD2N	222	100.43	101.47	0.28	1.93	169.19
HD2N	1461	131.48	102.44	0.22	1.82	108.84

Table 2 Random forest model performance

	Train MSE	Test MSE	Train R^2	Test R^2
Daytime	0.067	0.219	0.570	−0.039
Nighttime	0.067	0.310	0.625	0.052

4.2 Subjective Image Quality Assessment

The generated nighttime SVIs from the HD2N model were utilized to audit nighttime streetscape safety perceptions in the subjective IQA, yielding a mean opinion score (MOS) of 0.835. This indicates a high level of satisfaction with the image quality produced by the D2N model, reflecting the effectiveness of the generated images in conveying the intended safety perceptions. The within-group correlation coefficients provide further insights into the reliability of the subjective assessments. The moderate ICC values and high intra-rater reliability together reinforce the validity of the subjective assessment process, supporting its use to provide a well-rounded evaluation of the perceived safety of nighttime streetscapes.

4.3 Safety Perceptual Scores Prediction

The RF model uses streetscape indices to predict safety perception scores from TrueSkill. As shown in Table 2, both models perform well on training data but drop in performance on test data, especially the daytime model. The test Mean Squared Error (MSE) shows signs of overfitting. In Fig. 8, 'building', 'road', and 'sidewalk'emerged as the most influential features, indicating that structural elements of the streetscape play a critical role in daytime safety perception. In contrast, the features 'tree' and 'building' were identified as having the highest importance scores, suggesting a significant association of these features with safety perception at night.

4.4 D&N Safety Perception Discrepancy

Both models quantify how these features influence the difference in safety perception between night and day, as captured by two different scoring methods: the Likert scale and the TrueSkill score difference. The results are shown in Table 3. In model

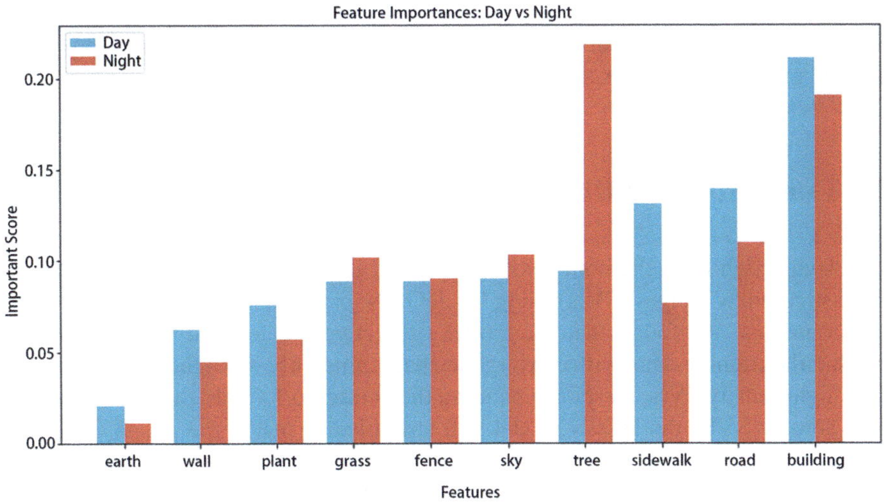

Fig. 8 Feature importances in random forest

Table 3 OLS regression results

	Model 0 Likert score		Model 1 TrueSkill score difference	
Variables	Std coef.	$p > \|t\|$	Std coef.	$p > \|t\|$
constant	−0.6128		0.1795	/
road	0.9166	**	0.1202	
fence	0.4944		−0.6970	
tree	−0.4761	**	−0.8970	***
plant	1.6509	+	0.0554	
building	0.1996		−0.0968	
grass	1.0390		−0.4387	
wall	0.5217		−3.4870	+
earth	0.7138		0.4356	
sidewalk	0.4789		0.0594	
R^2	0.167		0.148	

+ $p < 0.1$. **$p < 0.01$. ***$p < 0.001$

0, roads ($p < 0.01$) and trees ($p < 0.01$) have very significant effects on the Likert score for safety discrepancy. Roads positively correlate with a higher sense of nighttime safety, while trees have a negative impact, suggesting that well-maintained or visible roads enhance safety perceptions, and denser tree coverage might be perceived as less safe during nighttime. The presence of plants ($p < 0.1$) indicates a marginally significant positive effect on safety perception, possibly due to the aesthetic and barrier effects provided by greenery. In model 1, trees ($p < 0.001$) exhibit a highly significant negative impact on the TrueSkill score difference between nighttime and daytime, consistently indicating that trees may be perceived as reduced safety features during the night across different measurement scales. Wall

($p < 0.1$) has a marginally negative influence, possibly due to perceived barriers to natural surveillance. Model 0 ($R^2 = 0.167$) and Model 1 ($R^2 = 0.148$) both possess limited explanatory power.

5 Conclusion and Discussion

Our study highlights the complexities involved in transforming daytime to nighttime street view images (SVIs) using CycleGAN, particularly in high-density urban environments. We found that the model's effectiveness in such settings requires substantial training data, with convergence beginning only when the dataset exceeds 760 high-density SVIs. Despite expanding the dataset to 1,461 high-density and 222 low-density SVIs, the HD2N model still struggled to match the image quality of models trained on medium- or low-density data. This study reveals the necessity and feasibility of classifying different urban densities for D2N image translation, and also emphasizes the value of collecting D&N SVI on a larger scale and increasing the training dataset size for output accuracy.

Additionally, our research into safety score prediction using Random Forest models revealed significant overfitting issues, which means that the model has overlearned the features in the training data and cannot generalize well to new data. Future research can help verify and further analyze the causes of overfitting by collecting more safety perception scores from the questionnaire. Increasing the amount of data can not only improve the generalization ability of the model but also provide richer information to help identify which features have a more universal impact on safety perception. If overfitting still exists, it may be necessary to adjust the complexity of the model or introduce more methods to further optimize the model. Another possibility suggests that visual features alone may be insufficient for accurately modeling human safety perceptions, which are notably complex and vary significantly between daytime and nighttime.

Key findings include the variable influence of urban elements on safety perceptions. During the day, buildings, roads, and sidewalks are crucial for facilitating effective transportation and pedestrian flow, directly impacting safety perceptions. Conversely, at night, the focus shifts towards trees and building illumination, with trees negatively impacting safety due to potential obstructions and shadows, while well-lit buildings contribute to a sense of security and activity.

Our results align with Park and Garcia (2020), indicating that trees enhance safety perception during the day but pose safety risks at night due to visibility issues. The importance of structural elements like sidewalks also resonates with their findings. Based on the safety perception results, we can optimize the design and characteristics of streetscape features in different regions. Nighttime SVI can also assist crime prevention and environmental psychological intervention, and has good application prospects. However, our study also acknowledges the limitations. Morphological complexity, such as street lighting and land use, should be considered in future research.

Ethics Approval The article has obtained ethical approval from the University of Hong Kong's Human Research Ethics Committee (HREC) with the HREC reference number EA240182. The expiration date is May 5, 2027.

References

Adhikari, B., Hari, K. C., & Thapa, S. (2022). Night to day and day to night image transfer using generative adversarial network. *International Research Journal of Modernization in Engineering Technology and Science*.

Alzubaidi, S., & Soori, P. K. (2012). Study on energy efficient street lighting system design. *IEEE International Power Engineering and Optimization Conference, 2012*, 291–295. https://doi.org/10.1109/PEOCO.2012.6230877

Anoosheh, A., Sattler, T., Timofte, R., Pollefeys, M., & Gool, L. V. (2019). Night-to-day image translation for retrieval-based localization. In *2019 international conference on robotics and automation (ICRA)* (pp. 5958–5964). https://doi.org/10.1109/ICRA.2019.8794387

Barentine, J. C., Kundracik, F., Kocifaj, M., Sanders, J. C., Esquerdo, G. A., Dalton, A. M., Foott, B., Grauer, A., Tucker, S., & Kyba, C. C. M. (2020). Recovering the city street lighting fraction from skyglow measurements in a large-scale municipal dimming experiment. *Journal of Quantitative Spectroscopy and Radiative Transfer, 253*, 107120. https://doi.org/10.1016/j.jqsrt.2020.107120

Barentine, J. C., Walczak, K., Gyuk, G., Tarr, C., & Longcore, T. (2021). A case for a new satellite mission for remote sensing of night lights. *Remote Sensing, 13*(12), 2294. https://doi.org/10.3390/rs13122294

Basu, N., Oviedo-Trespalacios, O., King, M., Kamruzzaman, M., & Haque, M. M. (2023). What do pedestrians consider when choosing a route? The role of safety, security, and attractiveness perceptions and the built environment during day and night walking. *Cities, 143*, 104551. https://doi.org/10.1016/j.cities.2023.104551

Bennett, M. M., & Smith, L. C. (2017). Advances in using multitemporal night-time lights satellite imagery to detect, estimate, and monitor socioeconomic dynamics. *Remote Sensing of Environment, 192*, 176–197. https://doi.org/10.1016/j.rse.2017.01.005

Biljecki, F., & Ito, K. (2021). Street view imagery in urban analytics and GIS: A review. *Landscape and Urban Planning, 215*, 104217. https://doi.org/10.1016/j.landurbplan.2021.104217

Brands, J., & Schwanen, T. (2014). Experiencing and governing safety in the night-time economy: Nurturing the state of being carefree. *Emotion, Space and Society, 11*, 67–78. https://doi.org/10.1016/j.emospa.2013.08.004

Carr, N. (2001). An exploratory study of gendered di!Erences in young tourists perception of danger within London. *Tourism Management*.

Chalfin, A., Kaplan, J., & LaForest, M. (2020). Street light outages, public safety and crime displacement.

Chalfin, A., Hansen, B., Lerner, J., & Parker, L. (2022). Reducing crime through environmental design: Evidence from a randomized experiment of street lighting in new York City. *Journal of Quantitative Criminology, 38*(1), 127–157. https://doi.org/10.1007/s10940-020-09490-6

Chandler, J., Yang, L., & Ren, L. (2015). Procedural window lighting effects for real-time city rendering. In *Proceedings of the 19th symposium on interactive 3D graphics and games* (pp. 93–99). https://doi.org/10.1145/2699276.2699290

Chen, X., Li, G., Mehmood, M. S., Jin, A., Du, M., & Xue, Y. (2023). Using street view images to examine the impact of built environment on street property crimes in the Old District of CA City, China. *Discrete Dynamics in Nature and Society, 2023*, 1–15. https://doi.org/10.1155/2023/1470452

Chen, C., Liu, X., Qiu, T., & Liu, J. (2024). Evaluation on visual perception of urban street environment from the perspective of being friendly to nighttime running. *Landscape Architecture, 31*(4), 36–43. https://doi.org/10.3724/j.fjyl.202310160466

Conte, M. (2019). *Real-time rendering of cities at night*. https://papyrus.bib.umontreal.ca/xmlui/handle/1866/22119

Cui, Y., Shi, K., Jiang, L., Qiu, L., & Wu, S. (2021). Identifying and evaluating the nighttime economy in China using multisource data. *IEEE Geoscience and Remote Sensing Letters, 18*(11), 1906–1910. https://doi.org/10.1109/LGRS.2020.3010936

Cui, Q., Zhang, Y., Yang, G., Huang, Y., & Chen, Y. (2023). Analysing gender differences in the perceived safety from street view imagery. *International Journal of Applied Earth Observation and Geoinformation, 124*, 103537. https://doi.org/10.1016/j.jag.2023.103537

Dubey, A., Naik, N., Parikh, D., Raskar, R., & Hidalgo, C. A. (2016). Deep learning the city: Quantifying urban perception at a global scale. In B. Leibe, J. Matas, N. Sebe, & M. Welling (Eds.), *Computer vision—ECCV 2016* (pp. 196–212). Springer International Publishing. https://doi.org/10.1007/978-3-319-46448-0_12

Ershov, E., Savchik, A., Shepelev, D., Banić, N., Brown, M. S., Timofte, R., Koščević, K., Freeman, M., Tesalin, V., Bocharov, D., Semenkov, I., Subašic, M., Lončarić, S., Terekhin, A., Liu, S., Feng, C., Wang, H., Zhu, R., Li, Y., Mudenagudi, U. (2022). *NTIRE 2022 challenge on night photography rendering*. 1287–1300. https://openaccess.thecvf.com/content/CVPR2022W/NTIRE/html/Ershov_NTIRE_2022_Challenge_on_Night_Photography_Rendering_CVPRW_2022_paper.html

Ewing, R., & Handy, S. (2009). Measuring the unmeasurable: Urban design qualities related to walkability. *Journal of Urban Design, 14*(1), 65–84. https://doi.org/10.1080/13574800802451155

Fan, Y., Guthrie, A., & Levinson, D. (2016). Waiting time perceptions at transit stops and stations: Effects of basic amenities, gender, and security. *Transportation Research Part A: Policy and Practice, 88*, 251–264. https://doi.org/10.1016/j.tra.2016.04.012

Figueroa, R. L., Zeng-Treitler, Q., Kandula, S., & Ngo, L. H. (2012). Predicting sample size required for classification performance. *BMC medical informatics and decision making 12*(1) https://doi.org/10.1186/1472-6947-12-8

Foster, S., & Giles-Corti, B. (2008). The built environment, neighborhood crime and constrained physical activity: An exploration of inconsistent findings. *Preventive Medicine, 47*(3), 241–251. https://doi.org/10.1016/j.ypmed.2008.03.017

Fu, X., Jia, T., Zhang, X., Li, S., & Zhang, Y. (2019). Do street-level scene perceptions affect housing prices in Chinese megacities? An analysis using open access datasets and deep learning. *PLoS One, 14*(5), e0217505. https://doi.org/10.1371/journal.pone.0217505

Gallaway, T., Olsen, R. N., & Mitchell, D. M. (2010). The economics of global light pollution. *Ecological Economics, 69*(3), 658–665. https://doi.org/10.1016/j.ecolecon.2009.10.003

Gargiulo, I., Garcia, X., Benages-Albert, M., Martinez, J., Pfeffer, K., & Vall-Casas, P. (2020). Women's safety perception assessment in an urban stream corridor: Developing a safety map based on qualitative GIS. *Landscape and Urban Planning, 198*, 103779. https://doi.org/10.1016/j.landurbplan.2020.103779

Gong, F.-Y., Zeng, Z.-C., Zhang, F., Li, X., Ng, E., & Norford, L. K. (2018). Mapping sky, tree, and building view factors of street canyons in a high-density urban environment. *Building and Environment, 134*, 155–167. https://doi.org/10.1016/j.buildenv.2018.02.042

Gong, Z., Ma, Q., Kan, C., & Qi, Q. (2019). Classifying street spaces with street view images for a spatial indicator of urban functions. *Sustainability, 11*(22), Article 22. https://doi.org/10.3390/su11226424

Haldar, M. (2015). How much training data do you need? Medium. https://medium.com/@malay.haldar/how-much-training-data-do-you-need-da8ec091e956

Herbrich, R., Minka, T., & Graepel, T. (2006). TrueSkill™: a Bayesian skill rating system. Advances in neural information processing systems, 19.

Kang, Y., Abraham, J., Ceccato, V., Duarte, F., Gao, S., Ljungqvist, L., Zhang, F., Näsman, P., & Ratti, C. (2023). Assessing differences in safety perceptions using GeoAI and survey across neighbourhoods in Stockholm, Sweden. *Landscape and Urban Planning, 236*, 104768. https://doi.org/10.1016/j.landurbplan.2023.104768

Keane, C. (1998). Evaluating the influence of fear of crime as an environmental mobility restrictor on women's routine activities. *Environment and Behavior, 30*(1), 60–74. https://doi.org/10.1177/0013916598301003

Lee, E., & Kang, S. (2021). Daydriex: Translating nighttime scenes towards daytime driving experience at night. *Applied Sciences, 11*(5), Article 5. https://doi.org/10.3390/app11052013

Lee, M., Jackson, J., & Ellis, J. R. (2021). Everyday aesthetics, space, and the sensory: Fear of crime and affect in inner Sydney. *Open Qualitative Criminology*. https://doi.org/10.21428/51bae76e.19a822b4

Lee, J., Shiotsuka, D., Bang, G., Endo, Y., Nishimori, T., Nakao, K., & Kamijo, S. (2023). Day-to-night image translation via transfer learning to keep semantic information for driving simulator. *IATSS Research, 47*(2), 251–262. https://doi.org/10.1016/j.iatssr.2023.04.001

Levin, N. (2017). The impact of seasonal changes on observed nighttime brightness from 2014 to 2015 monthly VIIRS DNB composites. *Remote Sensing of Environment, 193*, 150–164. https://doi.org/10.1016/j.rse.2017.03.003

Li, Z., Zhou, X., Wang, X., & Guo, Z. (2013). Study on subjective and objective safety and application of expressway. *Procedia - Social and Behavioral Sciences, 96*, 1622–1630. https://doi.org/10.1016/j.sbspro.2013.08.184

Li, Y., Peng, L., Wu, C., & Zhang, J. (2022a). Street view imagery (SVI) in the built environment: A theoretical and systematic review. *Buildings, 12*(8), 1167. https://doi.org/10.3390/buildings12081167

Li, Z., Yi, S., & Ma, Z. (2022b). *Rendering nighttime image via cascaded color and brightness compensation* (arXiv:2204.08970). arXiv. http://arxiv.org/abs/2204.08970

Lin, L., & Moudon, A. V. (2010). Objective versus subjective measures of the built environment, which are most effective in capturing associations with walking? *Health & Place, 16*(2), 339–348. https://doi.org/10.1016/j.healthplace.2009.11.002

Lis, A., Zienowicz, M., Kukowska, D., Zalewska, K., Iwankowski, P., & Shestak, V. (2023). How to light up the night? The impact of city park lighting on visitors' sense of safety and preferences. *Urban Forestry & Urban Greening, 89*, 128124. https://doi.org/10.1016/j.ufug.2023.128124

Liu, J., Deng, Y., Wang, Y., Huang, H., Du, Q., & Ren, F. (2020). Urban nighttime leisure space mapping with nighttime light images and POI data. *Remote Sensing, 12*(3), Article 3. https://doi.org/10.3390/rs12030541

Liu, Z., Li, T., Ren, T., Chen, D., Li, W., & Qiu, W. (2024). Day-to-night street view image generation for 24-hour urban scene auditing using generative AI. *Journal of Imaging, 10*(5), 112. https://doi.org/10.3390/jimaging10050112

Ma, X., Ma, C., Wu, C., Xi, Y., Yang, R., Peng, N., Zhang, C., & Ren, F. (2021). Measuring human perceptions of streetscapes to better inform urban renewal: A perspective of scene semantic parsing. *Cities, 110*, 103086. https://doi.org/10.1016/j.cities.2020.103086

McCallum, I., Kyba, C. C. M., Bayas, J. C. L., Moltchanova, E., Cooper, M., Cuaresma, J. C., Pachauri, S., See, L., Danylo, O., Moorthy, I., Lesiv, M., Baugh, K., Elvidge, C. D., Hofer, M., & Fritz, S. (2022). Estimating global economic well-being with unlit settlements. *Nature Communications, 13*(1), 2459. https://doi.org/10.1038/s41467-022-30099-9

Microsoft. (2005). TrueSkillTM Ranking System. Microsoft Research. https://www.microsoft.com/en-us/research/project/trueskill-ranking-system/

Minka, T., Cleven, R., & Zaykov, Y. (2018). TrueSkill 2: An improved Bayesian skill rating system. https://www.microsoft.com/en-us/research/publication/trueskill-2-improved-bayesian-skill-rating-system/

Murray, A. T., & Feng, X. (2016). Public street lighting service standard assessment and achievement. *Socio-Economic Planning Sciences, 53*, 14–22. https://doi.org/10.1016/j.seps.2015.12.001

Naik, N., Philipoom, J., Raskar, R., & Hidalgo, C. (2014). Streetscore—Predicting the perceived safety of one million streetscapes. In *2014 IEEE conference on computer vision and pattern recognition workshops* (pp. 793–799). https://doi.org/10.1109/CVPRW.2014.121

Ngesan, M. R., Karim, H. A., & Zubir, S. S. (2013). Night time place identity: The relationship between urban public parks setting and human leisure behavior. 1281–1292. https://doi.org/10.2495/SC131092

Park, Y., & Garcia, M. (2020). Pedestrian safety perception and urban street settings. *International Journal of Sustainable Transportation, 14*(11), 860–871. https://doi.org/10.1080/15568318.2019.1641577

Phan, L., Fefferman, N., Hui, D., & Brugge, D. (2010). Impact of street crime on Boston Chinatown. *Local Environment, 15*(5), 481–491. https://doi.org/10.1080/13549831003767273

Plyushteva, A., & Boussauw, K. (2020). Does night-time public transport contribute to inclusive night mobility? Exploring Sofia's night bus network from a gender perspective. *Transport Policy, 87*, 41–50. https://doi.org/10.1016/j.tranpol.2020.01.002

Qiu, W., Zhang, Z., Liu, X., Li, W., Li, X., Xu, X., & Huang, X. (2022). Subjective or objective measures of street environment, which are more effective in explaining housing prices? *Landscape and Urban Planning, 221*, 104358. https://doi.org/10.1016/j.landurbplan.2022.104358

Rahm, J., Sternudd, C., & Johansson, M. (2021). "In the evening, I don't walk in the park": The interplay between street lighting and greenery in perceived safety. *Urban Design International, 26*(1), 42–52. https://doi.org/10.1057/s41289-020-00134-6

Saidi, O., Davenne, D., Lehorgne, C., & Duché, P. (2020). Effects of timing of moderate exercise in the evening on sleep and subsequent dietary intake in lean, young, healthy adults: Randomized crossover study. *European Journal of Applied Physiology, 120*(7), 1551–1562. https://doi.org/10.1007/s00421-020-04386-6

Salesses, P., Schechtner, K., & Hidalgo, C. A. (2013). The collaborative image of the city: Mapping the inequality of urban perception. *PLoS One, 8*(7), e68400. https://doi.org/10.1371/journal.pone.0068400

Sevtsuk, A., Basu, R., Li, X., & Kalvo, R. (2021). A big data approach to understanding pedestrian route choice preferences: Evidence from San Francisco. *Travel Behaviour and Society, 25*, 41–51. https://doi.org/10.1016/j.tbs.2021.05.010

Su, N., Li, W., & Qiu, W. (2023). Measuring the associations between eye-level urban design quality and on-street crime density around New York subway entrances. *Habitat International, 131*, 102728. https://doi.org/10.1016/j.habitatint.2022.102728

Svechkina, A., Trop, T., & Portnov, B. A. (2020). How much lighting is required to feel safe when walking through the streets at night? *Sustainability, 12*(8), 3133. https://doi.org/10.3390/su12083133

Syropoulos, S., Leidner, B., Mercado, E., Li, M., Cros, S., Gómez, A., Baka, A., Chekroun, P., & Rottman, J. (2024). How safe are we? Introducing the multidimensional model of perceived personal safety. *Personality and Individual Differences, 224*, 112640. https://doi.org/10.1016/j.paid.2024.112640

Truntsevsky, Y. V., Lukiny, I. I., Sumachev, A. V., & Kopytova, A. V. (2018). A smart city is a safe city: The current status of street crime and its victim prevention using a digital application. *MATEC Web of Conferences, 170*, 01067. https://doi.org/10.1051/matecconf/201817001067

Verma, D., Jana, A., & Ramamritham, K. (2020). Predicting human perception of the urban environment in a spatiotemporal urban setting using locally acquired street view images and audio clips. *Building and Environment, 186*, 107340. https://doi.org/10.1016/j.buildenv.2020.107340

Wadds, P. (2015). Crime, policing and (in)security: Press depictions of Sydney's night-time economy. *Current Issues in Criminal Justice, 27*(1), 95–112. https://doi.org/10.1080/10345329.2015.12036033

Wang, X., & Shi, W. (2023). Differences in users' activity characteristics and spatial patterns in neighborhood parks during the late afternoon and evening periods. *Land, 12*(7), 1438. https://doi.org/10.3390/land12071438

Wang, R., Liu, Y., Lu, Y., Zhang, J., Liu, P., Yao, Y., & Grekousis, G. (2019). Perceptions of built environment and health outcomes for older Chinese in Beijing: A big data approach with street view images and deep learning technique. *Computers, Environment and Urban Systems, 78*, 101386. https://doi.org/10.1016/j.compenvurbsys.2019.101386

Wu, F. (2023). Examining non-linear relationship between streetscape features and propensity of walking to school in Hong Kong using machine learning techniques. *Journal of Transport Geography*.

Yao, Y., Liang, Z., Yuan, Z., Liu, P., Bie, Y., Zhang, J., Wang, R., Wang, J., & Guan, Q. (2019). A human-machine adversarial scoring framework for urban perception assessment using street-view images. *International Journal of Geographical Information Science*. https://www.tandfonline.com/doi/abs/10.1080/13658816.2019.1643024

Yin, L., Cheng, Q., Wang, Z., & Shao, Z. (2015). 'Big data' for pedestrian volume: Exploring the use of Google street view images for pedestrian counts. *Applied Geography, 63*, 337–345. https://doi.org/10.1016/j.apgeog.2015.07.010

Zhang, F., Zhou, B., Liu, L., Liu, Y., Fung, H. H., Lin, H., & Ratti, C. (2018). Measuring human perceptions of a large-scale urban region using machine learning. *Landscape and Urban Planning, 180*, 148–160. https://doi.org/10.1016/j.landurbplan.2018.08.020

Zhao, H., Shi, J., Qi, X., Wang, X., & Jia, J. (2017). Pyramid scene parsing network. In *Proceedings of the IEEE conference on computer vision and pattern recognition* (pp. 2881–2890).

Zheng, Z., Wu, Y., Han, X., & Shi, J. (2020). ForkGAN: Seeing into the rainy night. *Computer Vision – ECCV, 2020*, 155–170. https://doi.org/10.1007/978-3-030-58580-8_10

Zhou, B., Zhao, H., Puig, X., Fidler, S., Barriuso, A., & Torralba, A. (2017). Scene parsing through ade20k dataset. In *Proceedings of the IEEE conference on computer vision and pattern recognition* (pp. 633–641).

Zhu, J.-Y., Park, T., Isola, P., & Efros, A. A. (2017). Unpaired image-to-image translation using cycle-consistent adversarial networks. In *2017 IEEE international conference on computer vision (ICCV)* (pp. 2242–2251). https://doi.org/10.1109/ICCV.2017.244

Spatial Variation Analysis of the Local Population Distribution During Urban Expansion and Contraction Processes

Kiichiro Kumagai, Kohei Uda, and Ayano Shiraishi

Abstract Japan is experiencing a rapid population decline and aging, necessitating new mechanisms for effective land-use planning and urban contraction. A critical issue is *urban spongification*, where vacant land and houses randomly appear, complicating collective redevelopment. To address this, we developed a spatial analysis method using local population data to identify partially low population-density areas as potential sites of *urban spongification*. This study examines spatial variations in local population distributions over the past 25 years, comparing data from 1995 to 2020. The number of low-density areas increases with population decline, accompanied by a decrease in their population densities. Using Geographical and Temporal Weighted Regression (GTWR), we analyse how these areas' economic impacts have evolved. Results indicate that since around 2000, the size of low-density areas has become increasingly significant in explaining land prices, with a clear negative influence on them over time.

Keywords Population decline · Local population dynamics · Urban spongification · Spatial analysis · GTWR

1 Background

While there have been many reported cases of population decline at the city level (Beauregard 2009; Audirac et al., 2012; Martinez-fernandez et al., 2012), Japan is facing a rapidly declining and ageing population on a national scale. Japan's

K. Kumagai (✉)
Department of Civil and Environmental Engineering, Setsunan University, Neyagawa, Osaka, Japan
e-mail: kumagai@civ.setsunan.ac.jp

K. Uda · A. Shiraishi
Division of Social Development Engineering, Graduate School of Science and Engineering, Setsunan University, Neyagawa, Japan

R. Goodspeed et al. (eds.), *Digital-Era Urban Transformations*, The Urban Book Series, https://doi.org/10.1007/978-3-031-98300-9_16

population decline, which began after peaking at around 128.5 million people in 2010, is expected to continue in the future. By 2065, the total population is anticipated to reach approximately 88 million, and it is expected to continue declining beyond this point. The proportion of elderly individuals is projected to increase, highlighting the severe demographic challenges the nation faces (National Institute of Population and Social Security Research, 2019).

The requirement of a new mechanism for effective comprehensive land-use planning and urban contraction (reverse urbanisation) has been discussed since the population decline began to be predicted (The Science Council of Japan, 2005). The Ministry of Land, Infrastructure, Transport and Tourism (2014) revised the Law on Special Measures for Urban Revitalisation and the Law on Revitalisation of Regional Public Transport in anticipation of the urban shrinkage to show a policy of guiding and concentrating the population in core areas established as attraction zones, e.g., Residence Attraction Districts and Urban Function Attraction Districts, over the medium to long term. On the other hand, the current situation is a mix of policy-driven urban contraction and traditional urban expansion because the suburbanisation of cities is still in progress. Research indicates that cities are likely to experience an unprecedented and multifaceted shrinking process in the future, diverging significantly from historical urban contraction patterns. It is therefore desirable to spatially analyse information on where and how changes have occurred up to now, and to use the analysis results for future urban monitoring.

Particularly noteworthy is the phenomenon known as *urban spongification,* where vacant land and empty houses appear in a seemingly random pattern within urban areas, creating a patchy urban fabric. The random nature of this phenomenon generally makes it difficult to redevelop cities collectively. Urban spongification can make it harder to sustain a stable or high population density (Ministry of Land, Infrastructure, Transport and Tourism, 2019). Previous research has developed a spatial analysis method, integrating spatial autocorrelation and overlay analysis based on local population data, to explore this phenomenon (Kumagai & Kameda, 2021; Kumagai et al., 2022). It was confirmed that the application of the method made it possible to detect the size and location of low population-density areas, as the potential areas of *urban spongification* occurrence (Kumagai & Kameda, 2021; Kumagai et al., 2022). In addition, it was clarified that the larger a low population-density area was, the lower its population density became.

There have been, however, few studies of the spatial variation of the local population distributions. Population dynamics are often discussed in conjunction with urban expansion and contraction, generally. Municipalities maintain population records, which they use as one of the bases for formulating their policies. Despite the various attraction zones being established by municipalities, the spatial variation of local population distribution around these zones is rarely discussed.

In this study, we examine the spatial variation in the local population distribution over the last 25 years. There has been both population growth and decline over the past quarter century in Japanese cities. First, we compare the size and location of the partially low population-density areas between 1995 and 2020. Second, we identify the economic impacts of the partially low population-density areas. We apply a hedonic approach with Ordinary Least Squares (OLS) and Geographically Weighted Regression (GWR) to the analysis of the relationship between the assessed values of land and geographical information in order to estimate the importance of the partially low population-density areas at different years. Finally, we discuss how the impact of the partially low population-density areas on the economy has shifted over the last 25 years, using the Geographical and Temporal Weighted Regression (GTWR) because the issue of population is not only to explain how shrinkage occurs, but also to study shrinkage as a process (Haase et al., 2014). Year-over-year comparisons provide robust evidence that the size of low-density areas of the local population has become increasingly significant in explaining land prices since around 2000, with a clear trend toward their negative impact on land prices.

2 Data and Methodology

2.1 Study Area

Osaka Prefecture was selected as the study area because its diverse urban and suburban characteristics provide a microcosm of Japan's demographic trends, making it an ideal case study for exploring spatial population variation in the context of urban shrinkage. Figure 1 shows the location of the study area. Osaka Prefecture is in the western part of Japan. The Osaka Prefecture contains Japan's second-largest city, Osaka City, surrounded by numerous satellite cities. The area of Osaka prefecture is divided into four city planning areas: Osaka City Planning area (OCP), North Osaka City Planning area (NCP), East Osaka City Planning area (ECP), and South Osaka City Planning area (SCP), because of their regional characteristics. OCP

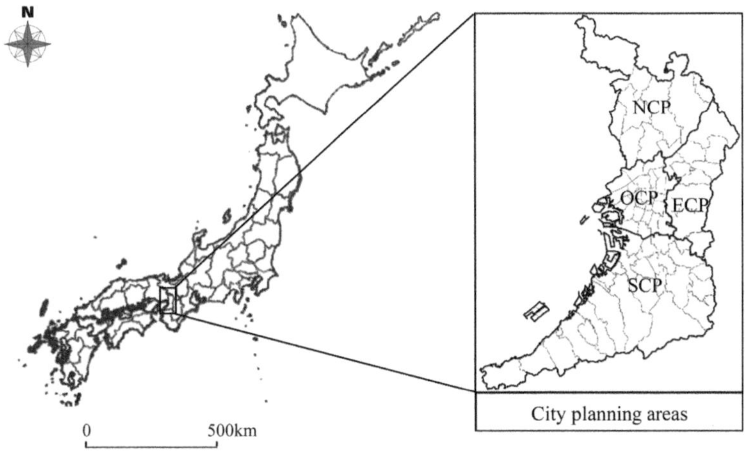

Fig. 1 Study area and its location. The enlarged view of the map shows the division of the city planning areas: OCP, NCP, ECP, and SCP

perfectly coincides with the municipal boundaries of Osaka City, whereas NCP has been gradually shaped over time by the city's continuous development. ECP includes densely built-up areas generated during the period of rapid economic growth after WWII. SCP has the largest area of the four city planning areas and contains not only urbanized areas but also a lot of agricultural fields and green areas. Hence, local populations of the study area are distributed in various spatial patterns, according to completely urbanized areas, moderately urbanized areas, and suburban areas.

2.2 Geographical Data

We used the basic unit block population data of the Population Census of Japan, obtained in 1995, 2000, 2010, 2015, and 2020, for the spatial analysis of demographic dynamics. The basic unit block population data are point-type data with the finest spatial spacing on the basis of the management of National Census investigators. While the population data are primarily recorded at the city block level, they are occasionally recorded at the building level (e.g., for a high-rise apartment) or across multiple blocks for the sake of survey efficiency. The blocks are also unevenly distributed due to variations in parcel sizes, road networks, and land use patterns. As a result, point data for the basic unit block has been shown to exhibit extremely high levels of spatial bias and dispersion, potentially leading to inaccurate representations of actual geographical trends in the spatial autocorrelation analysis used in this study. The size ranges from a minimum of 40 m^2 to a maximum of 15,304,216 m^2, with a lower quartile value of 2294 m^2, a median size of 3942 m^2, and an upper quartile value of 6718 m^2. In suburban areas, basic unit blocks tend to be larger due to the presence of agricultural and low-density residential zones. However, this study focuses on urban areas, where block sizes are generally smaller and clusters are closer to the median value. Given these considerations, we generated a 60 m-by-60 m mesh of data as local population data through the population of each basic unit block being summed and stored in each corresponding mesh. This grid size was chosen based on the median block area of 3942 m^2, ensuring accurate representation of local population distribution while enabling effective spatial analysis. This approach enhances data consistency and supports the raster-based methods described in this study. Figure 2a shows the population dynamics of the Osaka prefecture derived from the National Census. The period from 1995 to 2020 is both a post-monotonic population growth and a period of conversion from population growth to population decline. Figure 2b reveals the population fluctuations in the four city planning areas. The graphs of OCP and NCP exhibit similar trends, reflecting relatively stable or increasing population levels, whereas those of ECP and SCP show a clear downward trend, indicating population decline. We employed published value standard site data to discuss the impact of low-density areas of population on the economy. The published value of standard site data, obtained from National Land Numerical Information provided by the Ministry of Land, Infrastructure, Transport and Tourism, is GIS data on appraised, evaluated, and

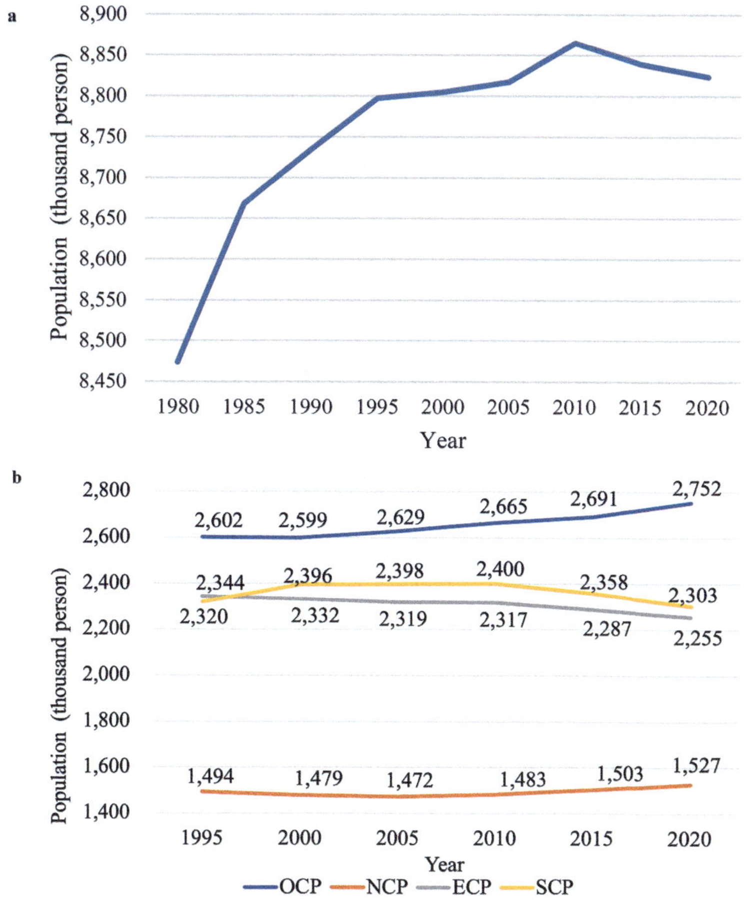

Fig. 2 Population dynamics in the study area. (**a**) Population of the Osaka prefecture from 1980 to 2020, (**b**) Population from 1995 to 2020 in the 4 city-planning areas

published land prices. The data consists of locations (points), published prices, current status of use, zoning, land area, and other data for a standard site throughout the country. The standard site is sampled from residential and commercial areas where land use, environment, land area, shape, and other factors are normal.

2.3 *Methodology*

2.3.1 Spatial Analysis of Local Population Distribution

A spatial analysis method applied in this study consists of a spatial autocorrelation analysis based on G statistics and a moving window analysis (Getis & Ord, 1992; Ord & Getis, 1995; Kumagai, 2011; Kumagai et al. 2021).

Application of G Statistics By applying the G statistic, larger clusters of local objects, i.e., larger local populations, can be detected along with the size of the cluster. If the local object is the local population, the G statistic can be expressed as

$$G_i(d) = \frac{\sum_i w_{ij}(d) h_j}{\sum_i h_i} \tag{1}$$

where G_i is G statistics, h_j is the local population located in the area, w_{ij} is a symmetric binary spatial weight matrix with ones for all links defined as being within distance d of a given i; all other links are coded zero, including the link of a point i to itself (Getis & Ord, 1992). If the null hypothesis is that the set of h_j values within d of location i is a random sample, we derive the Z value described in Eq. (2).

$$Z_i(d) = \frac{G_i(d) - E[G_i(d)]}{\sqrt{VarG_i(d)}} \tag{2}$$

Positive or negative spatial autocorrelation is obtained depending on whether the Z value is positively or negatively greater than a specific level of significance (Ord & Getis, 1995). The results of statistical tests at a significance level of 10% showed that the target regions were divided into three types of spatial autocorrelation: positive spatial autocorrelation, no spatial autocorrelation, and negative spatial autocorrelation. Positive spatial autocorrelation means that a clustering of larger local populations occurs within distance d of a given i, while negative spatial autocorrelation denotes a cold spot with sparse population. In this spatial analysis method, positive spatial autocorrelation is employed to identify areas of local population loss within urban population agglomerations. Then, the cluster size is usually described as a distance parameter, d. The maximum distance d (d_{max}) was obtained from the convergence of the differences of spatial correlation/non-spatial correlation areas between contiguous distances. For analysing the range of the distance parameter d, experiments were conducted by varying the value of d. By gradually changing the parameter and calculating the Z-value using moving window analysis, it was demonstrated that the fluctuation in the area difference between the positively spatially correlated region and the uncorrelated region converged. The convergence condition was defined as the point where the fluctuation amount became less than 0.5% of the total area, and the maximum value of the distance d (d_{max}) was determined accordingly (Kumagai & Kameda, 2021). In this study, d_{max} was found to be 750 m.

Definition of Spatial Index Meanwhile, to analyse phenomena such as 'Urban spongification', it is necessary to detect partial population-density decreases in densely populated areas. Then, in this study, a spatial analysis based on the occurrence of positive spatial autocorrelation regions along the distance parameter d was conducted. Figure 3 shows a conceptual diagram of the detection of the spatial features of local populations. In this spatial analysis method, the distance parameter d is treated as a variable, and its value is varied step by step to investigate whether or not high population values are concentrated within a radius d centred on the point. The value of d is determined when it is statistically determined that no high population values are concentrated within a radius d. Based on Fig. 3, the area being within distance d_{max} of point i, showing positive autocorrelation, generally covers the positive areas being within distance $d_{max\text{-}1}$ or less of i because G statistics are basically derived from the summation of local populations within d of i (see Fig. 3a, b). As the distance decreases from d_{max}, a change of statistics (from positive spatial autocorrelation to no spatial autocorrelation) happens in distance d_n at some points (see Fig. 3b, c). The distance d_n therefore uniquely detected and is defined as the Ambiguity of Spatial scale in a densely Populated area (ASP) since the null hypothesis that the set of local populations within d_n of location i is a random sample is not rejected even though positive spatial autocorrelations are consecutively shown in distance between d_{max} and d_{n+1} (Kumagai & Kameda, 2021). In other words, there could be low population density within distance d_n in spite of the fact that there are significantly larger

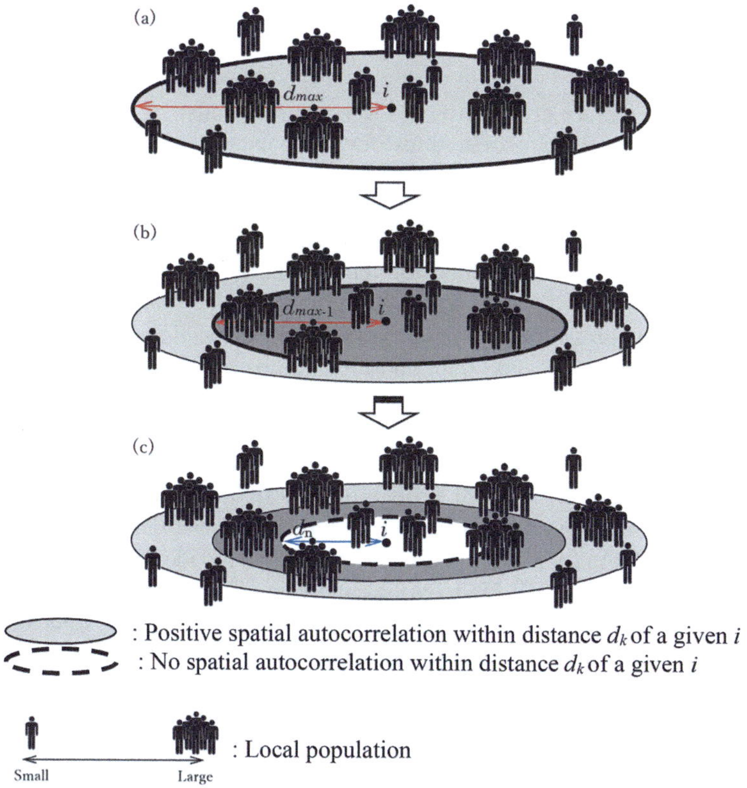

Fig. 3 Conceptual diagram with respect to the detection of the spatial features of local populations (Kumagai & Kameda, 2021). (**a**) Aa larger local population is clustered within distance d_{max} of a given i, (**b**) As distance d_k decreases ($d_{max\text{-}1}$), a larger local populations is still clustered within that range, (**c**) As distance d_k decreases to d_n, a smaller population appears within that range

local populations within d_{max} of point i. The ASP, as determined by d_n, is identified for each point i and stored in the mesh corresponding to i. In this study, the stepwise changes in the distance parameter d are carried out at 60 m intervals, as the population data is based on a 60 m mesh. Consequently, the ASP values are discrete and at 60 m intervals, e.g. 150 m, 210 m, etc. In the event that d_n is not detected, it can be deduced that a small local population area cannot be identified. This case is defined as ASP_0. Conversely, if no positive spatial autocorrelation appears at any spatial extent (e.g. in mountainous or large agricultural areas), it can be concluded that a small local population is widely distributed. This case is defined as ASP_{inf}.

Mapping of ASP It is required to detect the potential site where phenomenon such as 'Urban spongification' occurs. We thus applied the ASP calculation to moving window operations for iterating it on a mesh-by-mesh basis over the study area. Consequently, the ASP distributions were mapped with the spatial size of low population-density areas (Kumagai & Tokoi, 2023).

2.3.2 Statistical Analysis

We examine the feature of ASP from 2 viewpoints of statistical analysis. In the first approach, we apply the chi-squared test for independence to the distribution of the ASPs in the two time periods: 1995 and 2020, and then analyse the differences between ASPs using adjusted standardized residuals with Bonferroni correction (Snee, 1974; Miller, 1981; Bewick et al., 2004).

In the second approach, we focus on the population density within the ASP. We compare the population densities within the same ASP in 1995 and 2020. Specifically, we extract and aggregate the population densities at locations represented by the same ASP value in 1995 and 2020, and then compare their distributions. In other words, it is an attempt to understand the differences within areas of the same size, i.e. same ASP, at the two points in time. The Kolmogorov-Smirnov (K-S) test is used for this examination (Massey Jr, 1951). The K-S test is used to test whether two underlying one-dimensional probability distributions differ. The population density within the ASP is defined as PDA: Population Density in the ASP-designates area. We apply PDAs for 1995 and 2020 to the K-S test to identify differences under the condition of equal ASPs. This implies analysing changes in the features of partially low-population-density areas. If the distribution of PDA values shows a decreasing trend, it suggests that population decline is progressing, and urban spongification may be advancing further, even if ASP remains the same, i.e. even if the low-population area remains unchanged.

2.3.3 Hedonic Approach

We use a hedonic method with OLS and GWR to measure the relationship between the size of the partially low population density areas, namely the ASPs, and land prices. For the examination of the global relationship between independent and dependent variables, OLS is initially applied. After validating GWR with F_1, F_2, and F_3 (Leung et al., 2000), we apply GWR to the 1995 and 2020 data, respectively, for clarifying the role of the ASP in each time cross-section. Finally, GTWR is used for the analysis of the processes of spatiotemporal variations based on the relationship between land prices and geographic information e.g. local population distributions.

3 Results

3.1 *Maps of the Ambiguity of Spatial Scale in a Densely Populated Area (ASP)*

Figure 4 reveals the ASP Maps of 1995 and 2020. Coloured areas in Fig. 4 mean densely populated areas because positive spatial autocorrelations are shown within distance d_{max} in the areas. Gradations in colour from green to red denote the value of ASP, i.e. the extent to which the lower population is distributed around that point. ASP_0, which means that larger local populations exist from the narrowest range to the widest range (d_{max}), is coloured Iceberg (slightly darker light grey). Colourless areas indicate ASP_{inf}, showing that there is no spatial autocorrelation at any scale. Rephrased, it is the point at which the low-density range of the local population extends to infinity. In Fig. 4, it can be seen that the large ASPs, shown in red, are mainly distributed at the edges of the coloured areas. Those areas are mostly boundary areas with agricultural land. ASP may depend on the current status of land use in the surrounding area of the location. In addition, small green areas are scattered in the middle of the study area. Zoomed-in images and aerial photographs are also shown in Fig. 4. There seem to be various ASP changes: increase, decrease, and invariant between 1995 and 2020. As compared with the aerial photographs, development and redevelopment appears to be related to changes in the ASP (see dotted circles on enlarged images in Fig. 4). We can also recognise, however, that the changes of ASP occur in some areas where the aerial photographs do not show significant changes (see single-pointed circle on enlarged images in Fig. 4).

3.2 *Analysis Results in Time Cross Section: 1995 and 2020*

3.2.1 Statistical Analysis

We conducted the chi-square test for association based on the contingency table for ASP and year. The null hypothesis is that the distribution of ASP and year is independent. As a result, the p-values were below 0.001 in all areas: the whole study area, OCP, NCP, ECP, and SCP, so the null hypothesis was fully rejected. ASP patterns may therefore have changed over time. Table 1 indicates the results of testing the adjusted standardized residuals. These were calculated on the basis of adjusted

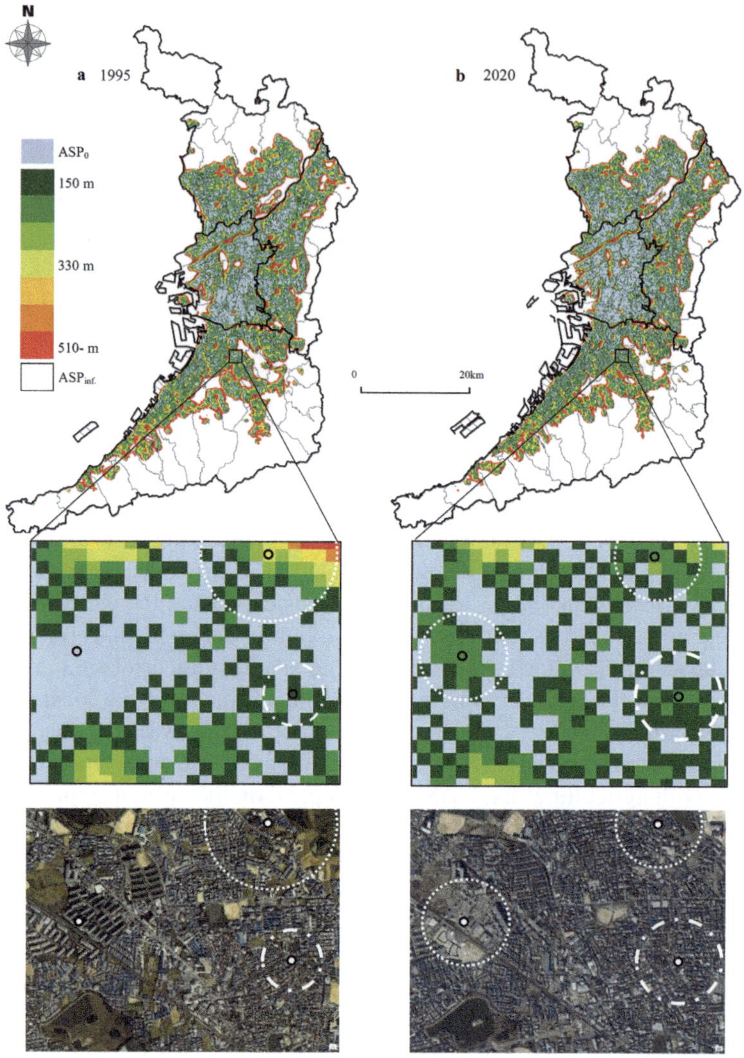

Fig. 4 (**a**) ASP map in 1995 with a partially enlarged map and an aerial photograph taken around 1984-1986, (**b**) ASP map in 2020 with a partially enlarged map and an aerial photograph taken in 2021. Dotted lines and single-pointed lines show the circles with radius ASP, which is indicated with the colour of centre pixels, covered with black small circles, in ASP images. The circle of ASP_0 does not appear

cutoffs derived from Bonferroni correction for testing the residuals of multiple cells in the contingency table (Snee, 1974; Miller, 1981; Bewick et al., 2004).

In the ASP 210 m–ASP 750 m and ASP_{inf} columns of Table 1a, the results show that the residuals in the 2020 row are significantly below the expected values. They can also be said to be clearly lower than the 1995 values. Meanwhile, there are quite different residuals between the city planning areas. It is shown that in SCP, the

Table 1 Results of testing the adjusted standardized residuals (**a**) Whole study area, (**b**) OCP, (**c**) NCP, (**d**) ECP, (**e**) SCP

(a) Whole study area

ASP (m)	ASP_0	150	210	270	330	390	450	510	570	630	690	750	$ASP_{inf.}$
1995	(−)***	(−)***	*	***	***	***	***	***	***	***	***	***	***
2020	***	***	(−)*	(−)***	(−)***	(−)***	(−)***	(−)***	(−)***	(−)***	(−)***	(−)***	(−)***

(b) OCP

ASP (m)	ASP_0	150	210	270	330	390	450	510	570	630	690	750	$ASP_{inf.}$
1995	(−)***	***	***	***	***	*	**						**
2020	***	(−)***	(−)***	(−)***	(−)***	(−)*	(−)**						(−)**

(c) NCP

ASP (m)	ASP_0	150	210	270	330	390	450	510	570	630	690	750	$ASP_{inf.}$
1995	(−)***		***	***	***	***	***	***	***	***	***		
2020	***		(−)***	(−)***	(−)***	(−)***	(−)***	(−)***	(−)***	(−)***	(−)***		

(d) ECP

ASP (m)	ASP_0	150	210	270	330	390	450	510	570	630	690	750	$ASP_{inf.}$
1995	(−)***	(−)***	*	***	***		**	***	*	***	**	***	
2020	***	***	(−)*	(−)***	(−)***		(−)**	(−)***	(−)*	(−)***	(−)**	(−)***	

(e) SCP

ASP (m)	ASP_0	150	210	270	330	390	450	510	570	630	690	750	$ASP_{inf.}$
1995	(−)***	(−)***	(−)***		*	***	***	***	***	***	***	***	***
2020	***	***	***		(−)*	(−)***	(−)***	(−)***	(−)***	(−)***	(−)***	(−)***	(−)***

*** – 0.001%, **0.001% – 0.01%, *0.01% – 0.05%

residuals in the 2020 row in the ASP 150 m and ASP 210 m columns of Table 1e are greater than those in the 1995 row. The occurrence of ASP appears to vary by city planning area, reflecting differences in population trends. In OCP and NCP, where the population is increasing, the number of low-density areas indicated by ASPs has decreased. Conversely, in SCP and ECP, where the population is declining, the occurrence of ASPs has increased, although these areas are small in size.

Table 2 reveals the results of K-S test for whether two underlying one-dimensional probability distributions, derived from the population density in ASP-designated areas (PDA) in 1995 and 2020, differ. We also applied the K-S test to the distributions of population density in townships as a comparison. In Table 2, the distribution of population density in most ASPs is shown to be different in 1995 and 2020. In the OCP, the distributions of PDA in ASP 150 m to ASP 570 m in 2020 is biased towards being smaller than the distribution in 1995, though the distribution of population density by township in 2020 is biased towards being larger than that of population density in 1995.

3.2.2 Results of OLS and GWR

The independent variables shown in the rows of Table 3a were applied to OLS and GWR. The variables of Geographic Information (GI) are acreage, shortest distance to a railway station, building coverage, floor area ratio, and other current status of the standard site. We basically used the variables of GI in all cases, which have been used in the land price model, showing a better goodness of fit (Shimizu & Nishimura, 2006). We also added variables of ASP to GI. The ASP values were applied separately to eight regions: commercial area, other area, and four city planning areas. We applied log-transformation to land values to address the positively skewed distribution of the data, a common characteristic in property value datasets. This transformation normalizes the data, ensuring linearity and reducing heteroscedasticity, which are crucial for the accuracy of OLS and GWR analyses. Additionally, ASP values were exponentially transformed to address the unique properties of ASP measurements, particularly ASP_{inf}, which represents areas where low-density population distributions extend infinitely. This transformation is described in Eq. (3).

$$x_{ASP} = 1 - e^{\left(-\frac{ASP}{d_{max}}\right)} \tag{3}$$

The transformation allows ASP values to be scaled from ASP_0 (denoting the most densely populated areas) to ASP_{inf} (representing areas of infinitely low density) within a range of 0–1. According to Eq. (3), ASP_0 becomes 0, ASP values between 150 m and 750 m are converted to a range between 0.181 and 0.632, and ASPinf is converted to 1. By preserving the ordinal relationship between ASP values and ensuring consistent comparability, this approach facilitates seamless integration into regression models.

The relationship between the variables was also examined through the calculation of the VIF. The results showed no multi-collinearity between the variables in Table 3a.

Table 2 Results of K-S test for the PDA distributions of every ASP in 1995 and 2020

		Z_{KS} value											
		OCP			NCP			ECP			SCP		
ASP representing the range of the PDA's calculations (m)	150	3.07	***	–	3.59	***	–	8.09	***	–	9.83	***	–
	210	4.07	***	–	5.95	***	–	6.02	***	–	7.28	***	–
	270	2.90	***	–	5.15	***	–	4.77	***	–	7.29	***	–
	330	3.97	***	–	4.86	***	–	5.12	***	–	7.46	***	–
	390	3.96	***	–	2.89	***	–	5.00	***	–	5.68	***	–
	450	2.45	***	–	3.85	***	–	3.63	***	–	6.64	***	–
	510	2.02	***	–	4.49	***	–	2.86	***	–	7.53	***	–
	570	1.69	**	–	3.14	***	–	2.61	***	–	5.22	***	–
	630	1.20			2.78	***	–	2.61	***		6.13	***	–
	690	0.56			2.29	***	–	2.56	***		6.28	***	–
	750	1.13			2.92	***	–	4.08	***	–	6.98	***	–
Population density at town level		2.52	***	+	0.86			2.67	***	–	2.17	***	

*** – 0.001%, ** 0.001% – 0.01%, * 0.01% – 0.05%

+/– distribution in 2020 is biased towards larger/smaller than the distribution in 1995

Table 3 Results of OLS and GWR in 1995 and 2020. The independent variables are shown in the rows of the table. (**a**) Results of OLS, (**b**) Results of GWR. The proportions of t-values of the variable in ASP are shown based on the significance level of 5%

(a)		1995 Estimate	Std.Error	t value	p value		2020 Estimate	Std.Error	t value	p value	
Intercept		5.36E+00	1.36E-01	39.516	<0.001	***	4.47E+00	5.15E-02	86.745	<0.001	***
Acreage [m^2]		−9.87E-06	1.90E-06	−5.196	<0.001	***	2.69E-06	1.52E-06	1.772	0.076544	.
Shortest distance to a railway station [m]		−3.17E-05	3.13E-06	−10.143	<0.001	***	−5.91E-05	3.90E-06	−15.15	<0.001	***
Building coverage [%]		6.38E-04	1.51E-04	4.229	<0.001	***	−3.50E-03	7.91E-04	−4.422	<0.001	***
Floor area ratio [%]		1.57E-03	4.01E-05	39.208	<0.001	***	1.84E-03	5.34E-05	34.417	<0.001	***
Industrial area [dummy]		4.82E-02	1.14E-02	4.219	<0.001	***	−4.77E-02	1.31E-02	−3.64	<0.001	***
Fire protection area [dummy]		1.46E-01	1.87E-02	7.848	<0.001	***	3.93E-02	1.18E-02	3.318	<0.001	***
Gas supply [dummy]		1.37E-01	1.27E-02	10.768	<0.001	***	1.43E-01	1.96E-02	7.302	<0.001	***
Sewerage [dummy]		7.80E-02	8.82E-03	8.84	<0.001	***	2.42E-01	2.24E-02	10.81	<0.001	***
ASP [m]	Commercial area										
	OCP	−1.22E-01	2.57E-02	−4.761	<0.001	***	3.89E-01	3.53E-02	11.011	<0.001	***
	NCP	−1.71E-01	3.51E-02	−4.856	<0.001	***	3.68E-01	4.23E-02	8.71	<0.001	***
	ECP	6.58E-02	3.28E-02	2.006	0.045	*	−2.09E-02	4.36E-02	−0.48	0.63151	
	SCP	−4.36E-02	3.37E-02	−1.293	0.196		1.41E-01	4.12E-02	3.429	<0.001	***
	Other area										
	OCP	1.14E-02	2.24E-02	0.508	6.11E-01		3.71E-01	2.88E-02	12.853	<0.001	***
	NCP	−1.32E-01	1.77E-02	−7.449	<0.001	***	4.21E-01	2.54E-02	16.548	<0.001	***
	ECP	5.15E-03	1.71E-02	0.302	0.763		9.28E-02	2.48E-02	3.742	<0.001	***
	SCP	−7.01E-02	1.66E-02	−4.223	<0.001	***	2.09E-01	2.58E-02	8.115	<0.001	***

(continued)

(a)	1995				2020			
	Estimate	Std.Error	t value	p value	Estimate	Std.Error	t value	p value
R^2	0.808				0.805			
AIC	−1836.37				−1088.25			
*** – 0.001%, **0.001% – 0.01%, * 0.01% – 0.05%,. 0.05% – 0.1%								

(b)	1995			2020		
	$t > 1.960$ ($p(>t) = 0.025$)	$1.960 >= \|t\|$	$-1.960 > t$ ($p(<t) = 0.025$)	$t > 1.960(p(>t) = 0.025)$	$1.960 >= \|t\|$	$-1.960 > t$ ($p(<t) = 0.025$)
ASP [m]						
Commercial area						
OCP	93.02	6.98	0.00	5.45	53.94	40.61
NCP	21.05	73.68	5.26	22.73	68.18	9.09
ECP	39.22	21.57	39.22	46.94	36.73	16.33
SCP	76.19	21.43	2.38	1.96	39.22	58.82
Other area						
OCP	100.00	0.00	0.00	29.80	52.94	17.25
NCP	77.68	22.32	0.00	44.90	52.72	2.38
ECP	18.48	36.95	44.57	81.32	18.11	0.57
SCP	75.99	21.11	2.90	24.46	61.77	13.76 (%)

The results of OLS are indicated in Table 3a, showing the global relationship between the independent and dependent variables. We can see that for 7 of 8 ASP variables, the p-values show significance in 2020, compared to 5 of 8 variables in 1995.

We then examined the validation of GWR using the statistics of F_1, F_2, and F_3 (Leung et al., 2000). The results were $F_1 = 0.749$ (p-value < 0.001), $F_2 = 11.028$ (p-value < 0.001) in 1995, and $F_1 = 0.738$ (p-value < 0.001), $F_2 = 11.524$ (p-value < 0.001) in 2020. The p-values of F_3 show significance in 15 out of 17 variables for 1995 and in 16 out of 17 variables for 2020. The validity of GWR model is thus clarified in this study.

In Table 3b, the proportions of t-values of the variable for ASP are indicated on the basis of the significance level of 5% derived from the GWR model. Applying the GWR resulted in $R^2 = 0.859$ in 1995 and $R^2 = 0.850$ in 2020, and AIC = −2443.694 in 1995 and AIC = −1643.554 in 2020. The results demonstrated an improvement in model fit compared to OLS. R^2 value increased approximately from 0.81 in the OLS model to 0.86 in the GWR model, indicating a better explanation of the variability in land prices. Also, AIC decreased significantly from −1,088 and −1,836 in the OLS model to −1,644 and −2,444 in the GWR model, further confirming the enhanced fit of the GWR model.

Table 3b shows that in OCP, ECP, and SCP of 1995, ASP of other areas mostly provides a positive impact on nearby land prices, while ASP for 2020 gives various impacts on nearby land prices. Comparisons by time cross-section suggest that the impact of ASP on land prices in 1995 and 2020 may be quite different.

4 Discussion

In this section, we discuss how the impact of the partially low population-density areas, ASPs, on the economy has shifted over the last 25 years. The GTWR model was applied to this analysis with consecutive ASPs and geographic data generated in 1995, 2000, 2005, 2010, 2015, and 2020. Before the application, we identified the appropriate parameter for the measurement of the extent of closeness in a spatio-temporal space, based on the convergence of R^2 and AIC (Huang et al., 2010). When the parameters were varied stepwise, the R^2 converged to a maximum value of 0.844, and the AIC converged to a minimum value of −11995.9.

Figure 5 indicates the fluctuations of t-value distributions for the last 25 years. In Fig. 5a–d, the t-values of ASP show negative impacts on land prices since 2000, though the *t*-in Fig. 5e has always had a negative impact on land prices over the past 25 years. The trend of t-values of ASPs can be recognised in common across the 4 urban planning areas.

Figure 6 reveals spatial distributions of *t*-values of ASP. The *t*-values for 1995 show a different distribution compared to other years. On the other hand,

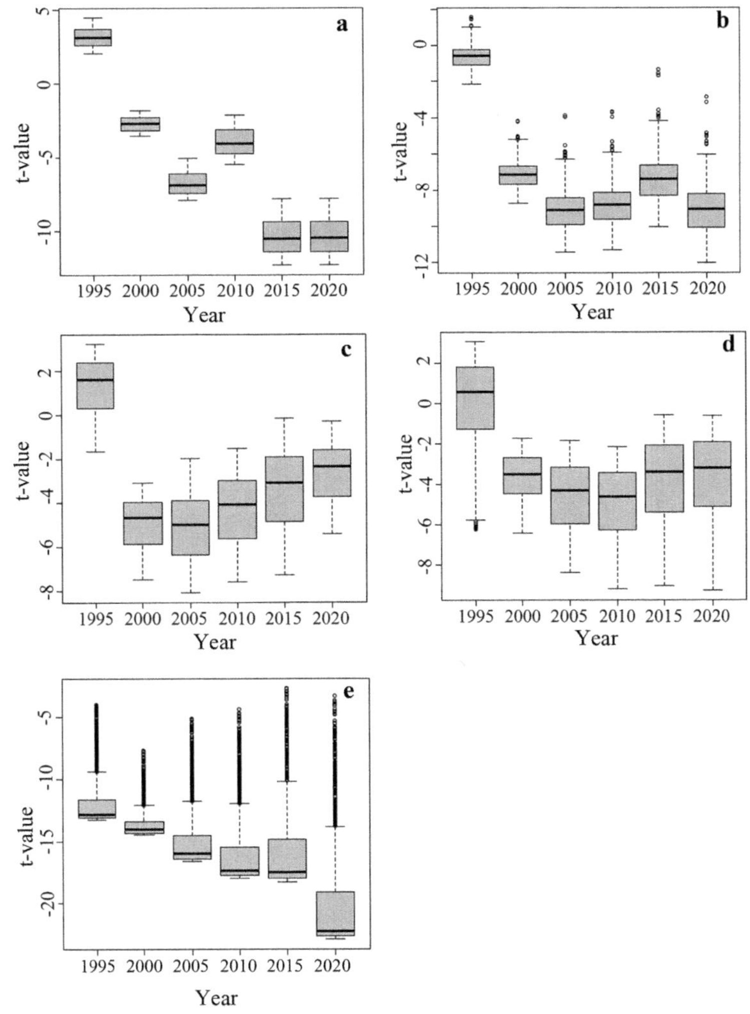

Fig. 5 Box plot of *t*-values derived from GTWR model. (**a**) Variable of ASP in OCP for other area, (**b**) Variable of ASP in NCP for other area, (**c**) Variable of ASP in ECP for other area, (**d**) Variable of ASP in SCP for other area. (**e**) Variable of shortest distance to a railway station as a comparison

between 2000 and 2020, the spatial distribution of *t*-values negatively affecting land prices is gradually becoming more regionally restricted. The size of the low-density area of the local population appears to have become significant in explaining land prices since around 2000, with a clear tendency to negatively affect land prices, although the total population of the study area peaked around 2010.

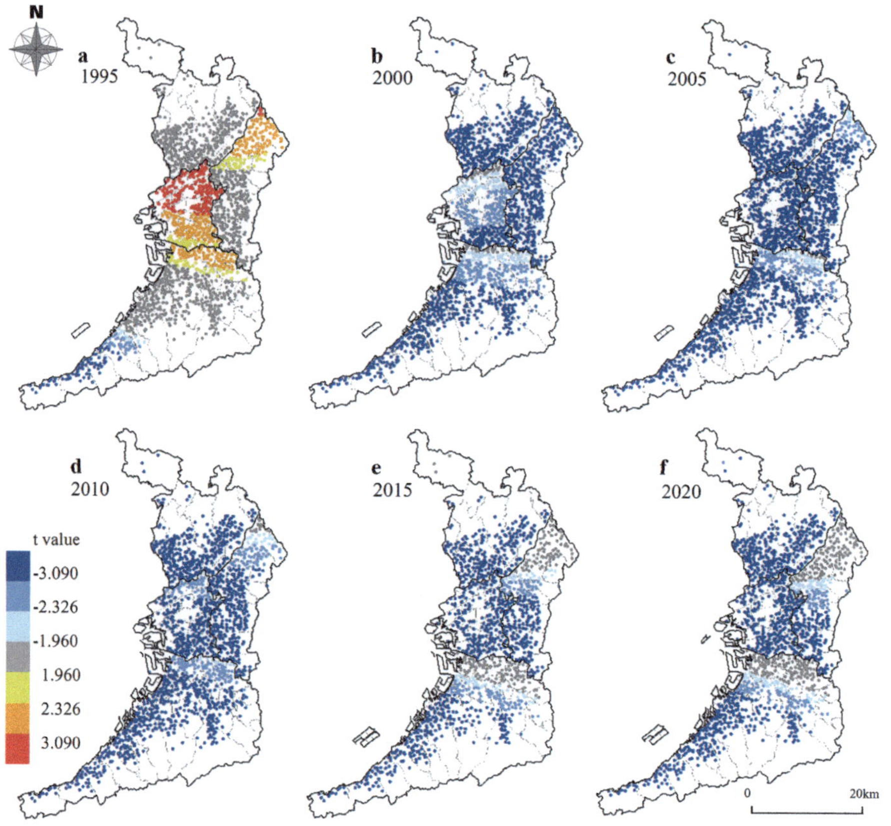

Fig. 6 *t*-value map of ASP, generated by GTWR model. (**a**) 1995, (**b**) 2000, (**c**) 2005, (**d**) 2010, (**e**) 2015, (**f**) 2020

5 Conclusions

In this study, we examined the spatial variation in the local population distribution over the last 25 years. Firstly, the size and location of the partially low population-density areas, defined as ASP, were compared between 1995 and 2020 from the viewpoints of time cross-section. Through the statistical analysis, the smaller size of the low-density area of the local population occurred more often within suburban city planning areas in 2020. The distribution of local population density designated by ASP in 2020 tended to be biased towards being smaller than the 1995 distribution. We identified the economic impacts of the partially low population-density areas. We applied a hedonic approach with OLS and GWR to the analysis of the relationship between the assessed values of land and geographical information in order to compare the importance of the partially low population-density areas between 1995 and 2020. The comparisons by time cross-section suggested that the impact of ASP on land prices in 1995 and 2020 might be quite different. Finally, we discussed how the impact of the partially low population-density areas on the economy has shifted over the last 25 years, using the GTWR. Although the total population of the study area peaked around 2010, the size of low-density areas of the local population has become more important in explaining land prices since approximately 2000. Furthermore, the analysis confirmed a clearer tendency for the size of these areas to exert a negative impact on land prices over time.

Acknowledgments The basic maps of the study area and aerial photographs are adopted from GSI tiles (https://maps.gsi.go.jp/development/ichiran.html) provided by the Geospatial Information Authority of Japan.

References

Audirac, I., Cunningham-Sabot, E., & Fol, S. (2012). Declining suburbs in Europe and Latin America. *International Journal of Urban and Regional Research, 36*, 226–244.

Beauregard, R. (2009). Urban population loss in historical perspective: United States, 1820–2000. *Environment and Planning A, 41*, 514–528.

Bewick, V., Cheek, L., & Ball, J. (2004). Statistics review 8: Qualitative data—tests of association. *Critical Care, 8*, 46–53.

Getis, A., & Ord, J. K. (1992). The analysis of spatial association by use of distance statistics. *Geographical Analysis, 24*(3), 189–206.

Haase, A., Rink, D., Grossmann, K., Bernt, M., & Mykhnenko, V. (2014). Conceptualizing urban shrinkage. *Environment and Planning A: Economy and Space, 46*(7), 1519–1534. https://doi.org/10.1068/a46269

Huang, B., Wu, B., & Barry, M. (2010). Geographically and temporally weighted regression for modeling spatio-temporal variation in house prices. *International Journal of Geographical Information Science, 24*(3), 383–401. https://doi.org/10.1080/13658810802672469

Kumagai, K. (2011). Verification of the analysis method for extracting the spatial continuity of the vegetation distribution on a regional scale. *Computers, Environment and Urban Systems, 35*, 399–407.

Kumagai, K., & Kameda, Y. (2021). Detection of the spatial variations of local populations from the viewpoint of urban structure analysis, ISPRS annals of the photogrammetry, remote sensing and spatial. *Information Sciences, V-4-2021*, 177–185. https://doi.org/10.5194/isprs-annals-V-4-2021-177-2021

Kumagai, K., Tokoi, H. (2023). Spatial variation of local population distribution in the urban shrinking process. Proceedings of the 18th International Conference on Computers in Urban Planning and Urban Management. https://doi.org/10.17605/OSF.IO/6YR5V

Kumagai, K., Tsukamoto, S., Tokoi, H., & Wada, M. (2022). Spatial variation analysis of local population data and local household-size data for urban management in a declining population. *ISPRS Annals of the Photogrammetry, Remote Sensing and Spatial Information Sciences, V-4-2022*, 283–291. https://doi.org/10.5194/isprs-annals-V-4-2022-283-2022

Leung, Y., Mei, C., & Zhang, W. X. (2000). Statistical tests for spatial nonstationarity based on the geographically weighted regression model. *Environment and Planning A, 32*, 9–32.

Martinez-fernandez, C., Audirac, I., Fol, S., & Cunningham-sabot, E. (2012). Shrinking cities: Urban challenges of globalization. *International Journal of Urban and Regional Research, 36*, 213–225. https://doi.org/10.1111/j.1468-2427.2011.01092

Massey, F. J., Jr. (1951). The Kolmogorov-Smirnov test for goodness of fit. *Journal of the American Statistical Association, 46*(253), 68–78.

Miller, R. G. (1981). *Simultaneous statistical inference*. Springer. https://doi.org/10.1007/978-1-4613-8122-8

Ministry of Land, Infrastructure, Transport and Tourism (2014). Promotion of urban renovation and compact cities. Retrieved September 23, 2024, from http://www.mlitgo.jp/common/000996976.pdf

Ministry of Land, Infrastructure, Transport and Tourism (2019). White paper on land, infrastructure, transport and tourism in Japan. Retrieved September 24, 2024, from https://www.mlit.go.jp/en/statistics/white-paper-mlit-2019.html.

National Institute of Population and Social Security Research (2019) Social security in japan (2019 edition). Retrieved September 26, 2024, from https://www.ipss.go.jp/s-info/e/pssj/pssj2019.pdf

Ord, J. K., & Getis, A. (1995). Local spatial autocorrelation statistics: Distributional issues and an application. *Geographical Analysis, 27*(4), 286–306.

Science Council of Japan (2005). For the future of big cities. Retrieved September 23, 2024, from https://www.scj.go.jp/ja/info/kohyo/pdf/kohyo-19-t1030-22.pdf

Shimizu, C., & Nishimura, K. G. (2006). Biases in appraisal land price information: The case of Japan. *Journal of Property Investment and Finance, 26*(2), 150–175.

Snee, R. D. (1974). Graphical display of two-way contingency tables. *The American Statistician, 28*(1), 9–12.

Do Good Schools Influence Property Prices? Disentangling School and Neighbourhood Effects with Empirical Evidence from Brighton

Josiah Tan and Adam Dennett

Abstract A perception that proximity to good schools pushes house prices up is common and one factor that contributed to a shift in the secondary-school admissions policy away from proximity-based criteria in Brighton, England, in 2008. This decoupling of proximity was expected to reduce perceived house price inflation near good schools. But is this link as strong as many believe, or are other factors more important? We examine 81,000 residential transactions in Brighton between 2000 and 2019, assessing the effect of schools alongside other contextual factors such as building attributes, accessibility to jobs, deprivation, and macro price trends, as well as the impact of the policy change in 2008. We find that after accounting for local neighbourhood confounders and a general upward drift in property prices, proximity to good schools has only a limited impact on house prices, with other local factors often playing an often more important role. The admissions policy reform has little discernible impact on prices.

Keywords Word · House prices · Schools · Accessibility · Hedonic models · Brighton and Hove

---

Josiah Tan completed the MSc in Urban Spatial Science in CASA in 2023. Adam Dennett is Professor of Urban Analytics in CASA.

J. Tan · A. Dennett (✉)
Bartlett Centre for Advanced Spatial Analysis (CASA), University College London, London, UK
e-mail: a.dennett@ucl.ac.uk

R. Goodspeed et al. (eds.), *Digital-Era Urban Transformations*, The Urban Book Series, https://doi.org/10.1007/978-3-031-98300-9_17

1 Introduction

In popular media and marketing narratives as well as the field of housing research, the theory that proximity to good schools makes housing more expensive is widely accepted as something approaching an unassailable fact. The causal pathways appear obvious to anyone with even a rudimentary understanding of economic theory—in a system where resource scarcity (good, free-to-access schools) can be high, and access to that limited resource is determined, frequently, by geographic residential proximity, then competition for that residential proximity will increase. As demand and competition rise and the route to winning is through paying more, then prices will naturally increase in this classical supply and demand scenario.

The literature from across the world is full of evidence that this theory holds in many different contexts (see, amongst other studies, Bonilla-Mejía et al., 2020; Chin & Foong, 2006; Fack & Grenet, 2010; Haurin & Brasington, 1996; Cheshire & Sheppard, 2004; Kane et al., 2005; Wen et al., 2017; Rosiers et al., 2001). However, despite the headlines of these papers boosting the narrative, the strength of the association can sometimes be quite weak (e.g. Wen et al., 2017) with other factors (such as accessibility to urban centres) playing an equal if not more important role.

Some of the work in this area has had the advantage of longitudinal data and clear policy shifts at particular times (see Bonilla-Mejía et al., 2020) which have allowed for natural experiments to be conducted with treatment and control groups to help determine the potential impact of policy changes and the size of the school effect, but these studies are in the minority. And in the case of the Bonilla-Mejía et al. paper specifically, while spatial catchment covariates were tested, they failed to account for other local spatial factors that might be playing an important role in the story.

In this paper, we address these issues through taking a more critical approach to the way the influence of schools on house prices has, to date, been conceived. Schools may well play a role in the decisions that some home-buyers make—indeed it can feature prominently in the marketing materials of estate agents—but can we say for certain that the presence of a good school is any more important, than access to employment opportunities or other neighbourhood factors, especially where good schools, good neighbourhoods and good employment opportunities might often overlap? We explore this question through a detailed analysis of house prices in Brighton, England. Brighton makes for a particularly interesting case study as in 2008, the admissions policy for its secondary schools was altered, if not primarily—as social mixing was a priority—partially in response to a perceived house price premium near some of the more popular secondary schools (Allen et al., 2013). We carry out a detailed spatial and temporal analysis which examines the sale prices of homes across the city between 2000 and 2019—crucially on either side of the admissions policy change—and relates these prices to school performance, proximity to those schools, and a basket of other potentially influential variables.

Our analysis is facilitated through the use of a detailed linked dataset created by Chi et al. (2020), which links Land Registry Price Paid data to detailed property attributes contained within the Department for Levelling Up, Housing and Communities (DLUHC) Energy Performance Certificate dataset. These data are further enhanced through attaching school performance and travel-time/accessibility data for the city centre and major railway stations, as well as contextual neighbourhood and deprivation data.

The main research question of this study is thus:

How strong is the effect of proximity to good schools on house prices in Brighton and did the 2008 secondary school admissions reform have any impact on prices?

Our findings in this paper are important for a number of reasons: There is of course academic merit in contributing to the evidence base in this popular area of housing research; however, the relevance of this work to sound policymaking, particularly around school admissions, is key. Any policy formulated using poor evidence or 'received wisdom' runs the risk of being a poor policy. Where policies are implemented that disconnect pupils from their local schools, then the resulting impacts on the social welfare of pupils or on the transport systems of cities need to be balanced against the positive impacts intended, such as increased social mixing and housing affordability. If some of these positive impacts do not exist, then the policy is on weaker foundations.

Brighton's admissions reforms were implemented to promote equity with a secondary benefit of combating the phenomenon of expensive housing near popular schools (Allen et al., 2013). Over a decade after these changes were made, we are now able to review the house price dimension of this story and evaluate the impacts of Brighton's admissions reforms, with potentially significant relevance for other geographical contexts. The rest of this paper proceeds with a review of the relevant contextual literature relating to house prices and school proximity, followed by an exploration of the local Brighton setting before detailing data used, methodology employed, and our results. We conclude with a discussion of our findings.

2 Review

2.1 Examining the Relationship between Schools and House Prices

A large body of international work exists which makes a causal link between proximity to good schools and higher house prices. In the United States (US), the link between residential choice and school choice has been established through qualitative (Lareau & Goyette, 2014) and quantitative studies in Boston, Massachusetts (Black, 1999) and Chicago (Bonilla-Mejía et al., 2020). In Paris, France, there is evidence that a standard deviation increase in public school performance raises housing prices by 1.4–2.4% (Fack & Grenet, 2010) while in Singapore where primary school admissions are dependent on proximity, distance to schools can explain a large amount of the variation in house price differences (Chin & Foong, 2006). In Japan, Kuroda (2018) demonstrated that proximity to better schools increased rents for family dwellings.

In England, proximity to state-funded schools is an established admission criterion (UK Government, 2023), and research to date suggests a straightforward link between this and house prices. Gibbons and Machin (2006) for example, showed that a 10% increase in children reaching their expected grades at Key Stage 2 (ages 7–11) increases house prices by 3%, and that each 100 m distance increase from a school lowers the premium by 8.4% relative to the original level. That said, it was also found that only those schools in the top performance deciles had a positive influence on house prices—most schools depress house prices in their vicinity as 'average schools are not desirable local amenities' (Stephen Gibbons & Machin, 2006, p. 90).

While a large body of quantitative evidence has emerged in this area, many of the studies feature methodological limitations. For example, data samples that are either relatively small (Black, 1999; Cheshire & Sheppard, 2004; Steve Gibbons & Machin, 2003; Leech & Campos, 2003), do not account for property-level characteristics (Stephen Gibbons & Machin, 2006), or cover a short period of time (Cheshire & Sheppard, 2004; Steve Gibbons & Machin, 2003; Leech & Campos, 2003). The outcome of any of these limitations is that the inferential certainty of the results is called into question.

An established methodology for examining the factors that influence house prices is the hedonic regression model. Based on the Rosen model (Rosen, 1974), it theorises that house prices demonstrate consumer choice as an optimal balance of bundled attributes. With the appropriate data inputs, the hedonic price model can be a powerful explanatory tool and illustrate that property prices are a predictable function of property-level and neighbourhood-level attributes. A major challenge with this model, however, is dealing with omitted variable and endogeneity problems (Gibbons & Machin, 2008), which could lead to upwardly biased estimates if confounding variables are not recognised (Black, 1999). In other words, should the effects of schools on house prices be confounded by other, unaccounted for, neighbourhood factors, then their influence may be exaggerated.

Sometimes, absolute distance is less important than being within a school's catchment boundary, and some research designs that use boundary discontinuities are well-documented in the existing literature. These focus on house-price differentials between houses separated by catchment area boundaries (Cheshire & Sheppard, 2004; Davidoff & Leigh, 2007; Fack & Grenet, 2010; Stephen Gibbons & Machin, 2006). These studies also mostly report a 2–4% increase in house prices for a one-standard-deviation increase in school performance for a given catchment. However, this approach does not account for varying neighbourhood-level attributes within school catchment boundaries.

Where distinct changes in policy or school availability have occurred at particular times points, another approach is to use 'natural experiments' which look at before and after 'treatment' effects. In our context, the 'treatment' could be the opening or closing of schools or changes in admissions policies, with the effect measured in house price changes. This approach was used in the US by Reback (2005) and Kane et al. (2005) that capitalised on school policy changes, and could be similarly applied to Brighton's admissions reforms where one approach could be to explore comparable housing and other contextual data have been collated for local authorities where admissions policies have not been altered like Brighton's. Directly relevant to this work, Shah (2018) studies the effect of Brighton's reforms on house prices for a small subset of repeat sales and found that a 10% increase in the GCSE pass rate of a secondary school increased house prices in that sample by 2.38%. Shah analysed treatment and control samples affected by Brighton's 2008 reforms using a difference-in-difference approach. However, the study has limitations as property-level attributes are not accounted for, and only a limited sample of repeat sales data of the same houses pre- and post-reform between the years 1995–2017 is used. Furthermore, their definition of treatment and control groups could be questioned as their control groups were still subject to the same policy changes as the treatment groups.

2.2 *Other Contextual Influences on House Prices*

While proximity to good schools is important, it is not the only consideration for home buyers. Housing is often characterised as a 'bundled good' in economic discourse (Goodman, 1978) as home buyers pay for the property's intrinsic attributes and its neighbourhood-level utilities such as school quality, transport links, and accessibility to the city centre—a proxy for employment opportunities and other services. A classically economically rational home buyer (Huu Phe & Wakely, 2000) would be aiming to optimise their housing investment with respect to other amenities such as public goods, proximity to urban centres, and other spatial factors (Alonso, 1964; Cho, 2001). Services like rail connectivity (Du & Mulley, 2007; Stephen Gibbons & Machin, 2005), distance to city centre (McCord et al., 2014b), and travel time to London (Chi et al., 2020), are key considerations in residential choice in the UK context. Other neighbourhood factors that could potentially be considered include local rates of crime (McIlhatton et al., 2016) and access to green spaces (McCord et al., 2014a).

2.3 *School Admissions in Brighton*

Before the school admissions reform was announced in 2007, Brighton's admissions criteria were similar to most English Local Authorities (LAs). Parents expressed their preferences for any state school, and schools admitted students until their published capacity was reached. Schools and students were then matched using a priority-matching mechanism that considered only the first choice, then using oversubscription criteria to rank and allocate remaining students where necessary (Allen et al., 2013). Under the UK national School Admissions Code, 2007, distance between home and school is cited as a 'clear and objective criterion' for oversubscription that ensures students do not have overly long journeys to and from the school (School Admissions Code, 2007)—and indeed is used as the primary tie-breaker in many local authorities. However, the code was also cognisant of how this could advantage families who can afford properties near the school, so a range of other suitable decision criteria are listed—including random (lottery) allocation. While the national code serves as a guide, the decision on which criteria are applied rests with local admissions authorities.

Brighton's authorities recognised the problem of fair access to popular secondary schools (Eastwood & Turvey, 2008) and consequently, in 2008, introduced a lottery system, alongside new catchment areas[1] to its secondary school admissions process (distance was retained as a criterion for primary schools). The reform was designed to improve equitable access to secondary schools and stop the creation of 'golden halos' of housing premiums near oversubscribed schools (Allen et al., 2013, p.152). Interviews with parents in the city (Robinson et al., 2016) confirmed that this perception of higher house prices in the proximity of popular schools was well established.

In the post-reform admissions process, a lottery was used first to select applicants within the catchment areas and then, if places were still available, applied once more for applicants living outside the catchment area who had expressed a preference for that school. While this meant that families living in the catchment areas for less popular secondary schools lacked choice, the policy reform transformed the chances of families getting a place at their preferred school within a catchment, especially those within dual catchment areas. Dual catchment areas (e.g. Blatchington Mill and Hove Park, Dorothy Stringer and Varndean) give children in the catchment an equal probability of admission, with a place at either school virtually guaranteed. This means that the composition of student populations in both single and dual school catchments would be changed post-reform—with those within the catchment but living a relatively long way from the school given equal probability of admission to those living next door.

While the Dorothy Stringer/Varndean (DS/V) catchment area is relatively affluent with both schools being popular, the Blatchington Mill/ Hove Park catchment area is interesting because Blatchington Mill performed significantly better than Hove Park in the post-16 General Certificate of Secondary Education (GCSE) examination, with Hove Park having a higher proportion of students eligible for free school meals (FSM)—a marker of social deprivation within the school cohorts (Allen et al., 2013). The reform, thus, somewhat delinks proximity to a popular school from the probability of access. In the Blatchington Mill/ Hove Park catchment area, houses previously guaranteed a place by virtue of proximity now have to enter a lottery with all the families in the catchment area. It is also important

[1] https://applyforaschool.brighton-hove.gov.uk/p/schoolsmap

to note that the equitable access to popular schools in Brighton also now depends on the drawing of catchment areas, which are devised by the Local Authority. Nonetheless, proximity still matters as being within a catchment area with good schools (e.g. DS/V) virtually guarantees a place at either school, just that it should have a weaker effect.

3 Study Design

To investigate the school effect on property prices and to try and deal as much as we can with the issues of endogeneity, potentially causing problems in other studies, we use an established hedonic regression modelling framework with a comprehensive selection of building-level and neighbourhood-level variables, with some additional adjustments for spatial and temporal elements. We incorporate both structural/building-level attributes alongside locational attributes as advocated by Orford (2002), but unique in this study relative to others is the detailed linkage of school quality via realistic walking and public transport journey times along the road network to home locations—this allows for a much more accurate representation of proximity/accessibility than crude distance buffers. We are also cognisant of the perceptions of school quality changing over time, so every sale is linked to the perceived quality of the nearest primary and secondary school when the sale occurred. Also, conscious that grouping variables like school catchment areas might be unwittingly capturing general area effects, we include alternative area groupings to control for erroneous associations. To explore the impact of the 2008 admissions policy change, the final model splits the data into two panels of pre- and post-reform sales.

4 Data

4.1 House Prices and Structural Attributes

The spine of our research dataset is the new linked House Price (HP) dataset for England and Wales (1995–2021), created by Chi et al. (2022). This dataset covers more than 18 million transactions from the Land Registry Price Paid dataset (LR-PPD) between 1995 and 2021 and links these transactions with key building attribute data from the Department for Levelling Up, Housing and Communities' (DLUHC) Energy Performance Certificates (EPCs) dataset. From this we select seven structural variables including price paid, construction year, year of sale, property type, floor area, number of rooms, energy efficiency rating and postcode, for 81,652 sales transactions in Brighton between 2000 and 2020, spatially joining them with our own neighbourhood, accessibility and school quality perception variables.

The full research dataset and associated code are available on an online repository. All data were obtained from open and publicly available repositories, including but not limited to: the London Datastore, Ofsted Online Reports, Department for Education (DfE) Archives, and the Office for National Statistics (ONS).

4.2 *Locational Attributes: Local Schools Data*

Perceptions of school quality amongst homebuyers are subjective, but using information from a range of school-related datasets, we hope to derive a categorical composite 4-level indicator of perceived school quality, using the Ofsted terminology of: Outstanding, Good, Satisfactory, and Poor. Ofsted ratings are only factored in from the year 2007, as archived data was patchy until then. Ratings persist between inspections, so a school rated 'Outstanding' in 2007 maintains that rating until a new inspection is carried out. Past Ofsted reports for Academies (a new class of state school, partially funded from private investment and exempt from local authority control) are also wiped 5 years after academisation (a controversial process occasional forced upon under-performing state schools), which means useful data was lost for schools such as Portslade Aldridge and Brighton Aldridge. This denies the historicity of their school quality, as they might have been rated 'good' or better in many of the years prior, but only deemed inadequate for a short period before academisation. Hence, we adopt a 'business-as-usual' approach, continuing the last recorded rating where data is lacking, assuming that school ratings, the demography of their cohorts, and local perceptions do not change rapidly over short periods of time. An example of this is using the previous year's data for St Joseph's Primary School and Balfour Primary School, when they had no data for specific years.

The DfE Archives were a rich resource, with data available on key school information (school names and postcodes), Key Stages 2 and 4 results, Free School Meals (FSM), and pupil destinations. We recognise that including FSM in a perceptions variable could erroneously link deprivation with quality, so we give it a low weighting in the mix. However, the link between FSM and attainment levels might be relevant in perceptions. Key Stage 2 (KS2) results were available from 2003, while Key Stage 4 (KS4) results (GCSEs) were available from 2000, and FSM and Pupil Destination (for secondary school) information was available from 2012. The availability of datasets shapes the ranking of schools, with variables only being factored in and weighted from the year they are available. The process of ranking and categorising perceived school quality is summarised in the appendix.

4.3 *Other Locational Attributes and Proximity Calculations*

To calculate proximity/accessibility to schools and other relevant services, the ONS Postcode Directory (ONSPD) was used to add latitude and longitude coordinates to the postcodes in the HP and schools datasets. Coordinates for the Brighton, Hove, and Preston Park rail stations were obtained using their coordinates on Google Maps. These stations were chosen as they are the key stations that are connected to London via the Thameslink/Southern Rail/Gatwick Express network (Thameslink, 2023) and service Brighton's substantial London commuter population—particularly those 'employed full-time in manager, director and senior official, professional and associate professional and technical occupations'[2] which have the highest salaries and which may add to the pool of available property purchasing capital in the city. A general catch-all for other services and local employment services was the city centre of Brighton. The point used as the city centre reference was obtained by calculating the centroid of the Brighton and Hove Regional Centre polygon as determined by the Consumer Data Research Centre's (CDRC) dataset on retail centres in the UK (Consumer Data Research Centre, 2022).

Accessibility to these important services and opportunities was calculated using the Rapid Realistic Routing with R5 (r5r) package (Pereira et al., 2021). This uses the road network from OpenStreetMap alongside public transport schedules and elevation to generate realistic travel times between all residential postcodes, schools, railway stations, and the city centre. These were calculated for both walking and taking the bus. The bus schedules used are from 2021, and while these may not accurately represent bus routes and timetables for the full 2000–2019 period, in the UK, bus routes and schedules do not undergo drastic changes over time.

Bands of 5 min, up to 30 min were generated for walk times, while for bus times it was up to 45 min, with a maximum of 20 min walking. Transactions outside these bounds were labelled 'Not near' whichever service. Variables representing walk times to the nearest rail stations, and primary and secondary schools (regardless of quality) were also generated for all postcodes in Brighton to provide insights into the baseline effects of schools on property prices and a comparable variable for rail stations. For proximity to schools and rail stations, walk times up to 30 min where with anything beyond that classed as 'not near'. Bus time to city centre was used in preference to car travel times as the 2021 Census found 37.4% of Brighton households had no cars or vans—much higher than the English average of 23.5% (ONS, 2023)—which means walking or public transport are the main forms of transport in the city.

To explore local neighbourhood effects, we used the 2019 Index of Multiple Deprivation (IMD) ranks and deciles, as well as 2011 Output Area Classification supergroups, school catchment areas, and electoral ward boundaries. With the IMD, we do not account for changes in relative deprivation and demographic profiles over time and just use the latest iteration. While it might have been preferable to try and match deprivation profiles temporally to house sales, studying the evolution of the Index of Multiple Deprivation over time from 2010 via the CDRC Map Maker, the general drift within Brighton is towards being less deprived, with relative geographic positions remaining quite constant between 2010 and 2019. If anything, any associations between higher price and affluence or lower price and deprivation might be underplayed at the beginning of the study period.

[2] See—https://www.commute-flow.net/classification.html#g9—for a map demonstrating how closely linked Brighton is to London for 'Techs & the City Types' commuters.

Given the potential importance of secondary school catchment areas post reform, we include these as area dummy variables in the analysis. To deal with the possibility of erroneous neighbourhood effects being ascribed to catchment boundaries, we also include electoral wards as a comparison grouping (or random effect using statistical terminology) variable.

4.4 Data Pre-Processing

Constructing the dataset involved three main steps: the first being the categorisation of schools according to their perceived quality; the second generating variables for geographical proximity of the city centre, rail stations and schools perceived as 'outstanding/good'; the third merging all of these new derived variables with the transactions in the HP dataset and cleaning (removing incomplete records or extreme values) before analysis.

All variables were combined into one dataset, with code adapted from (Chi et al., 2021) being used to clean the dataset of missing or overly high or low sale price values. The top and bottom 5% of transactions in terms of prices for each year (2000–2019) were also excluded from the dataset to account for outliers, which are common in house-price datasets. More information on the data cleaning process can be found in the supplementary material.

5 Explaining House Prices: the Hedonic Price Model

The general hedonic price model (Freeman, 1979; Rosen, 1974) we use can be written as follows:

$$P_{it} = f\left(X_{ht}, X_{gt}, X_s, e_t\right)$$

where:

P_{it}= price of transaction i in period t
X_{ht}= a vector of property-level characteristics at the time of transaction
X_{gt}= a vector of neighbourhood-level contextual variables at the time of transaction
X_s= a vector of socio-economic characteristics of the neighbourhood assumed to be persistent over time.

For P_{it}, the element of time is factored in by using the year of transaction as a categorical variable to control for year-on-year variations in price, and coefficient changes are interpreted as percentage change in price from the reference year (Fleming & Nellis, 1985). We could have attempted to standardise prices across years using national information on inflation, but in retaining year as a dummy variable, we are able to compare these annual inflationary effects with those of the other variables. X_s does not have a time element to it, as we just use 2019 IMD data as an approximation of relative deprivation/affluence within the city.

5.1 Exploratory Analysis

The map above (Fig. 1) illustrates the spatial distribution of the house prices analysed in this report, aggregated to Lower Level Super Output Areas (LSOAs) in 2019 (the geographical variation in house prices does not change much year-by-year). Higher prices tend to be seen in the North and East of the City Centre, with the LSOAs with the highest prices near Preston Park and Hove rail stations (Fig. 2c) and also near in-demand schools such as Blatchington Mill, Dorothy Stringer, and Varndean. Proximity to good primary schools (Fig. 2a) seems centred near the city centre and rail stations, with good secondary schools being located more towards the west (Fig. 2b). However, Whitehawk and Mouselcoomb (red areas eastward of Brighton station are deprived in terms of both IMD and a lack of good schools (Fig. 2c).

The dataset contains data on price per square metre, but our experiments found that using price per square metre as the dependent variable resulted poorer models fits compared to raw price, and size moved to the explanatory side of the model. Price per square metre also made some variables such as number of rooms difficult to interpret. Examining the frequency distribution of price (Fig. 1), we can observe that the data are right-skewed—a long-tail of fewer, more expensive properties shifting the mean value to the right of the modal property price. To correct this skew, we transform it by taking the natural logarithm (Box & Cox, 1964) and use that as

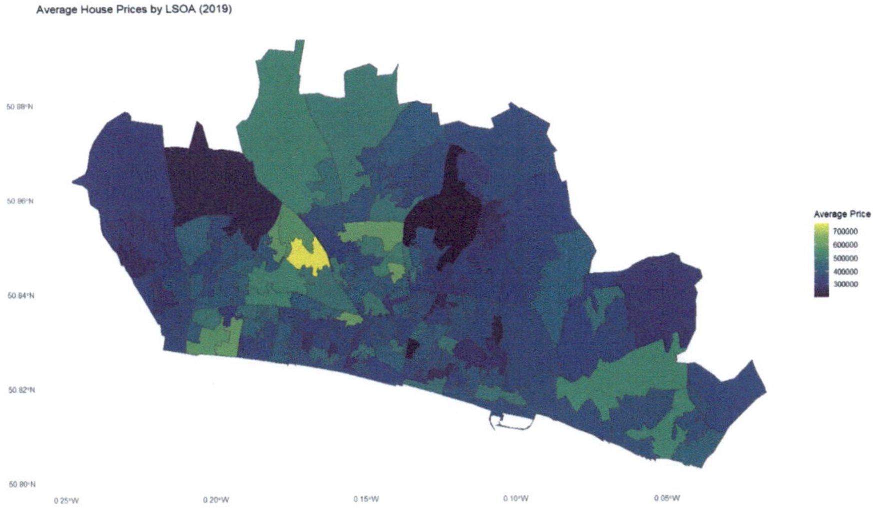

Fig. 1 Map showing average house prices in Brighton and Hove by LSOA in 2019

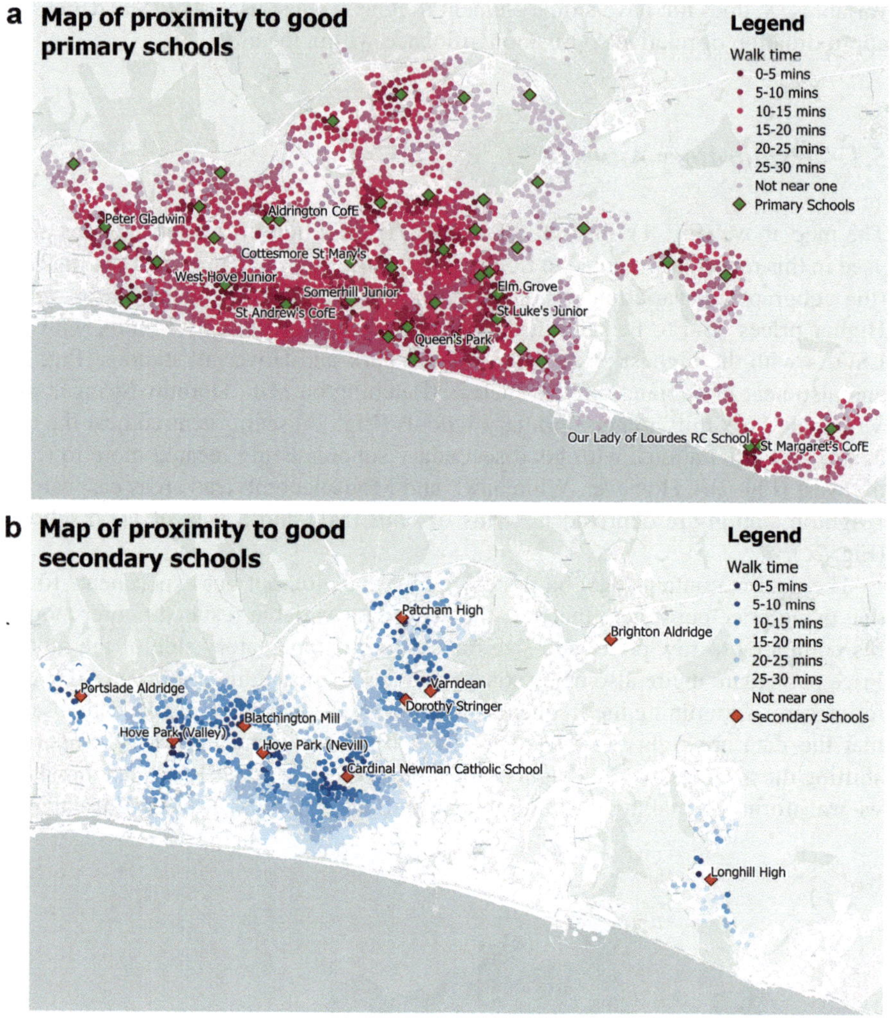

Fig. 2 Maps of proximity to good primary (**a**), and secondary schools (**b**) (selected good schools labelled), and IMD deciles (**c**)

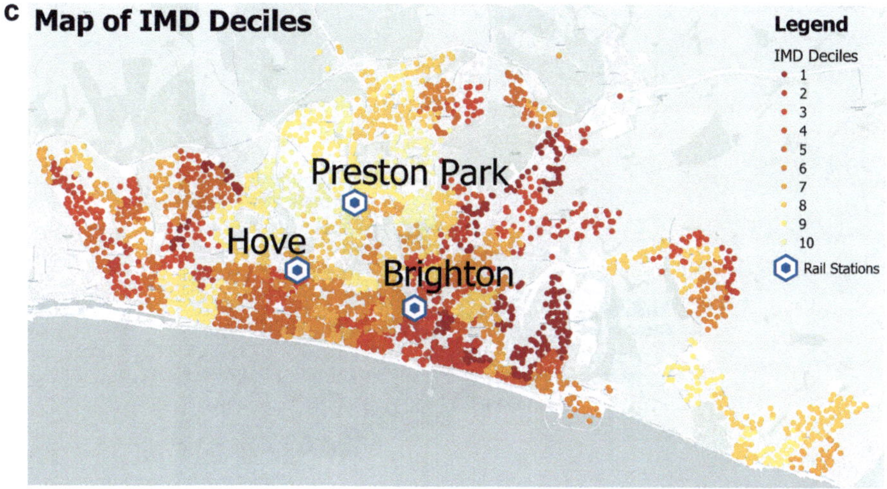

Fig. 17.2 (continued)

the dependent variable (Fig. 3). The explanatory variables, total floor area and walk time to the nearest train station, are transformed in the same way to correct a similar positive skew. The distributions of each explanatory variable are illustrated through histograms (continuous variables) and barplots (categorical variables) are shown in Figs. 7, 8 and 9 in the Appendix.

The correlation matrix (Fig. 4) reveals the correlations between all explanatory variables. The coefficients for price, total floor area, and number of rooms are moderately high, while some groups of variables, such as those related to the accessibility of rail or schools, are only moderately correlated. Electoral wards are highly correlated with rail and secondary school variables, but this is understandable as it is a categorical geographical variable that covers the entire study area.

Generalised VIF (GVIF) values were used to check for multicollinearity since there are both numerical and categorical variables. $\mathrm{GVIF}^{1/2 \cdot D\mathrm{f}}$, where D_f is the number of coefficients in the subset (or categories), is equivalent to the Variance Inflation Factor (VIF) to check for multicollinearity (Fox & Monette, 1992). The $GVIF^{1/2 \cdot Df}$ values were all below 5 (see Table 4), which is an acceptable threshold for multicollinearity (Hair, 2009). Some level of correlation between certain categorical variables might be inevitable as they are unavoidably geographically linked.

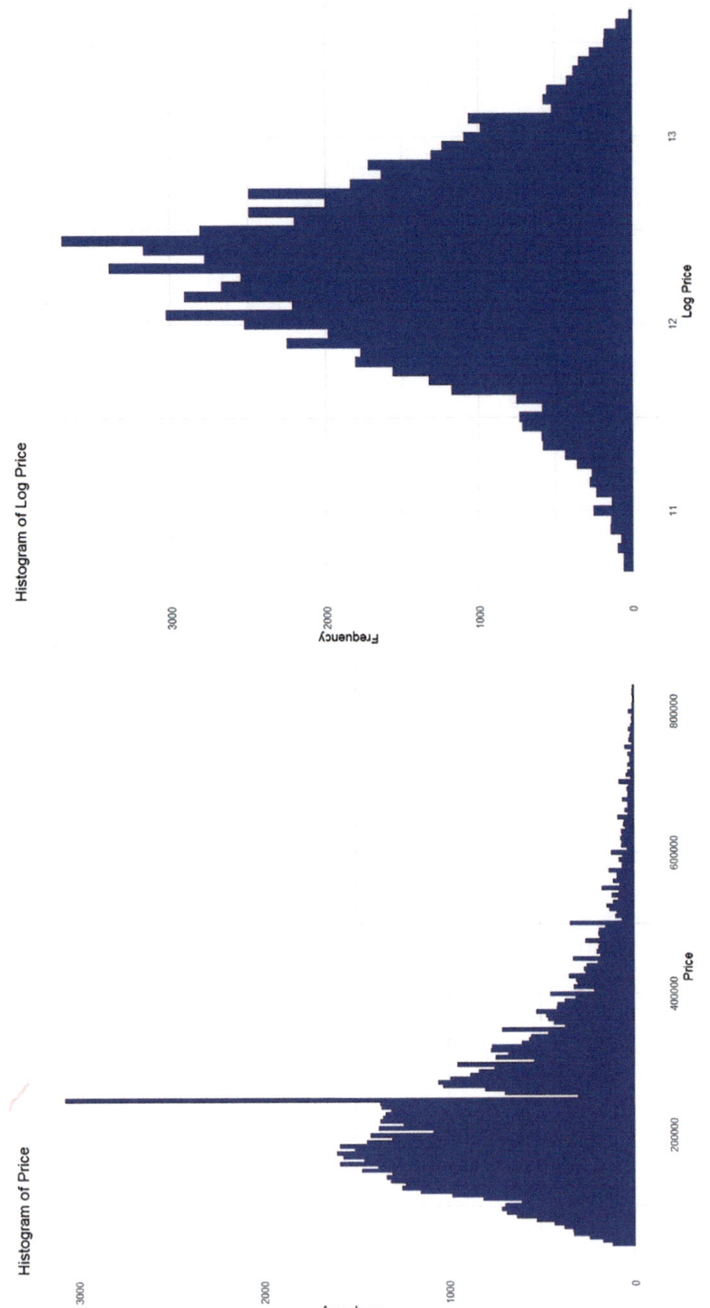

Fig. 3 Distribution of dependent variable before–after natural log transformation

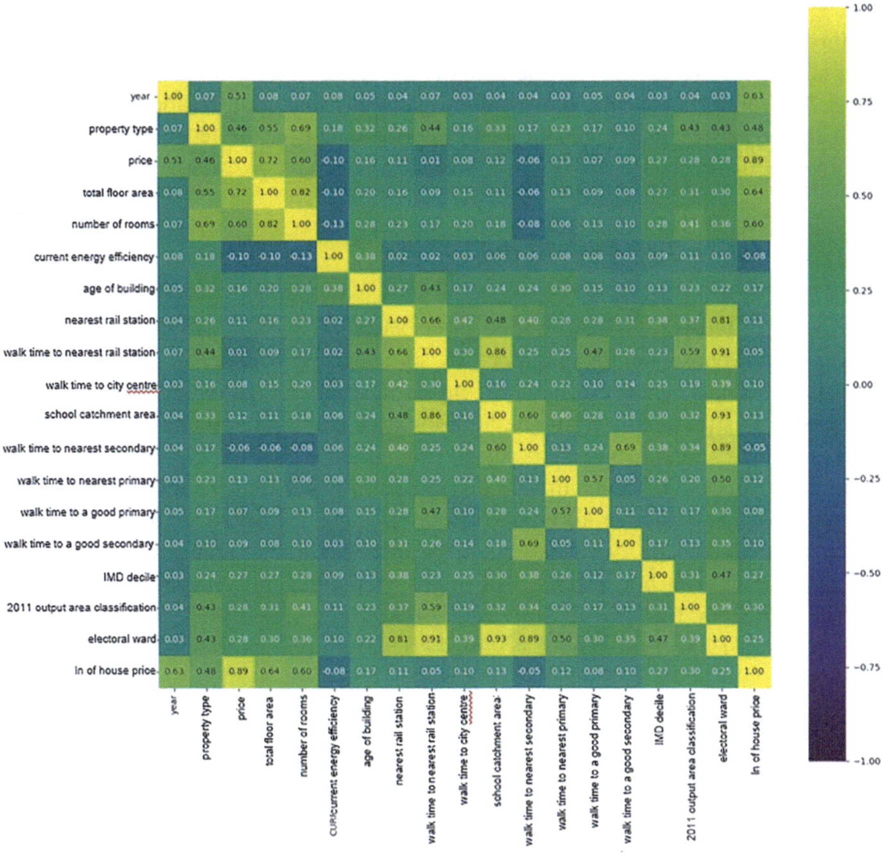

Fig. 4 Correlation matrix for the dataset

6 Results

The final hedonic price model models (the natural log of) house price paid, as a function of property type, total floor area, proximity to services (including schools and employment through mainline rail and the town centre), and 2019 IMD deciles. The model is first run on the whole dataset spanning the years 2000–2019, then, to observe differences before and after the admissions policy change in 2008, the data are split into two panels; the first containing transactions from 2000 to 2007, second spanning 2008 to 2019. We were unable to carry out a classical difference-in-differences analysis as all houses within the Brighton and Hove study area were subject to the policy change, so defining clear treatment and control groups was not feasible.

The coefficient associated with each variable in the model is an implicit indicator of importance, but where these are dependent on the original scale of measurement, their standardised *t*-values provide insights into their relative importance with each other. The model 1 baseline model investigates just the contribution that building-level characteristics and year sold make to price variations in the city.

6.1 Model 1: Baseline Model

In model 1 (Table 1), 83.1% of the price variation paid in Brighton across the 19 year study period can be explained by sale year, property type, total floor area, number of rooms, energy efficiency, and construction age band, with the most important physical characteristic being floor area. However, relentless year-on-year price increases (except for a short period immediately after the global financial crisis in 2008) since 2000 mean that when a property is sold has the most influence on the price, over and above all other factors. Detached properties are 22.7% more expensive than flats, on average, while a 1% increase in floor area is associated with a 0.57% increase in house price, and more rooms generally mean higher prices. Energy efficiency has a negligible (and slightly negative) effect, while buildings built between 1900–1990 have a negative effect on price as compared to buildings built after.

6.2 Model 2: Baseline + Contextual Variables

In model 2, we compare three variants of a model containing property-level *and* additional contextual variables. Over repeated iterations of experimentation, three contextual variables proved the most interesting: secondary school catchment areas, proximity to good schools in the year sold, and electoral wards. These variables, plus the Index of Multiple Deprivation in 2019, were added to the baseline model (Table 2). The additional contextual variables are helpful in explaining house prices, with the adjusted R^2 increasing from 83.1% to >86%.

Taking catchment areas, proximity to schools, and electoral wards separately, they all add a similar amount of explanatory power to the model. Combining them offers no perceptible advantage in explanatory power, but we do see a noticeable confounding effect on school catchment areas from electoral wards. This suggests it is not being in a catchment per se that can affect house prices, rather that some neighbourhoods are more preferable than others, and these can be home to particular schools and catchments that then overlap. The importance of a house being in the two most important catchment areas (Blatchington and Hove, and Dorothy Stringer and Varndean) virtually disappears when electoral wards/neighbourhoods are included. On their own, electoral wards also add more to the adjusted R^2 than catchment areas. Good schools do appear to be part of this mix in local area effects that

Table 1 Baseline model 1

	Full dataset; property-level	
Variable	Coefficient	T-value
Constant	9.002**	591.83
Year (2000)		
2001	0.176**	40.43
2002	0.371**	85.29
2003	0.493**	110.49
2004	0.587**	131.21
2005	0.629**	136.30
2006	0.708**	162.45
2007	0.825**	186.32
2008	0.777**	136.49
2009	0.725**	143.07
2010	0.848**	175.26
2011	0.847**	176.63
2012	0.878**	182.95
2013	0.911**	194.68
2014	1.01**	224.37
2015	1.10**	236.78
2016	1.18**	249.13
2017	1.23**	254.68
2018	1.238**	247.65
2019	1.223**	242.21
Property type (flat/Maisonette)		
Detached	0.205**	56.02
Semi-detached	0.091**	31.78
Terrace	0.076**	33.57
Number of rooms	0.021**	21.08
Log total floor area	0.571**	167.61
Energy efficiency	−0.0002**	−2.83
Construction age		
Before 1900	0.030**	3.95
1900–1929	−0.012	−1.61
1930–1949	−0.080**	−10.32
1950–1966	−0.158**	−20.39
1967–1975	−0.119**	−14.98
1976–1982	−0.096**	−10.62
1983–1990	−0.043**	−4.90
1991–1995	0.035**	3.57
1996–2002	0.128**	14.68
2003–2006	0.145**	15.22
2007 onwards	0.088**	7.80
Sample Size	73,559	
Adjusted R^2	0.831	
F-statistic	9540**	

**$p < 0.05$–

Table 2 Model 2: baseline + contextual variables model

	(a) Main model with school catchment areas	(b) Main model with proximity to schools	(c) Main model with electoral wards	(d) Main model with all variables of interest
Variable (reference)	Coefficients (*t*-value)			
Constant	8.882** (536.13)	8.903** (543.04)	9.139** (516.03)	9.085** (367.60)
Year (2000)				
2001	0.179** (45.78)	0.180** (45.63)	0.179** (46.25)	0.180** (46.55)
2002	0.374** (95.70)	0.374** (95.07)	0.374** (96.73)	0.374** (97.03)
2003	0.498** (124.19)	0.500** (123.76)	0.499** (125.67)	0.500** (126.11)
2004	0.591** (146.96)	0.592** (146.07)	0.591** (148.55)	0.592** (149.09)
2005	0.631** (152.24)	0.633** (151.71)	0.630** (153.70)	0.632** (154.32)
2006	0.712** (181.83)	0.713** (180.88)	0.711** (183.49)	0.712** (184.00)
2007	0.835** (210.02)	0.834** (208.31)	0.835** (211.89)	0.835** (212.44)
2008	0.786** (153.77)	0.786** (152.67)	0.786** (155.35)	0.787** (155.78)
2009	0.728** (160.01)	0.729** (159.08)	0.726** (161.34)	0.727** (161.70)
2010	0.847** (196.76)	0.846** (193.42)	0.846** (196.75)	0.846** (197.23)
2011	0.847** (196.76)	0.846** (195.01)	0.847** (198.51)	0.846** (198.89)
2012	0.877** (203.63)	0.876** (201.66)	0.875** (205.30)	0.875** (205.36)
2013	0.914** (217.33)	0.912** (215.40)	0.913** (219.38)	0.912** (219.62)
2014	1.015** (250.41)	1.016** (248.84)	1.014** (252.72)	1.014** (253.16)
2015	1.103** (265.40)	1.103** (263.47)	1.102** (267.98)	1.103** (268.41)
2016	1.187** (279.96)	1.187** (277.94)	1.186** (282.61)	1.186** (283.28)
2017	1.236** (284.92)	1.234** (282.52)	1.235** (287.53)	1.234** (288.08)
2018	1.242** (276.58)	1.238** (273.88)	1.241** (279.21)	1.241** (279.75)
2019	1.229** (270.88)	1.228** (268.83)	1.230** (273.84)	1.230** (274.47)

Table 2 (continued)

	(a) Main model with school catchment areas	(b) Main model with proximity to schools	(c) Main model with electoral wards	(d) Main model with all variables of interest
Property type (flat/ maisonette)				
Detached	0.299** (79.58)	0.301** (80.58)	0.307** (81.77)	0.310** (82.65)
Semi-detached	0.216** (73.34)	0.213** (72.20)	0.226** (76.43)	0.229** (77.20)
Terrace	0.175** (77.33)	0.169** (74.14)	0.192** (82.91)	0.192** (83.04)
Number of rooms	0.028** (30.84)	0.027** (29.85)	0.029** (32.46)	0.029** (32.55)
Log total floor area	0.497** (158.08)	0.504** (159.46)	0.490** (156.56)	0.488** (156.30)
Energy efficiency	0.0001** (2.28)	0.0001** (2.14)	0.0003** (4.03)	0.0003** (4.24)
Construction age (NA)				
Before 1900	−0.013 (−1.88)	−0.011 (−1.52)	−0.012 (−1.75)	−0.012 (−1.84)
1900–1929	−0.025** (−3.62)	−0.027** (−3.79)	−0.022** (−3.23)	−0.023** (−3.32)
1930–1949	−0.051** (−7.30)	−0.051** (−7.18)	−0.052** (−7.39)	−0.050** (−7.15)
1950–1966	−0.097** (−13.88)	−0.091** (−12.98)	−0.094** (−13.60)	−0.091** (−13.20)
1967–1975	−0.087** (−12.16)	−0.083** (−11.59)	−0.087** (−12.26)	−0.086** (−12.11)
1976–1982	−0.065** (−8.00)	−0.066** (−8.01)	−0.067** (−8.26)	−0.066** (−8.20)
1983–1990	−0.033** (−4.22)	−0.034** (−4.23)	−0.032** (−4.09)	−0.033** (−4.18)
1991–1995	0.031** (3.53)	0.018** (1.99)	0.021** (2.43)	0.025** (2.83)
1996–2002	0.099** (12.47)	0.084** (10.52)	0.090** (11.43)	0.091** (11.49)
2003–2006	0.136** (15.84)	0.129** (14.92)	0.135** (15.80)	0.139** (16.25)
2007 onwards	0.086** (8.44)	0.081** (7.91)	0.079** (7.80)	0.084** (8.30)
Bus time to city centre (not near)				
0–5 min	0.326** (13.77)	0.352** (14.79)	0.306** (12.86)	0.229** (11.19)
5–10 min	0.237** (34.37)	0.285** (41.81)	0.243** (31.57)	0.216** (27.50)

(continued)

Table 2 (continued)

	(a) Main model with school catchment areas	(b) Main model with proximity to schools	(c) Main model with electoral wards	(d) Main model with all variables of interest
10–15 min	0.181** (35.07)	0.232** (47.19)	0.188** (30.21)	0.159** (24.35)
15–20 min	0.148** (31.82)	0.183** (40.08)	0.152** (28.31)	0.125** (21.87)
20–25 min	0.092** (20.31)	0.119** (26.93)	0.098** (19.79)	0.075** (14.62)
25–30 min	0.079** (19.18)	0.102** (25.20)	0.076** (16.84)	0.057** (12.19)
30–35 min	0.035** (9.55)	0.044** (12.30)	0.042** (10.43)	0.031** (7.57)
35–40 min	−0.011** (−3.11)	−0.011** (−3.26)	0.004 (1.19)	−0.005 (−1.41)
40–45 min	−0.018** (−5.43)	−0.014** (−4.09)	−0.008** (−2.23)	−0.011** (−3.15)
Rail station (none)				
Brighton	−0.020** (−7.10)	−0.009** (−3.20)	−0.039** (−12.10)	−0.027** (−8.01)
Hove	−0.011** (−3.30)	0.003 (1.01)	−0.056** (−13.98)	−0.047** (−11.57)
Preston Park	0.038** (9.39)	0.038** (9.66)	−0.009** (−1.96)	−0.016** (−3.08)
Log walk time to nearest rail	−0.027** (−13.66)	−0.033** (−18.82)	−0.045** (−17.89)	−0.038** (−14.36)
Walk time to nearest secondary school	0.0004** (6.16)	0.0004** (6.21)	0.0004** (4.04)	0.0007** (6.36)
Walk time to nearest primary school	0.005** (35.60)	0.007** (41.90)	0.004** (29.74)	0.005** (30.11)
IMD deciles (decile 1)				
Decile 2	0.098** (20.22)	0.073** (15.00)	0.107** (20.89)	0.100** (19.17)
Decile 3	0.132** (28.75)	0.115** (25.08)	0.119** (24.67)	0.116** (23.18)
Decile 4	0.137** (29.37)	0.138** (29.53)	0.144** (30.72)	0.138** (28.82)
Decile 5	0.145** (30.59)	0.147** (31.22)	0.123** (23.85)	0.122** (23.06)
Decile 6	0.162** (37.45)	0.164** (38.00)	0.161** (34.51)	0.155** (32.10)
Decile 7	0.183** (41.86)	0.169** (38.59)	0.174** (38.45)	0.169** (36.23)
Decile 8	0.194** (42.95)	0.186** (41.24)	0.191** (40.04)	0.183** (36.84)
Decile 9	0.231** (46.66)	0.235** (47.32)	0.210** (38.96)	0.204** (36.70)

Table 2 (continued)

	(a) Main model with school catchment areas	(b) Main model with proximity to schools	(c) Main model with electoral wards	(d) Main model with all variables of interest
Decile 10	0.290** (51.92)	0.288** (51.22)	0.249** (41.08)	0.241** (38.49)
Catchment areas (Aldridge)				
Blatchington and Hove	0.153** (33.58)			0.026 (1.58)
Stringer and Varndean	0.118** (28.87)			0.016 (1.04)
Longhill	0.085** (17.85)			−0.060** (−3.58)
Patcham	0.082** (16.56)			−0.039** (−2.48)
Portslade	0.034** (6.94)			−0.122** (−6.87)
Proximity to good secondary school in year sold				
0–5 min		0.064** (4.81)		0.030** (2.20)
5–10 min		0.064** (10.93)		0.037** (5.85)
10–15 min		0.032** (8.21)		0.016** (3.81)
15–20 min		0.023** (7.28)		0.008** (2.23)
20–25 min		0.016** (5.69)		−0.003 (−1.04)
25–30 min		0.008** (3.15)		−0.009** (−3.63)
Proximity to good primary school in year sold				
0–5 min		0.097** (19.88)		0.052** (10.37)
5–10 min		0.057** (15.75)		0.020** (5.14)
10–15 min		0.041** (12.63)		0.007** (2.14)
15–20 min		0.042** (13.29)		0.012** (3.44)
20–25 min		0.044** (12.70)		0.020** (5.41)
25–30 min		0.025** (6.58)		0.012** (3.10)
Electoral wards (Goldsmid)				
Brunswick and Adelaide			0.003 (0.50)	0.005 (0.99)
Central Hove			0.046** (10.08)	0.046** (9.65)
East Brighton			−0.054** (−6.84)	−0.049** (−5.01)
Hangleton and Knoll			−0.016** (−2.44)	−0.045** (−6.67)
Hanover and Elm Grove			−0.089** (−15.59)	−0.091** (−11.81)

(continued)

Table 2 (continued)

	(a) Main model with school catchment areas	(b) Main model with proximity to schools	(c) Main model with electoral wards	(d) Main model with all variables of interest
Hollingdean and Stanmer			−0.116** (−17.92)	−0.105** (−12.20)
Hove Park			0.018** (3.07)	−0.008 (−1.30)
Moulsecoomb and Bevendean			−0.171** (−25.29)	−0.162** (−9.34)
North Portslade			−0.119** (−16.56)	0.005 (0.79)
Patcham			−0.064** (−10.31)	−0.021** (−2.03)
Preston Park			0.019** (3.72)	0.022** (2.85)
Queen's Park			−0.052** (−8.26)	−0.052** (−6.04)
Regency			−0.001 (−0.24)	−0.002 (−0.34)
Rottingdean Coastal			−0.007 (−1.02)	0.036** (3.45)
South Portslade			−0.128** (−18.76)	NA (NA)
St. Peter's and North Laine			−0.058** (−11.53)	−0.053** (−7.92)
Westbourne			0.046** (9.27)	0.042** (8.07)
Wish			−0.012** (−2.12)	−0.023** (−3.89)
Withdean			−0.013** (−2.01)	0.004 (0.38)
Woodingdean			−0.130** (−17.74)	−0.072** (−6.64)
Sample Size	73,559	73,559	73,559	73,559
Adjusted R^2	0.864	0.862	0.867	0.868
F-statistic	6980**	6230**	5840**	4920**

**$p < 0.05$ – Notes: T-values are in parentheses

improve the desirability of housing, but the neighbourhoods they are in appear to be even more important. The best model for Brighton includes proximity to good schools in the year sold, electoral wards, but not catchment areas.

6.3 *Model 3: The Pre- and Post-Admissions Reform Panel Model*

This study then runs the optimised model for the whole period alongside pre- and post-reform periods. Pre-reform refers to the years 2000–2007, and post-reform refers to 2008–2019.

6.4 General Observations

6.4.1 Model Assumptions

This optimised model was then checked to see if it fits the assumptions for linear regression, which are that errors are independent and normally distributed, and if there is equal error variance (Harris, 2016). The residual plot (Fig. 5) shows that the variance is largely consistent with no distinct patterning. The Q-Q plot and histogram show that the residuals are close to being normally distributed, although there is evidence for positive kurtosis, meaning there are more observations in the tails of the distribution than a true normal distribution would expect. These plots are after dropping the top and bottom 5% of transactions in terms of price, so the upper tail of the Q-Q plots would have been more pronounced without the filtering; although none of this invalidates the analysis.

6.4.2 Analysis Observations

The general trends in the directions and magnitudes of the effects of the explanatory variables are similar both pre- and post-admissions reform. The model with electoral ward dummy variables (Model 3a—almost identical to Model 2c) seems to explain house prices well, with a high adjusted R^2 of 86.7% for the whole period and 80.0% and 81.6% for pre- and post-reform periods, respectively. Year-on-year variation is important, with 2009 house prices being −6.0% lower than 2008, on average, being explained by delayed effects of the Global Financial Crisis in 2008. Property-level effects are also important, with semi-detached and terrace properties being on average 23.0% and 19.6% more expensive than flats, respectively (Table 1). This increase in coefficients and *t*-values, even after including contextual variables, confirms what all home-buyers already know—that property-level variables are key determinants of prices, and more important than contextual variables. No matter where they are, bigger houses are almost always more expensive than smaller ones, but this is often ignored when studies fail to adjust for dwelling mix in their analysis.

Across the three regressions (Table 3), IMD deciles have relatively high *t*-values, suggesting that ambient levels of deprivation around properties matter. Decile 2 (second most-deprived) is associated with a 10.5–12.1% increase in price relative to Decile 1, with Decile 10 (least deprived) associated with a 25.7–28.1% increase. The other variable for socio-economic context, OAC supergroups, was confounded by IMD data and hence dropped from the model.

Looking at locational factors, proximity to the city centre (and its associated employment and service opportunities) has a clear positive effect on prices, with houses within a 5-min bus journey of the centre being associated with a 27.4–36.6% increase in prices (Table 3). This positive effect decreases as houses become more distant from the city centre, turning into a slight negative effect of 0.7% for the whole period when houses are 40–45 min away from the city centre relative to houses not near the city centre at all. As walk-time from a rail station increases by

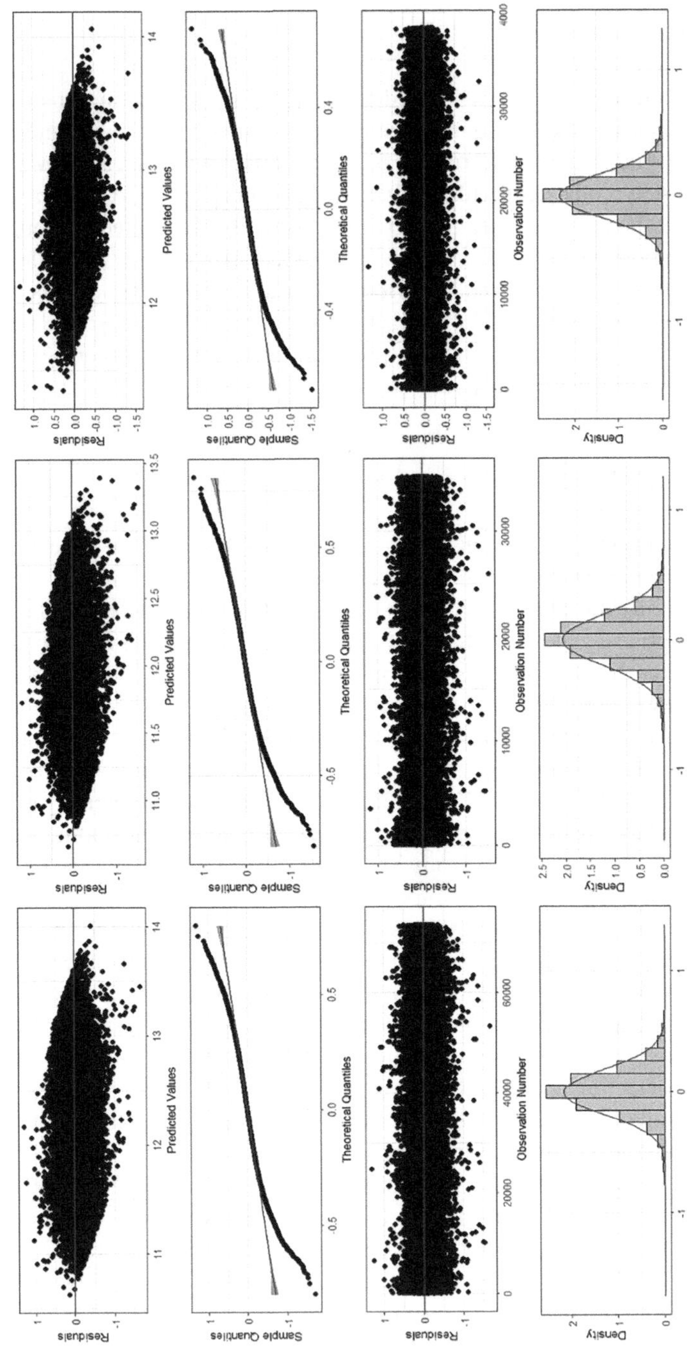

Fig. 5 Residual, Q-Q, Index and Histogram plots (from top to bottom) of models in Table 3 (L: full period, C: pre-reform, R: post-reform)

Table 3 Model 3: whole period (a) and pre (b)-post (c)-reform comparison

	(a) Whole period	(b) Pre-reform	(c) Post-reform
Variable (Reference)	Coefficients (*t*-value)		
Constant	9.103** (491.80)	9.289** (337.26)	9.715** (400.34)
Year (2000/2008)			
2001	0.180** (46.51)	0.180** (44.37)	
2007	0.835** (212.20)	0.833** (201.10)	
2009	0.726** (161.48)		−0.060** (−11.91)
2019	1.230** (274.03)		0.439** (86.88)
Property type (flat/ maisonette)			
Detached	0.307** (81.95)	0.271** (43.51)	0.336** (73.93)
Semi-detached	0.226** (76.46)	0.207** (45.72)	0.246** (65.21)
Terrace	0.192** (82.70)	0.179** (51.67)	0.206** (68.33)
Number of rooms	0.029** (32.51)	0.025** (18.14)	0.032** (28.35)
Log total floor area	0.489** (156.47)	0.465** (99.99)	0.514** (126.13)
Energy efficiency	0.0003** (4.05)	−0.00008 (−0.76)	0.0007** (8.57)
Construction age (NA)			
Before 1900–1990	−0.022** (−3.17) to −0.093** (−13.36)	−0.023** (−2.04) to −0.102** (−9.89)	−0.030** (−3.28) to −0.083** (−8.98)
1991–2007 onwards	0.020** (2.34) to 0.136** (15.92)	0.055** (4.30) to 0.186** (14.77)	0.029** (2.72) to 0.085 (7.66)
Bus time to city centre (not near)			
0–5 min	0.283** (11.87)	0.242** (7.67)	0.312** (8.57)
25–30 min	0.067** (14.72)	0.036** (4.90)	0.087** (15.35)
40–45 min	−0.007** (−1.98)	−0.019** (−3.31)	−0.002 (−0.45)
Rail station (none)			
Brighton	−0.030** (−9.09)	−0.019** (−3.94)	−0.038** (−8.95)
Hove	−0.054** (−13.30)	−0.051** (−8.41)	−0.052** (−10.07)
Preston Park	−0.007 (−1.36)	−0.011 (−1.31)	−0.002 (−0.34)
Log walk time to nearest rail	−0.044** (−17.12)	−0.048** (−12.48)	−0.041** (−12.04)
Walk time to nearest secondary school	0.0007** (6.49)	0.0008** (4.74)	0.0003** (2.05)
Walk time to nearest primary school	0.005** (29.37)	0.005** (19.21)	0.005** (22.49)
IMD Deciles (Decile 1)			
Decile 2	0.104** (20.34)	0.114** (14.90)	0.096** (14.46)
Decile 10	0.241 (39.21)	0.248** (25.61)	0.229** (29.79)
Proximity to good secondary school in year sold			
0–5 min	0.034** (2.55)	0.030 (1.20)	0.027 (1.81)
5–10 min	0.043** (6.74)	0.027** (2.51)	0.040** (5.09)
20–25 min	0.0009 (0.32)	−0.016** (−3.56)	0.005 (1.38)

(continued)

Table 3 (continued)

	(a) Whole period	(b) Pre-reform	(c) Post-reform
25–30 min	−0.007** (−2.65)	−0.012** (−3.30)	−0.005 (−1.67)
Proximity to good primary school in year sold			
0–5 min	0.055** (10.83)	0.054** (6.63)	0.062** (9.63)
5–10 min	0.023** (5.96)	0.017** (2.64)	0.040** (8.10)
20–25 min	0.026** (7.38)	0.019** (3.33)	0.041** (8.91)
25–30 min	0.018** (4.66)	0.010 (1.76)	0.028** (5.82)
Electoral wards			
	−0.165** (−23.58) to 0.052** (10.30)	−0.151** (−13.91) to 0.045** (5.75)	−0.178** (−19.90) to 0.051** (7.83)
Sample size	73,559	35,256	38,303
Adjusted R^2	0.867	0.800	0.816
F-statistic	5110**	1720**	1980**

**$p < 0.05$. Notes: T-values are in parentheses. Range of values presented for electoral wards are smallest and largest range of significant values. Full results available in Table 5 in the Appendix

1%, prices fall by 4.4% (Table 3)—a logical effect given the association with access to higher-paid jobs in London. However, this contrasts with the categorical variable for houses near Brighton, Hove, and Preston Park rail stations. For the whole period, Brighton and Hove stations are respectively associated with a −3.0% and −5.4% effect on house prices relative to houses not near stations, while Preston Park's coefficient is statistically insignificant.

Zooming in on walk-time to the nearest secondary school (regardless of quality), prices *increase* by 0.03–0.08% as walk-time increases (Table 3), and this is counter to the expected direction of the relationship. Speculatively, we might hypothesise an association with increased noise and traffic associated with the schools being a disamenity to home buyers; however, a more likely interpretation school proximity is only important for a sub-section of home-buyers, and for those that it is, there may not be value in locating near an *average* school. With the inclusion of electoral ward dummy variables, the coefficients for proximity to a *good* secondary school in year sold are statistically significant and positive up to a walk-time of 20 min away.

Over the whole study period, walking proximity to a good secondary school does increase prices, but this should be contextualised by acknowledging that the effect is considerably smaller than walking proximity to the mainline railway stations. Furthermore, when decomposed into pre- and post-reform datasets, there is limited statistical significance for the school proximity variable pre-reform, with houses between 5–10 min distance being 2.7% more expensive (Table 3), but statistically insignificant for the other distance bands and a negative effect when more than 20 min away. Post-reform, proximity only has a statistically significant positive effect on houses 5–10 min (Table 3) and 10–15 min away (Appendix 4). To be clear, this is the opposite effect to the one expected by bringing in the lottery for oversubscription—with the small proximity effect *increasing* after the link between admissions and distance was abolished.

Interestingly, the primary school effect is larger than the secondary school effect. There is a clear positive and statistically significant effect for proximity to good primary schools in the year sold. Being within 5 min of walking distance from a good primary school is associated with a 5.5% increase in prices relative to being more than 30 min away from one. This positive effect gradually decreases to 1.8% for the 25–30 min distance band for 2000–2019. These causal pathways for this could be associated with distance proximity still mattering as an admission criterion at the Primary level, but also with the fact that at Primary School age, parents will frequently accompany their children to school, adding to their daily duties in a way that does not occur at secondary school. Combined, these two factors may make proximity a desirable quality that does not exist for secondary schools. Another result of note, but one without causal explanation, is that this effect appears stronger after the secondary schools admissions reform, with the effect of the 0–5 min distance band increasing from 5.4% to 6.2% post-reform, and the 25–30 min band associated with a 2.8% increase as compared to a statistically insignificant effect pre-reform.

Electoral ward effects vary considerably, but with some effects of note. For example, those with the largest negative coefficients (relative to the reference central Ward of Goldsmid to the West of the Main railway station and spanning the Brighton/Hove border) were Hanover and Elm Grove and Hollingdean and Stanmer—these areas are notable for high levels of 'Studentification' (see Sage et al., 2012 for a detailed account), where there is an influx of university students living in these areas.

Mapping the residuals from the final model (Fig. 6) shows us the properties that either cost far more (positive residuals) or far less (negative residuals) than what the model predicts. The negative outliers are mostly transactions before 2010, in IMD deciles above 5 (more deprived), and more than 30 min away from the city centre,

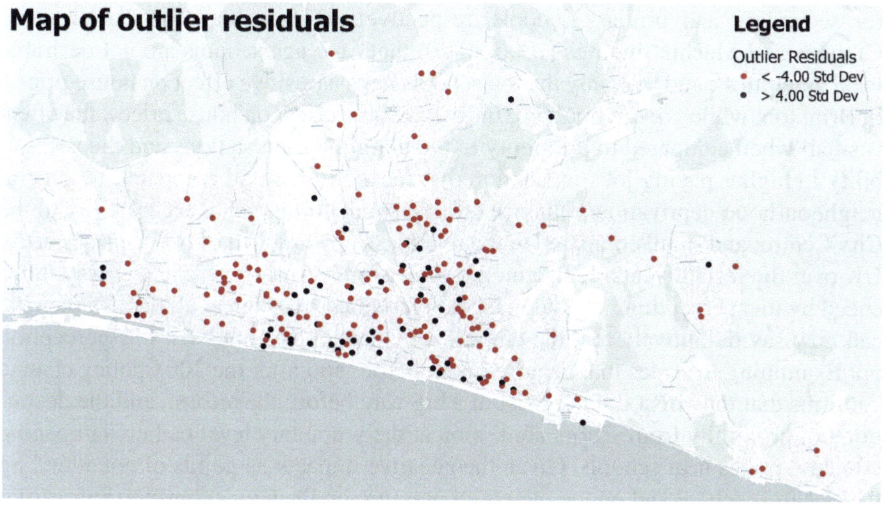

Fig. 6 Map of outlier residuals (model 3a)

although a number do occur more centrally. Some of these extreme negative residuals are transactions in the central and Hanover areas, where the 'studentification' and 'touristification' in the city are concentrated. Noise and anti-social behaviour from both activities (Sage et al., 2012; Prosser, 2022) can be higher in these areas, with the proliferation of short-term holiday let homes fuelled by websites like AirBnB, potentially dampening the sale values of neighbouring properties. As for the few positive outliers, most are also before 2010, but they also exhibit a clear spatial patterning—mainly concentrated in a thin band which runs East to West along the seafront. This is clearly a 'sea view' or accessibility to the beach and seafront premium, which was not captured as a variable in the model.

7 Discussion

This study sought to shed new light on the established perception that proximity to schools pushes up house prices. Our case study area in Brighton is a particularly interesting study due to the change in secondary school over-subscription criteria in 2008, which many believed would impact a perceived 'golden halo' effect around some of the city's best schools. We were able to explore the variables influencing house prices in the city at a finer spatial and temporal resolution than has been attempted in other studies, accounting for dwelling characteristics alongside detailed measurements of accessibility to different services, including schools, at the postcode level and accounting for temporal changes in school quality. In doing so, we could assess both the level of the influence that schools have on prices, both through direct spatial association and through any changes to this association after the proximity priority to school place allocation was removed.

We found that contrary to expectations, *increasing* the walk-time to the nearest school (ignoring its perceived quality) *increases* house prices by 0.07% and 0.5% for secondary and primary schools, respectively. This echoes the earlier noted Gibbons and Machin findings (2006, p. 90) that 'average schools are not desirable local amenities', and that only the top schools have a positive effect on house prices. In Brighton, while *good* school proximity has some effect on house prices, the effect is small when compared to proximity to the mainline train stations and the accessibility to higher paying jobs in London that these offer; small compared to general neighbourhood deprivation/affluence effects; small compared to accessibility to the City Centre; and small compared to the relentless upward drift of house prices in the UK over the last 20 years. If Brighton's admissions reforms were in any way influenced by this perception of 'golden halos' recounted by Allen et al. (2013), then we can now say definitively that the empirical evidence does not back this perception up. Examining the sales that occurred both before and after the 2008 policy change confirms that the effect did not exist in a big way before the reform, and the decoupling of proximity from school admission at the secondary level had no dampening effect on prices near schools. Given the negative impacts on pupils of not attending their nearest school and on the increased pressure on the City's transport infrastructure for those having to travel long distances to access their education, without a compelling house price case underpinning the reform, the policy itself may need revisiting. This empirical study may also invite policymakers to reconsider the reality of policy reforms, that desired policy outcomes may not necessarily come to fruition, and there should be scope to refine policies further following any review of new evidence.

In this analysis, we see the importance of time and how this might confound our perceptions. Rather than standardising our dependent price variable to account for on-going inflationary trends in the housing market—for example by selecting a base year and adjusting by the consumer price index (CPI) or the retail price index (RPI)—we chose to account for inherent inflation in the system by including a dummy variable for the sale year. This had the benefit of allowing us to compare standardised coefficients across variables and see to what extent simply buying a house later would have, compared to being close to a school or any other hedonic variable. This is important as many of the perceptions of school proximity price inflation have been made against a backdrop of a long-term trend of house price inflation. Anyone living in a city and observing price inflation might be forgiven for attaching any number of plausible reasons to that increase, with macro trends often unlikely to make their list. However, in Brighton at least, we now have some strong empirical evidence that proximity to good schools plays only a small role in house prices and is less important than other contextual factors.

7.1 *Study Limitations*

It could always be argued that we could have taken the modelling analysis further than we have to try and account for even more of the unexplained variation in the model—the inclusion of a coastal location variable or attempts to match more closely school quality to when house purchase offers were made rather than taking the date at when entries make it onto the land registry database. It is also true that our metric for determining perceived school quality, which was based primarily on Ofsted ranking and exam performance, could easily be critiqued—particularly where some measure of 'value added' (outcomes relative to cohort composition) might be a more nuanced measure and one which some parents may prioritise.

Another direction we could have explored in more detail is spatial or temporal non-stationarity (where relationships may not hold in the same way over space or time), particularly with differences between school catchment areas—something which we could have achieved through using the catchments or sale years as additional panels or levels in the data, adopting a multi-level or linear mixed-effects framework. These options are still open for future research—and something we experimented with in earlier iterations of this research. At this time we found that given our primary objective was to ascertain how strong the effect of good school proximity is on house prices in Brighton and whether the 2008 secondary school admissions reform had any impact on prices; and little evidence of a strong link was found at any point; we can be satisfied that the analysis we carried out in its present form and the methods we used worked, were easy to interpret, and allowed us to draw useful conclusions.

We have not in this study attempted to account for the deeper structural, social, and spatial forces interacting to reproduce geographies of educational advantage and housing market affordability that clearly intersect in ways that can be profoundly problematic when discussing equitable access to both. To an extent this was beyond the scope of what we were trying to achieve, however within our current analysis framework, interacting the index of multiple deprivation with good school proximity variables might have revealed more about how deprivation and distance factors can work both together and in opposition to one-another in different parts of the city to have variegated influences on price.

8 Conclusions

Referring back to our primary research question, we can conclude that the influence of good schools on house prices should not be overstated compared with other contextual factors, at least in Brighton and Hove. Because the effect of proximity to secondary schools was already small, changes to the secondary school admissions policy in 2008 had very little impact, contrary to some expectations. Breaking the explicit link between proximity and access to good schools in Brighton did not lead to houses near those schools commanding less of a premium.

To unearth this finding, we studied the spatial and temporal dimensions of the housing market context at a level of detail rare in the existing literature. Our use of the r5r algorithm in estimating detailed walking and public transport journey times from homes to schools, railway stations, and the city centre meant we were able to generate a highly nuanced representation of proximity to these services in the city. Examining Ofsted and DfES data on school performance over a twenty-year period meant that we were able to attach plausible quality perceptions to schools at the time when houses in the dataset were sold, giving a level of temporal resolution to the analysis also rarely seen.

In incorporating detailed time and dwelling attribute data to our price analysis—thanks to the pioneering work of Chi et al. (2022) and their unique linked dataset—and bringing in both deprivation and important catchment and confounding neighbourhood/electoral ward random effects, we were able to fully disentangle different type of influence on the sale prices of properties sold. It would have been very easy to claim a clear school catchment area effect had we not fully tested other spatial groupings at the Ward level and discovered that the catchment effect disappears when we do. It helps prove that the neighbourhood context of a school is important but not necessarily a result of the school on its own—this challenges work that has been carried out in other UK studies and shows that while Brighton might be a unique case, merit may be found in adopting this approach in studies elsewhere.

9 Materials

The methods used in this paper are contained in the following Github repository: https://github.com/jtej22/CASA0010-Dissertation

Appendix

Process for Deriving Perceived school Quality Scores

For primary schools, more weight was given to KS2 academic results and Ofsted ratings where data was available. KS2 results were used from 2003–2006, with KS2 results and Ofsted ratings weighted at 50% each from 2007–2011, and 45% each from 2012–2019 with percentage of students on FSM weighted at 10%.

For secondary schools: from 2003–2006 categories were similarly dependent on just KS4 results, but from 2007–2011 rankings were derived from an equal weighting between academic results (KS4) and Ofsted ratings. From 2012–2019 they were weighted as 30% each for academic results, Ofsted ratings and pupil destination, with 10% given to percentage of students on FSM. The metric for academic results is the percentage of 15 year old pupils achieving 5 or more grades A*-C (GCSE/GNVQ) and similar metrics as it is a common indicator of secondary school performance (Cheshire & Sheppard, 2004; Glen & Nellis, 2010), with the equivalent metric after the 2017 curriculum change (Ofqual, 2017) being used too. For pupil destination, the total percentage of students from each school going to a Sixth Form School or College is computed and ranked, with the highest percentage school ranked 1. This assumes that parents link progression to post-16 education with academic quality which could be contested, but we still think a justifiable metric when raw exam results are also used.

After accounting for all the variables in both primary and secondary school datasets, all schools were ranked and classified into one of four bands (Outstanding, Good, Satisfactory and Poor), with each band comprising their respective quantiles. Secondary schools are given one of three bands (Outstanding, Good, Poor) as there only 9 state secondary schools within the city (Brighton & Hove City Council, 2023). The Cardinal Newman Catholic School (CNCS) has autonomy over its admissions and prioritises Catholic children (CNCS, 2023), but after this priority, other priority categories still apply (Brighton & Hove City Council, 2023) and on this basis we still include it in our analysis. Hove Park School also has 2 campuses, Valley campus and Nevil campus, and both locations were used to determine proximity to good schools. It was assumed that both locations would have been used for admissions, as is the policy for King's School in Brighton. Finally, the primary school rankings for 2003 are extended to cover 2000–2002 as well on the assumption that school quality does not change rapidly. While a time lag between school ranking and transaction year could have been included, it is also difficult to estimate an ideal time lag, hence the same years were used as an approximation.

Variable Frequency Distributions

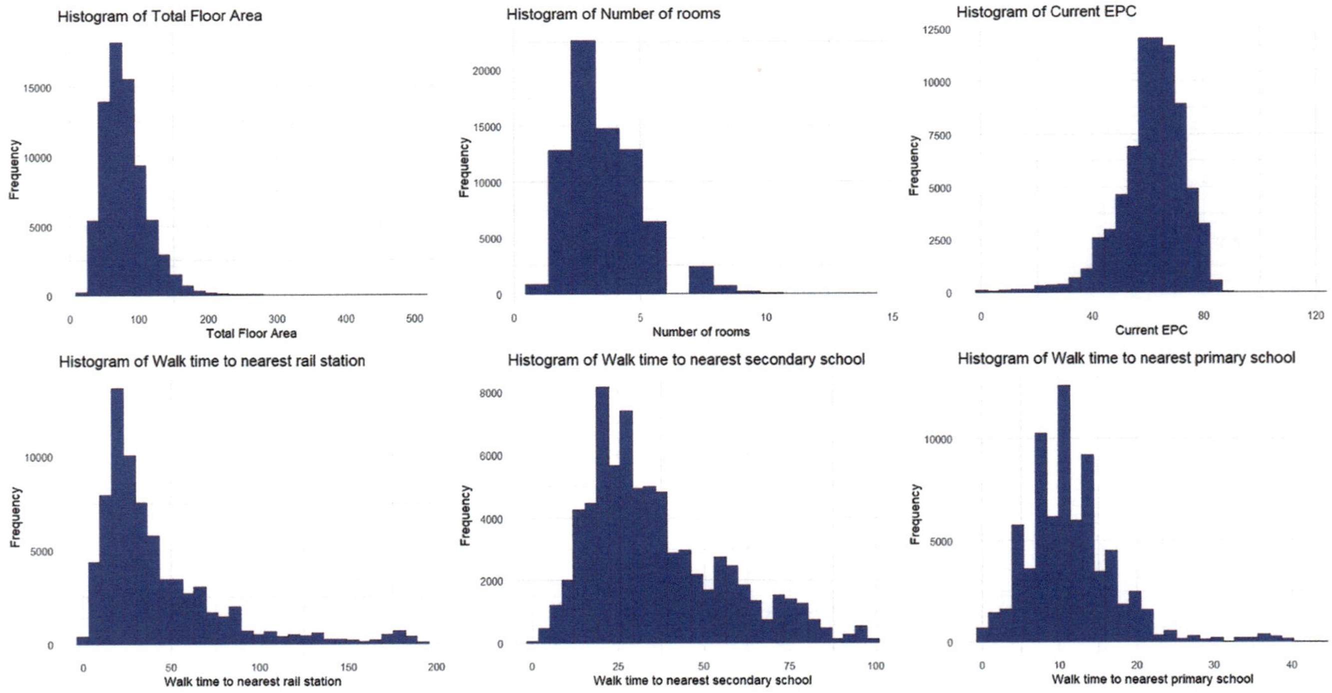

Fig. 7 Histograms of continuous independent variables

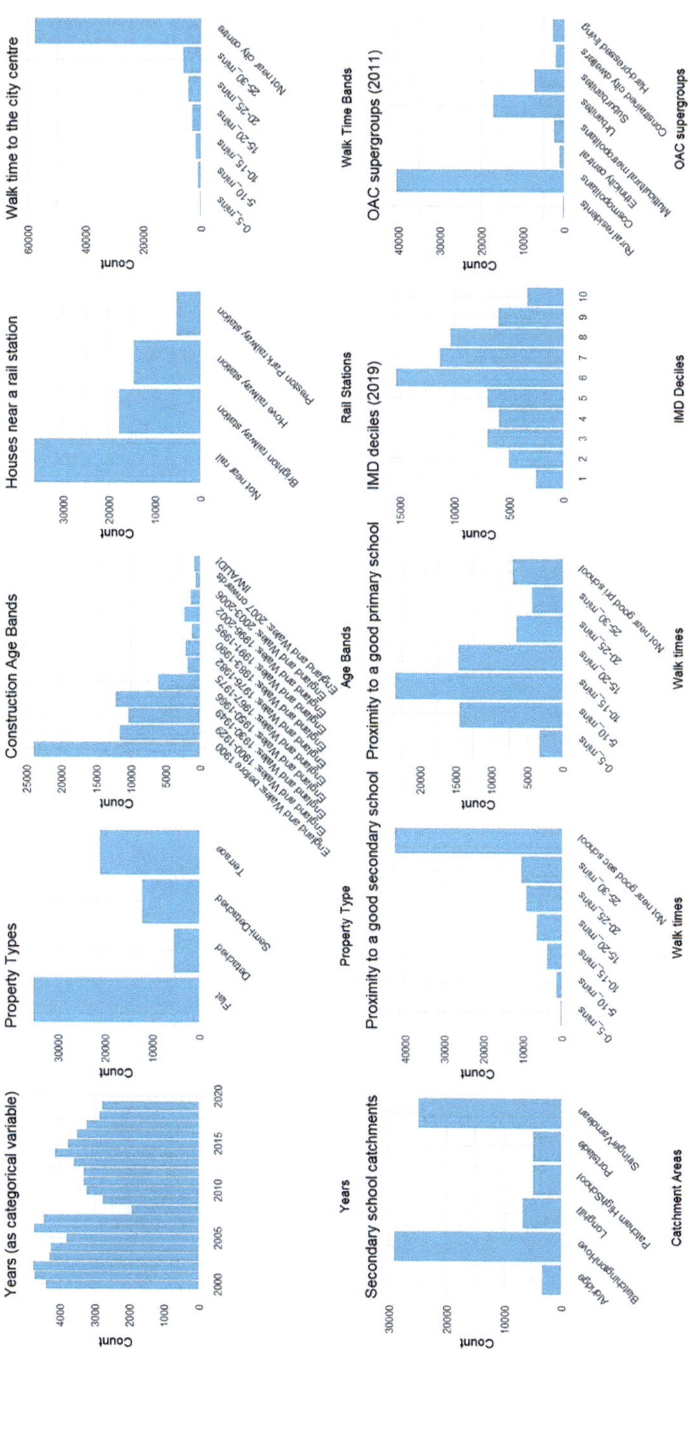

Fig. 8 Barplots of categorical independent variables

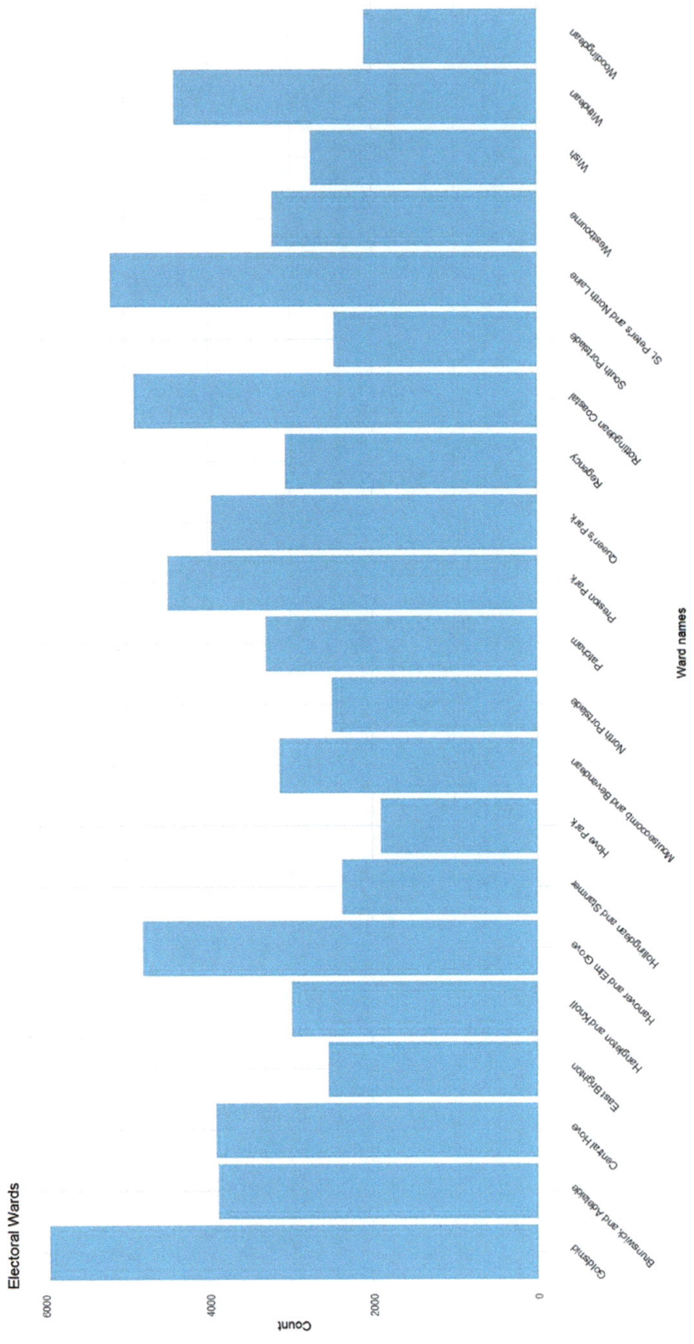

Fig. 9 Barplots of Brighton's electoral wards

GVIF Values

Table 4 GVIF values of variables

	GVIF	D_f	$GVIF^{1/(2 \cdot Df)}$
Years (as factors)	1.120	19	1.003
Property type	4.962	3	1.306
Number of rooms	3.691	1	1.921
Construction age band	4.394	12	1.064
Energy efficiency	1.216	1	1.103
Log Total floor area	3.051	1	1.747
Bus time band to city centre	79.282	9	1.275
Nearest rail station name	56.541	3	1.959
Log Walk time to nearest rail	8.599	1	2.932
Walk time to nearest primary school	2.274	1	1.501
Walk time to nearest secondary school	8.660	1	2.943
IMD decile	29.066	9	1.206
Proximity to good primary school (year sold)	5.552	6	1.154
Proximity to good secondary school (year sold)	4.156	6	1.126
Electoral wards	68,969	20	1.321

Full Regression Results

Table 5 Whole period and pre-post-reform regressions full results

	Whole period	Pre-reform	Post-reform
Variable (reference)	Coefficients (*t*-value)		
Constant	9.103** (491.80)	9.289** (337.26)	9.715** (400.34)
Year (2000/2008)			
2001	0.180** (46.51)	0.180** (44.37)	
2002	0.374** (96.97)	0.375** (92.55)	
2003	0.500** (126.05)	0.499** (119.58)	
2004	0.592** (148.82)	0.591** (141.22)	
2005	0.632** (154.22)	0.632** (146.66)	
2006	0.712** (183.89)	0.710** (174.22)	
2007	0.835** (212.20)	0.833** (201.10)	
2008	0.786** (155.57)		*(Reference)*
2009	0.726** (161.48)		−0.060** (−11.91)
2010	0.846** (196.97)		0.059** (11.91)

(continued)

Table 5 (continued)

	Whole period	Pre-reform	Post-reform
Variable (reference)	Coefficients (*t*-value)		
2011	0.846** (198.53)		0.058** (11.89)
2012	0.874** (205.02)		0.086** (17.53)
2013	0.912** (219.33)		0.124** (25.80)
2014	1.014** (252.86)		0.227** (48.26)
2015	1.102** (268.00)		0.315** (66.04)
2016	1.186** (282.83)		0.399** (82.49)
2017	1.234** (287.71)		0.445** (90.65)
2018	1.240** (279.35)		0.451** (89.89)
2019	1.230** (274.03)		0.439** (86.88)
Property Type (Flat/ Maisonette)			
Detached	0.307** (81.95)	0.271** (43.51)	0.336** (73.93)
Semi-detached	0.226** (76.46)	0.207** (45.72)	0.246** (65.21)
Terrace	0.192** (82.70)	0.179** (51.67)	0.206** (68.33)
Number of rooms	0.029** (32.51)	0.025** (18.14)	0.032** (28.35)
Log Total floor area	0.489** (156.47)	0.465** (99.99)	0.514** (126.13)
Energy Efficiency	0.0003** (4.05)	−0.00008 (−0.76)	0.0007** (8.57)
Construction Age (NA)			
Before 1900	−0.011 (−1.58)	−0.028** (−2.83)	0.009 (1.02)
1900–1929	−0.022** (−3.17)	−0.048** (−4.70)	0.005 (0.51)
1930–1949	−0.051** (−7.34)	−0.072** (−6.96)	−0.030** (−3.28)
1950–1966	−0.093** (−13.36)	−0.102** (−9.89)	−0.079** (−8.76)
1967–1975	−0.086** (−12.16)	−0.086** (−8.23)	−0.083** (−8.98)
1976–1982	−0.067** (−8.31)	−0.060** (−4.96)	−0.073** (−6.97)
1983–1990	−0.033** (−4.21)	−0.023** (−2.04)	−0.050** (−4.80)
1991–1995	0.020** (2.34)	0.055** (4.30)	−0.021 (−1.82)
1996–2002	0.086** (10.92)	0.126** (11.05)	0.029** (2.72)
2003–2006	0.136** (15.92)	0.186** (14.77)	0.085** (7.66)
2007 onwards	0.080** (7.86)	0.109** (5.41)	0.082** (6.95)
Bus time to city centre (Not near)			
0–5 min	0.283** (11.87)	0.242** (7.67)	0.312** (8.57)
5–10 min	0.234** (30.25)	0.216** (18.61)	0.226** (22.11)
10–15 min	0.179** (28.57)	0.145** (14.95)	0.199** (25.03)

Table 5 (continued)

	Whole period	Pre-reform	Post-reform
Variable (reference)	Coefficients (*t*-value)		
15–20 min	0.146** (26.77)	0.116** (13.56)	0.164** (23.97)
20–25 min	0.090** (17.93)	0.058** (7.32)	0.110** (17.60)
25–30 min	0.067** (14.72)	0.036** (4.90)	0.087** (15.35)
30–35 min	0.037** (9.12)	0.026** (3.86)	0.044** (8.84)
35–40 min	0.0007 (0.21)	−0.003 (−0.53)	0.002 (0.57)
40–45 min	−0.007** (−1.98)	−0.019** (−3.31)	−0.002 (−0.45)
Rail Station (None)			
Brighton	−0.030** (−9.09)	−0.019** (−3.94)	−0.038** (−8.95)
Hove	−0.054** (−13.30)	−0.051** (−8.41)	−0.052** (−10.07)
Preston Park	−0.007 (−1.36)	−0.011 (−1.31)	−0.002 (−0.34)
Log Walk time to nearest rail	−0.044** (−17.12)	−0.048** (−12.48)	−0.041** (−12.04)
Walk time to nearest secondary school	0.0007** (6.49)	0.0008** (4.74)	0.0003** (2.05)
Walk time to nearest primary school	0.005** (29.37)	0.005** (19.21)	0.005** (22.49)
IMD Deciles (Decile 1)			
Decile 2	0.104** (20.34)	0.114** (14.90)	0.096** (14.46)
Decile 3	0.116** (23.95)	0.128** (17.63)	0.108** (17.21)
Decile 4	0.141** (30.04)	0.150** (21.22)	0.133** (22.08)
Decile 5	0.122** (23.66)	0.137** (17.48)	0.109** (16.63)
Decile 6	0.158** (33.77)	0.167** (23.70)	0.150** (24.92)
Decile 7	0.171** (37.64)	0.171** (24.85)	0.172** (29.52)
Decile 8	0.185** (38.17)	0.193** (26.14)	0.175** (28.32)
Decile 9	0.203** (37.36)	0.215** (25.54)	0.187** (27.17
Decile 10	0.241 (39.21)	0.248** (25.61)	0.229** (29.79)
Proximity to good secondary school in year sold			
0–5 min	0.034** (2.55)	0.030 (1.20)	0.027 (1.81)
5–10 min	0.043** (6.74)	0.027** (2.51)	0.040** (5.09)
10–15 min	0.021** (4.96)	0.006 (0.89)	0.019** (3.36)
15–20 min	0.012** (3.63)	0.004 (0.73)	0.007 (1.65)
20–25 min	0.0009 (0.32)	−0.016** (−3.56)	0.005 (1.38)
25–30 min	−0.007** (−2.65)	−0.012** (−3.30)	−0.005 (−1.67)
Proximity to good primary school in year sold			
0–5 min	0.055** (10.83)	0.054** (6.63)	0.062** (9.63)

(continued)

Table 5 (continued)

	Whole period	Pre-reform	Post-reform
Variable (reference)	Coefficients (*t*-value)		
5–10 min	0.023** (5.96)	0.017** (2.64)	0.040** (8.10)
10–15 min	0.012** (3.52)	0.004 (0.74)	0.029** (6.52)
15–20 min	0.017** (5.12)	0.016** (2.93)	0.030** (6.96)
20–25 min	0.026** (7.38)	0.019** (3.33)	0.041** (8.91)
25–30 min	0.018** (4.66)	0.010 (1.76)	0.028** (5.82)
Electoral Wards			
Brunswick and Adelaide	0.006 (1.22)	0.006 (0.82)	0.004 (0.57)
Central Hove	0.050** (10.74)	0.044** (6.22)	0.049** (7.98)
East Brighton	−0.063** (−7.91)	−0.055** (−4.47)	−0.062** (−6.04)
Hangleton and Knoll	−0.026** (−3.98)	−0.002 (−0.15)	−0.051** (−6.09)
Hanover and Elm Grove	−0.094** (−16.39)	−0.086** (−9.96)	−0.099** (−13.34)
Hollingdean and Stanmer	−0.110** (−16.95)	−0.114** (−11.39)	−0.110** (−13.24)
Hove Park	0.009 (1.47)	0.008 (0.85)	0.003 (0.40)
Moulsecoomb and Bevendean	−0.165** (−23.58)	−0.151** (−13.91)	−0.178** (−19.90)
North Portslade	−0.114** (−15.68)	−0.090** (−7.83)	−0.145** (−15.83)
Patcham	−0.064** (−10.18)	−0.049** (−4.74)	−0.084** (−10.63)
Preston Park	0.019** (3.54)	−0.005 (−0.70)	0.035** (5.14)
Queen's Park	−0.060** (−9.32)	−0.067** (−6.91)	−0.049** (−5.94)
Regency	−0.002 (−0.37)	−0.033** (−3.80)	0.026** (3.23)
Rottingdean Coastal	−0.015** (−2.05)	0.023** (2.08)	−0.038** (−4.18)
South Portslade	−0.129** (−18.78)	−0.124** (−11.65)	−0.145** (−16.71)
St. Peter's and North Laine	−0.060** (−11.63)	−0.076** (−9.94)	−0.050** (−7.26)
Westbourne	0.052** (10.30)	0.045** (5.75)	0.051** (7.83)
Wish	−0.007 (−1.28)	−0.023** (−2.61)	−0.004 (−0.49)
Withdean	−0.014** (−2.05)	−0.18 (−1.70)	−0.016 (−1.85)
Woodingdean	−0.130** (−17.58)	−0.089** (−7.64)	−0.162** (−17.56)
Sample Size	73,559	35,256	38,303
Adjusted R^2	0.867	0.800	0.816
F-statistic	5110**	1720**	1980**

References

Allen, R., Burgess, S., & McKenna, L. (2013). The short-run impact of using lotteries for school admissions: Early results from Brighton and Hove's reforms. *Transactions of the Institute of British Geographers, 38*, 149–166. https://doi.org/10.1111/j.1475-5661.2012.00511.x

Alonso, W. (1964). The historic and the structural theories of urban form: Their implications for urban renewal. *Land Economics, 40*(2), 227–231.

Black, S. E. (1999). Do better schools matter? Parental valuation of elementary education. *The Quarterly Journal of Economics, 114*(2), 577–599.

Bonilla-Mejía, L., Lopez, E., & McMillen, D. (2020). House prices and school choice: Evidence from Chicago's magnet schools' proximity lottery. *Journal of Regional Science, 60*(1), 33–55.

Box, G. E. P., & Cox, D. R. (1964). An analysis of transformations. *Journal of the Royal Statistical Society: Series B (Methodological), 26*(2), 211–243.

Brighton & Hove City Council (2023). Schools portal—Map overview. Retrieved August 4, 2023, from https://applyforaschool.brighton-hove.gov.uk/p/schoolsmap

Cheshire, P., & Sheppard, S. (2004). Capitalising the value of free schools: The impact of supply characteristics and uncertainty*. *The Economic Journal, 114*(499), F397–F424.

Chi et al., (2021). https://journals.uclpress.co.uk/ucloe/article/id/1332/

Chi, B., Dennett, A., Morphet, R., & Hutchinson, C. (2020). Exploring local authority travel time to London effects on spatio-temporal pattern of local authority house prices variation in England, [Working / discussion paper]. Retrieved April 5, 2023, from https://www.ucl.ac.uk/bartlett/casa/publications/2020/apr/casa-working-paper-218

Chi, B., Dennett, A., Oléron-Evans, T., & Morphet, R. (2022). House price per square metre in England and Wales, 1995–2021, UK Data Service. https://doi.org/10.5255/UKDA-SN-855033.

Chin, H. C., & Foong, K. W. (2006). Influence of school accessibility on housing values. *Journal of Urban Planning and Development, 132*(3), 120–129.

Cho, C.-J. (2001). Amenities and urban residential structure: An amenity-embedded model of residential choice*. *Papers in Regional Science, 80*(4), 483–498.

CNCS (2023). Admissions|Cardinal Newman Catholic School. Retrieved August 4, 2023, from https://www.cncs.co.uk/admissions

Consumer Data Research Centre (2022). Retail centre boundaries, (Version 1) [Data set] (Consumer Data Research Centre).

Davidoff, I., & Leigh, A. (2007). How much do public schools really cost? Estimating the relationship between house prices and school quality, SSRN Scholarly Paper, (Rochester, NY).

Du, H., & Mulley, C. (2007). The short-term land value impacts of urban rail transit: Quantitative evidence from Sunderland, UK. *Land Use Policy, 24*(1), 223–233.

Eastwood, R., & Turvey, K. (2008). Equal opportunities or loaded dice? The 2007 admissions code after the Brighton and Hove adjudication. *Research Intelligence, 102*, 20–21.

Fack, G., & Grenet, J. (2010). When do better schools raise housing prices? Evidence from Paris public and private schools. *Journal of Public Economics, 94*(1), 59–77.

Fleming, M. C., & Nellis, J. G. (1985). The application of hedonic indexing methods: A study of house prices in the United Kingdom. *Statistical Journal of the United Nations Economic Commission for Europe, 3*(3), 249–270.

Fox, J., & Monette, G. (1992). Generalized collinearity diagnostics. *Journal of the American Statistical Association, 87*(417), 178–183.

Freeman, A. M. (1979). The hedonic price approach to measuring demand for neighborhood characteristics, pp. 191–217 (Elsevier).

Gibbons, S., & Machin, S. (2003). Valuing English primary schools. *Journal of Urban Economics, 53*(2), 197–219.

Gibbons, S., & Machin, S. (2005). Valuing rail access using transport innovations. *Journal of Urban Economics, 57*(1), 148–169.

Gibbons, S., & Machin, S. (2006). Paying for primary schools: Admission constraints, school popularity or congestion? *The Economic Journal, 116*(510), C77–C92.

Gibbons, S., & Machin, S. (2008). Valuing school quality, better transport, and lower crime: Evidence from house prices. *Oxford Review of Economic Policy, 24*(1), 99–119.

Glen, J., & Nellis, J. (2010). 'The price you pay': The impact of state-funded secondary school performance on residential property values in England. *Panoeconomicus, 57*(4), 405–428.

Goodman, A. C. (1978). Hedonic prices, price indices and housing markets. *Journal of Urban Economics, 5*(4), 471–484.

Hair, J. (2009). *Multivariate Data Analysis*. Faculty and Research Publications.

Harris, R. (2016). *Quantitative geography: The basics*. SAGE Publications Ltd.

Haurin, D. R., & Brasington, D. (1996). School quality and real house prices: Inter- and Intrametropolitan effects. *Journal of Housing Economics, 5*(4), 351–368.

Huu Phe, H., & Wakely, P. (2000). Status, quality and the other trade-off: Towards a new theory of urban residential location. *Urban Studies, 37*(1), 7–35.

Kane, T. J., Staiger, D., & Riegg, S. (2005). School quality, neighborhoods and housing prices: The impacts of school desegregation, SSRN Scholarly Paper, (Rochester, NY).

Kuroda, Y. (2018). The effect of school quality on housing rents: Evidence from Matsue city in Japan. *Journal of the Japanese and International Economies, 50*, 16–25.

Lareau, A., & Goyette, K. (2014). *Choosing homes, choosing schools Russell Sage Foundation*. Russell Sage Foundation.

Leech, D., & Campos, E. (2003). Is comprehensive education really free?: A case-study of the effects of secondary school admissions policies on house prices in one local area. *Journal of the Royal Statistical Society Series A (Statistics in Society), 166*(1), 135–154.

McCord, J., McCord, M., McCluskey, W., Davis, P. T., McIlhatton, D., & Haran, M. (2014a). Effect of public green space on residential property values in Belfast metropolitan area. *Journal of Financial Management of Property and Construction, 19*(2), 117–137.

McCord, M., Davis, P. T., Haran, M., McIlhatton, D., & McCord, J. (2014b). Understanding rental prices in the UK: A comparative application of spatial modelling approaches. *International Journal of Housing Markets and Analysis, 7*(1), 98–128.

McIlhatton, D., McGreal, W., Taltavul de la Paz, P., & Adair, A. (2016). Impact of crime on spatial analysis of house prices: Evidence from a UK city. *International Journal of Housing Markets and Analysis, 9*(4), 627–647.

Ofqual (2017). GCSE maths grade boundaries. Retrieved August 8, 2023, from https://ofqual.blog.gov.uk/2017/11/03/gcse-maths-grade-boundaries/

ONS (2023). https://www.ons.gov.uk/census/maps/choropleth/housing/number-of-cars-or-vans/number-of-cars-3a/no-cars-or-vans-in-household?lad=E06000043

Orford, S. (2002). Valuing locational externalities: A GIS and multilevel modelling approach. *Environment and Planning B: Planning and Design, 29*(1), 105–127. https://doi.org/10.1068/b2780

Pereira, R. H. M., Saraiva, M., Herszenhut, D., Braga, C. K. V., & Conway, M. W. (2021). r5r: Rapid realistic routing on multimodal transport networks with R5 in R, Findings. https://doi.org/10.32866/001c.21262.

Prosser, B. M. (2022). Listening to urban seaside gentrification: Living with displacement injustices on the UK south coast.

Reback, R. (2005). House prices and the provision of local public services: Capitalization under school choice programs. *Journal of Urban Economics, 57*(2), 275–301.

Robinson, C., Castanheira, P., & Turvey, K. (2016). *Stakeholders' perspectives on the secondary school admissions procedures in Brighton and Hove*. Education Research Centre, University of Brighton. https://research.edgehill.ac.uk/files/21774447/bhcc20final20report2026.02.16.pdf

Rosen, S. (1974). Hedonic prices and implicit markets: Product differentiation in pure competition. *Journal of Political Economy, 82*(1), 34–55.

Rosiers, F. D., Lagana, A., & Theriault, M. (2001). Size and proximity effects of primary schools on surrounding house values. *Journal of Property Research, 18*(2), 149–168.

Sage, J., Smith, D., & Hubbard, P. (2012). The rapidity of studentification and population change: There goes the (student)hood. *Population, Space and Place, 18*(5), 597–613.

School Admissions Code (2007). King's printer of acts of parliament.

Shah, S. (2018). The price of free education: Extracting the school quality premium in housing using Brighton and Hove's school admission reforms, Christ's College, University of Cambridge Economics Dissertation.

Thameslink (2023). Our train journey routes | Our train destinations | Thameslink. Retrieved August 8, 2023 from https://www.thameslinkrailway.com/destinations-and-offers/where-we-travel-to/our-routes

UK Government (2023). School admissions. Retrieved May 29, 2023, from https://www.gov.uk/schools-admissions/admissions-criteria

Wen, H., Xiao, Y., & Zhang, L. (2017). School district, education quality, and housing price: Evidence from a natural experiment in Hangzhou. *China, Cities, 66*, 72–80.

Index

R. Goodspeed et al. (eds.), *Digital-Era Urban Transformations*, The Urban Book Series, https://doi.org/10.1007/978-3-031-98300-9

GPSR Compliance

The European Union's (EU) General Product Safety Regulation (GPSR) is a set of rules that requires consumer products to be safe and our obligations to ensure this.

If you have any concerns about our products, you can contact us on ProductSafety@springernature.com

In case Publisher is established outside the EU, the EU authorized representative is:

Springer Nature Customer Service Center GmbH
Europaplatz 3
69115 Heidelberg, Germany

Batch number: 10755329

Printed by Printforce, the Netherlands